Application of Short-Term Bioassays in the Fractionation and Analysis of Complex Environmental Mixtures

Environmental Science Research

Application of Short-Term Bioassays in the Fractionation and Analysis of Complex Environmental Mixtures

Edited by

Michael D. Waters
Stephen Nesnow
Joellen L. Huisingh
Shahbeg S. Sandhu
and
Larry Claxton

U.S. Environmental Protection Agency
Research Triangle Park, North Carolina

PLENUM PRESS · NEW YORK AND LONDON

Library of Congress Cataloging in Publication Data

Symposium on Application of Short-term Bioassays in the Fractionation and
  Analysis of Complex Environmental Mixtures, Williamsburg, Va., 1978.
  Application of short-term bioassays in the fractionation and analysis of com-
plex environmental mixtures.
  (Environmental science research; 15)
  Includes index.
  1. Biological assay—Congresses. 2. Mutagenicity testing—Congresses. 3. Environ-
mental chemistry—Technique—Congresses. I. Waters, Michael D. II. United States.
Environmental Protection Agency. III. Title.
QH541.15.B54S95 1978   574.2'4'028   79-22240
ISBN 978-1-4684-3613-6   ISBN 978-1-4684-3611-2 (eBook)
DOI 10.1007/978-1-4684-3611-2

Proceedings of the Symposium on Application of Short-Term Bioassays
in the Fractionation and Analysis of Complex Environmental Mixtures,
held at Williamsburg, Virginia, February 21–23, 1978.

This report has been reviewed by the Health Effects Research
Laboratory, U.S. Environmental Protection Agency, and approved
for publication. Mention of trade names or commercial products
does not constitute endorsement or recommendation for use.

Softcover reprint of the hardcover 1st edition 1979

Plenum Press, New York
A Division of Plenum Publishing Corporation
227 West 17th Street, New York, N.Y. 10011

Foreword

In recent years the development and utilization of short-term bioassays which detect cytotoxic, mutagenic, and carcinogenic activities of environmental chemicals has increased dramatically. The wide acceptance of the bacterial mutagenesis bioassay using Salmonella typhimurium developed by Bruce Ames has been a catalyst in the expanded growth of the discipline of genetic toxicology. The Ames test, a direct measure of mutagenesis and an indicator of presumptive carcinogenesis, was originally developed as a research tool and subsequently has demonstrated its usefulness in identifying potential genotoxicants. Initially, studies concentrated on the testing of pure agents. More recently, this bioassay system has been applied to the evaluation of complex environmental mixtures. Short-term bioassays using other end points such as cytotoxicity and oncogenic transformation are now being applied to the evaluation of these complex mixtures isolated from the environment.

The coupling of new analytical tools such as high pressure liquid chromatography with sensitive short-term bioassays, has created new areas of intensive investigation whose inception was not possible prior to this time. This interdisciplinary approach which places the engineer, analytical chemist, biochemist, and genetic toxicologist together in joint pursuit of common goals has produced a milieu of exciting collaborative research.

This volume is the proceedings of the "Application of Short-term Bioassays in the Fractionation and Analysis of Complex Environmental Mixtures," a symposium convened at

Williamsburg, Virginia, February 21-23, 1978. This symposium was sponsored by the U.S. Environmental Protection Agency, Office of Energy Minerals and Industry, Washington, DC, and Office of Health and Ecological Effects, Health Effects Research Laboratory, Biochemistry Branch, Research Triangle Park, NC. The symposium consisted of 24 formal presentations that amplify the three major topics discussed during the symposium: an overview of short-term bioassay systems; current methodology involving the collection and chemical analysis of environmental samples; and current research involving the use of short-term bioassays in the fractionation and analysis of complex environmental mixtures. The purpose of this symposium was to present the state-of-the-art techniques in bioassay and chemical analysis as applied to complex mixtures and to foster continued advancement of this important area. Complex mixtures discussed include ambient air and water, waste water, drinking water, shale oil, synthetic fuels, automobile exhaust, diesel particulate, coal fly ash, cigarette smoke condensates, and food products.

It is our hope that this volume will serve as a reference to catalyze and encourage further research in this field.

Michael D. Waters, Ph.D.
Stephen Nesnow, Ph.D.

Acknowledgment

We would like to thank Gerald Rausa, Office of Energy Minerals and Industry, for his advice, encouragement, and support of this program.

We would also like to express our appreciation to Wendy A. Martin, Peter A. Murphy, and David F. Wright of Kappa Systems, Inc., for their efforts and cooperation in coordinating the symposium and in editing the proceedings.

Contents

SECTION 2:
COLLECTION AND CHEMICAL ANALYSIS
OF ENVIRONMENTAL SAMPLES

SECTION 3:
CURRENT RESEARCH

SECTION 1

SHORT-TERM
BIOASSAY SYSTEMS—
AN OVERVIEW

THE USE OF MICROBIAL ASSAY SYSTEMS IN THE DETECTION OF ENVIRONMENTAL MUTAGENS IN COMPLEX MIXTURES

Herbert S. Rosenkranz, Elena C. McCoy,
Monica Anders, William T. Speck, and
David Bickers
Department of Microbiology
New York Medical College
Valhalla, New York

INTRODUCTION

Microbial mutagenicity procedures are widely used as prescreens for the detection of environmental agents (10,11) endowed with genetic activity. Because of the remarkably good correlation between mutagenicity in bacteria and the potential for causing cancer in animals (13,29,30,58,59,63, 76,78,91), these short-term assay procedures are also being used to identify potential environmental carcinogens. Our laboratory has participated in several studies dealing with the development, validation, and evaluation of a number of these short-term assay procedures which have included the following: the Salmonella (7), E. coli WP2 uvrA (17,42) and the multipurpose E. coli (66) mutagenicity assays, the host-mediated assay using Salmonella (51), the DNA repair-deficient E. coli (pol A$^+$/pol A$_1$$^-$) system (83,87), the Saccharomyces cerevisiae mitotic recombination assay (104), and the prophage λ inductest (67).

It is our opinion that of all the systems available to date the Salmonella mutagenicity assay procedure is the most versatile and is adaptable to a number of experimental situations that reflect environmental situations. Moreover, we have found that the reliability of the Salmonella assay can be improved when it is coupled to a second assay system which measures changes other than mutagenic events, such as modifications of the cellular DNA as determined with the pol A$^+$/pol A$_1$$^-$ E. coli system (31,32,83,84).

Certainly, with respect to the concern of the present symposium, that of the analysis of complex mixtures, the Salmonella mutagenicity assay procedure has demonstrated its usefulness in the detection of mutagenic activity present in complex mixtures: particulate air pollutants (21,92,94), biological fluids (36,52), excreta [urines (22,26,33,52,90, 103), and feces (18)], cigarette smoke condensates (48,49, 65), plant extracts, effluents from synthetic technologies (34,35,77), drinking water (53,86), pyrolysis products, and cooking smoke condensates (70).

THE SALMONELLA MUTAGENICITY ASSAY

The Salmonella mutagenicity assay developed by Ames and his associates (3,4,7) is quite simple; it involves placing the mutagen at the center of a petri dish containing a modified minimal medium that is seeded with bacteria (S. typhimurium) unable to grow because of a deficiency in their histidine biosynthetic pathway. Revertants to histidine-independence are seen as colonies in a ring around the area in which the agent has been deposited. After testing a large number of mutants, Ames selected several classes of strains with low spontaneous reversion rates and high sensitivity to various agents. Each of these types represents a different class of mutants, i.e., a) strains that detect mutagens that cause base-substitutions; they revert either by direct mutations or by suppressor mutations; b) strains capable of detecting frameshift mutations.

In addition, some of the tester strains carry the uvr B mutation which increases their susceptibility to several classes of mutagens. Furthermore, to increase their permeability to large molecules, a deficiency in cell envelope lipopolysaccharides has been introduced (rfa mutation, strains TA1535, TA1537, TA1538).

Recently (7,60), the sensitivity of the tester strains was increased further by introducing into the strains a plasmid which appears to participate in a type of error-prone repair (strains TA98, TA100).

Typical results obtained with this spot-test procedure are shown in Figure 1, which illustrates the response of the bacteria to a concentration gradient of a test agent (2-bromoethanol) which induces mutations of the base-substitution type.

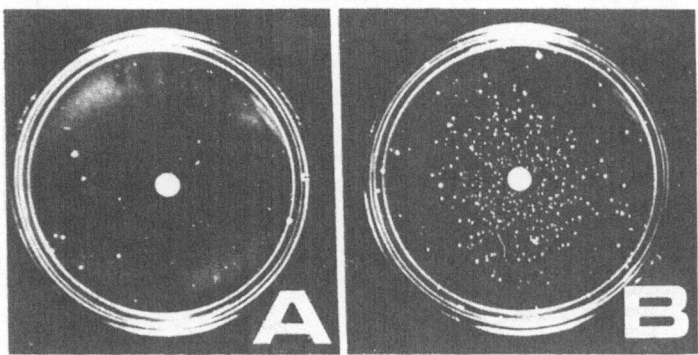

Figure 1. Mutagenicity of 2-bromoethanol for Salmonella typhimurium. Minimal plates containing a trace of histidine were inoculated with either (A) strain TA1538 or (B) strain TA1535, the indicator strains for frameshift and base substitution mutations, respectively. A paper disc impregnated with 1 µl of 2-bromoethanol was deposited on the surface of the agar plates. The plates were incubated in the dark at 37°C for 54 hours and then examined for the appearance of mutants (histidine-independent colonies). Note the presence of mutants in a zone surrounding the disc in the plate inoculated with S. typhimurium strain TA1535 (Figure 1B) but not TA1538 (Figure 1A) which indicates that this chemical induces mutations of the base-substitution but not of the frameshift type.

Ames and his associates refined their procedure further to allow better quantitation (6,7). In this procedure known amounts of the test substances are incorporated with the tester strain into the soft agar overlay and revertants to histidine prototrophy are scored after 2 days incubation at 37°C. This permits the determination of dose-response curves and allows certain conclusions based upon structure-activity relationships. Finally, the procedure permits comparisons between mutagenic and carcinogenic potencies (58,59, 63,85,93,97).

Results obtained with this incorporation procedure are shown in Figure 2, which deals with the mutagenicity of a group of halogenated propanols related to the mutagenic flame retardant tris (2,3-dibromopropyl) phosphate.

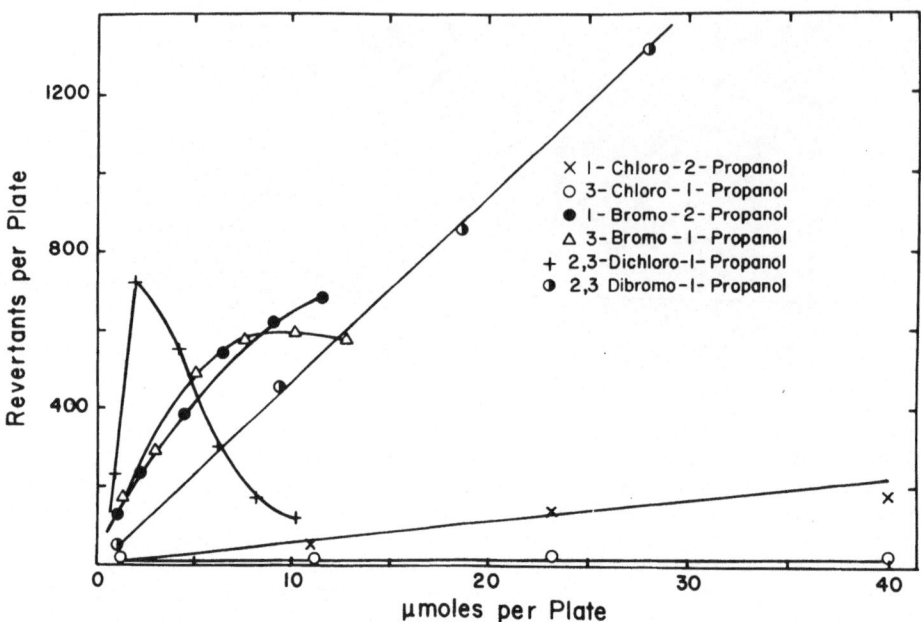

Figure 2. Effect of dose on the mutagenic response of
Salmonella typhimurium TA1535 to mono- and di-halopropanols
related to the flame retardant tris (2,3-dibromopropyl)
phosphate. In this procedure the test agent as well as the
bacteria are incorporated into the agar overlay. Note the
different shapes of the dose-response curves. Of the chemicals tested, 2,3-dichloro-1-propanol exhibited the highest
specific mutagenic activity; however, levels in excess of 2
µmoles per plate were toxic. 3-chloro-1-propanol was devoid
of measurable mutagenicity.

Metabolic Activation

Frequently agents that are carcinogenic for mammals are
not mutagenic for bacteria. This may be due to the fact that
these substances require metabolic activation by mammalian
enzymes. Such activation is beyond the metabolic capability
of microorganisms. To overcome the inability of bacteria to

metabolize procarcinogens to their ultimate active form, a
number of procedures have been devised wherein rat liver
extracts together with co-factors are incorporated into the
assay mixtures (5,37,39,54,87). These modifications have
been most useful as they have permitted the demonstration of
the mutagenicity of substances such as 2-aminofluorene, 4-
aminobiphenyl, 6-aminochrysene and others (5,6,59, Rosenkranz
and Poirier, in preparation). Recently, Ames and his associates showed that the enzyme activation of some promutagens
was greatly enhanced if the animals were pretreated with
Aroclor (7,49), a polychlorinated biphenyl which acts as
an inducer of hepatic enzymes (mixed function oxygenases).

Some Precautions and Procedural Modifications

Although the procedures for the Salmonella mutagenicity
assay have been described in great detail (7), there are a
number of precautions that should be observed to avoid arti-
factual results:

1. Phenotype Effects. Using the standard experimental
procedures, it was found on occasion (84, and unpublished
results) that a certain agent at a specific concentration
(e.g., 3-amino-1,2,4-triazole) promoted the growth of "normal
size" colonies, which, upon further testing (by replating on
minimal medium devoid of histidine and of the test agent),
were found to have retained their histidine-auxotrophic
character and would not, therefore, be classified as mutants
(i.e., revertants to histidine-independence). This effect,
which was also seen with low levels of streptomycin and 5-
fluorouracil, appears to reflect phenotypic changes resulting
from mistakes in translation or transcription (20,38,41).

In addition, it was found with some fairly bactericidal
(non-mutagenic) test agents that the mutagenicity plates contained large numbers of pinpoint colonies, which, upon further
testing, were found to be still histidine-requiring. This
phenomenon appears to reflect cross-feeding between material
(presumably histidine) released by killed cells and surviving
bacteria.

For standard testing, in order to eliminate such artifacts, we routinely test a number of colonies (a minimum of
five) from each assay plate for histidine-independence by
plating on minimal medium. (The minimal medium must contain
biotin as the tester strains are biotin-requiring) (see also
ref. 88).

2.   Volatile and Labile Test Agents. Agents that are
volatile (e.g., methyl iodide, methylene chloride) or heat-
labile (e.g., N-nitrosoethylurea) cannot be incorporated into
the molten (45°C) agar overlay (Table 1). Such substances
can, however, be detected by the spot-test procedure. As a
matter of fact, dose-response curves using discs impregnated
with graded amounts of substances with boiling points near
42°C have been obtained (e.g., haloalkanes, see ref. 16,61)
(Figure 3).

For substances with boiling points at or below 37°C, we
have devised two simple spot tests:

a)   Discs impregnated with the test agent are placed
      onto the plates which are then kept at approximately
      4°C for 2 hours, whereupon the discs are removed and
      the plates incubated in the usual manner, or

Table 1

Effect of Testing Procedure on Yield of Mutants

Substance	Amount		Strain	Revertants per Plate	
				On Disc	In Agar*
Methyl iodide	2.5	μmoles	TA1535	675	45
Methylene chloride	5	μmoles	TA1535	980	31
N-Nitrosoethylurea	250	μg	TA1535	3552	35
Methyl methanesulfonate	10	μl	TA1535	439	153
N-Acetoxy-N2-fluorenylacetamide	25	μg	TA1538	19	675
1-Phenyl-3,3-dimethyltriazene	250	μg	TA1535	23	2840

* Test agent was incorporated into the soft agar layer at 45°C.

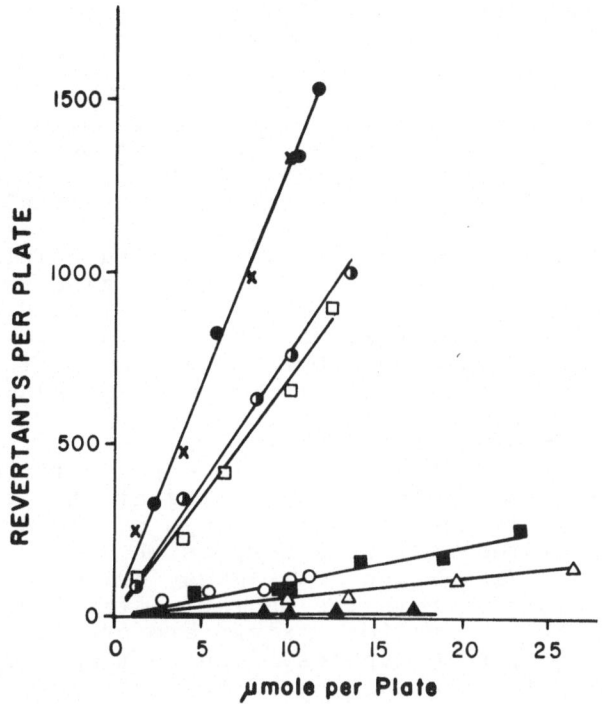

Figure 3. Effect of haloalkane concentration on mutagenicity
of S. typhimurium. In this experiment, discs containing
graded amounts of test agent were deposited on the surface of
the agar. ● = 1,2-dibromoethane; x = 1,5-dibromopentane;
◑ = 1,2-dibromo-2-methylpropane; □ = 1-bromo-2-chloroethane;
■ = 1,1,2,2-tetrachloroethane; o = 1,1-dibromoethane; △ = 1,2-
dichloroethane; ▲ = 1,1,2,2-tetrabromoethane.

   b)   The discs impregnated with the volatile agent are
         placed on the plates, which are then put into Bio
         Bags (Marion Scientific Co., Kansas City, Mo.) that
         are heat-sealed and incubated at 37°C in the usual
         manner.

In each instance specially modified incubators are used so
that volatile substances are removed and do not present a
potential hazard.

Of course, for quantitative study of the mutagenicity
of volatile substances or gases, more precise procedures are
used such as a "caisson Lwoff" with appropriate manometric
controls. This apparatus is opened in a chemical hood.

More recently, substances with boiling points in the
range of 35°—44°C have also been shown to give reproducible
results (61 and unpublished results) when tested by the pre-
incubation modification (69,102) of the standard Salmonella
assay. This modification is especially useful for direct-
acting mutagens or the mutagenic metabolites generated by
the activation mixture which are so unstable or labile that
they may react preferentially with the excess of soft agar
present in the standard assay. This appears to be especially
true of dialkylnitrosamines whose mutagenicity cannot be
readily detected in the standard assays but which exhibit
activity when tested in liquid suspension (54). To overcome
this deficiency, Yahagi and colleagues (69,102) have modified
the standard assay wherein the bacteria, the test chemical,
and the metabolic activation mixture are pre-incubated at
room temperature for 20 minutes. The soft agar is added, the
mixture poured onto the surface of the agar plates, then
incubated in the usual manner. This procedure is also useful
for volatile substances, provided the tubes are stoppered.

Results with volatile chemicals which were scored as
negative in the standard assay but which were positive by
the pre-incubation modification are illustrated in Table 2.

    3. Non-diffusible Test Agents. Just as volatile and
heat-labile substances do not give reliable results in the
quantitative plate test, substances which because of poor
solubility, size, or charge do not diffuse rapidly in agar,
will not give reliable tests in the spot-test procedure
(N-acetoxy-N2-fluorenylacetamide, 1-phenyl-3,3-dimethyltria-
zene, Table 1 and Figure 4). For these reasons, we routinely
test all chemicals by the spot-test and the quantitative
incorporation procedure. In addition, substances which are
negative by these methods are also tested by the pre-incubation modification (see above).

Table 2

Effect of Procedure on the Mutagenicity of Labile Chemicals

Compound	Amount per Plate	Revertants per Plate* Standard	Pre-Incubation
Dimethylcarbamyl chloride	0	22	29
	0.3 μg	18	24
	1.0 μg	22	27
	3.3 μg	19	25
	10 μg	23	56
	33 μg	33	138
	100 μg	57	517
	333 μg	98	644
Allyl chloride	0	14	20
	1 μl	13	42
	2 μl	19	56
	5 μl	15	82
	10 μl	15	60
Sodium azide	0.5 μg	206	200

*Strain TA1535 was used in all experiments.

4. Testing of Agents Which Form Oxygen-Sensitive Intermediates. It should also be mentioned that certain substances are metabolized by the S-9 fraction or by the bacteria to intermediates that may be very sensitive to oxygen. The mutagenicity of such substances may be demonstrated only when the incubation is carried out for a time under anaerobic conditions (e.g., azathioprine, (6-[1-methyl-4-nitroimidozol-5-yl)]thiopurine) (89), 6-nitrosopurine, 2-nitrofluorene and 2-nitronapthalene (Table 3). On the other hand, direct-acting chemicals (ethyl methanesulfonate, 1,2-epoxylbutane, etc., Table 3), which are not metabolized to oxygen-sensitive intermediates are not affected by the period of anaerobiosis.

5. Activation of Chemicals by Skin Enzymes. Although undoubtedly liver enzymes are very active in the biotransformation of cancer-causing chemicals and indeed the use of microsomal preparations derived from rat livers has greatly extended the usefulness of microbial mutagenicity assay systems, it is frequently not realized that entry of environmental agents that may be hazardous may be through the skin. It

Table 3

Effect of Anaerobiosis on Mutagenicity
for Salmonella typhimurium

Additions	Strain	µg per Plate	Revertants per Plate Aerobic	Anaerobic
Ethyl methanesulfonate	TA100	7	5000	5000
2-Bromoethanol	TA100	5.5	690	710
1,2-Epoxybutane	TA100	14	390	410
Propyleneimine	TA100	1.4	7000	7000
Azathioprine	TA100	0	109	107
		25		114
		100		239
		250	165	375
		400		553
		500	149	659
6-Nitrosopurine	TA1535	0	28	21
		10	24	33
		25	28	34
		100	28	61
		250	44	79
2-Nitrofluorene	TA1538	5	700	1220
2-Nitronaphthalene	TA1535	100	480	816

Anaerobiosis was achieved by placing the petri plates into
Gas Pak jars (BBL, Cockeysville, MD) which were incubated
(37°C) in the dark for 14 hours whereupon the plates were
removed from the jar and incubated aerobically for an
additional 34 hours.

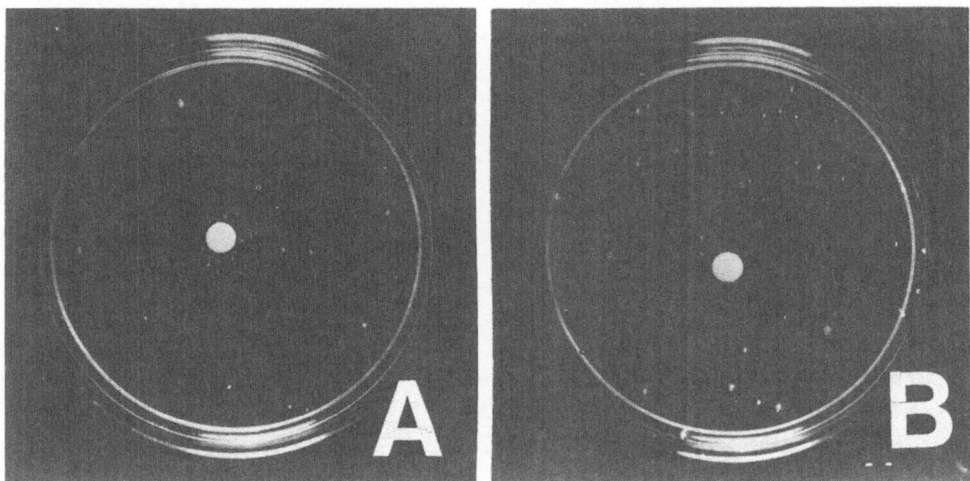

Figure 4. Mutagenicity of N-acetoxy-N2-fluorenylacetamide.
Each plate received a disc impregnated with 250 μg of the
test agent. Note the zone of colonies surrounding the disc
in the plate containing TA1538 (Figure 4A) but not in TA1535
(Figure 4B). This indicates that acetoxyl-fluorenylacetamide
causes frameshift mutations. The fact that the mutants are
close to the disc reflects the slow diffusion of the test
agent.

was, therefore, thought to be of interest to investigate the
ability of microsomal preparations derived from newborn mice
to activate such agents.

The previously widely-used flame retardant tris (2,3-
dibromopropyl) phosphate (TBPP) was shown to be mutagenic
for Salmonella typhimurium (15,75). It was found further
that the genetic activity of TBPP was greatly enhanced in
the presence of rat liver microsomes (75) especially if
these were derived from Aroclor-induced animals (Figure 5).
Recently, it was demonstrated that ingestion of TBPP by
experimental animals resulted in the induction of tumors
(1,12,71,95). Although these results are very significant,

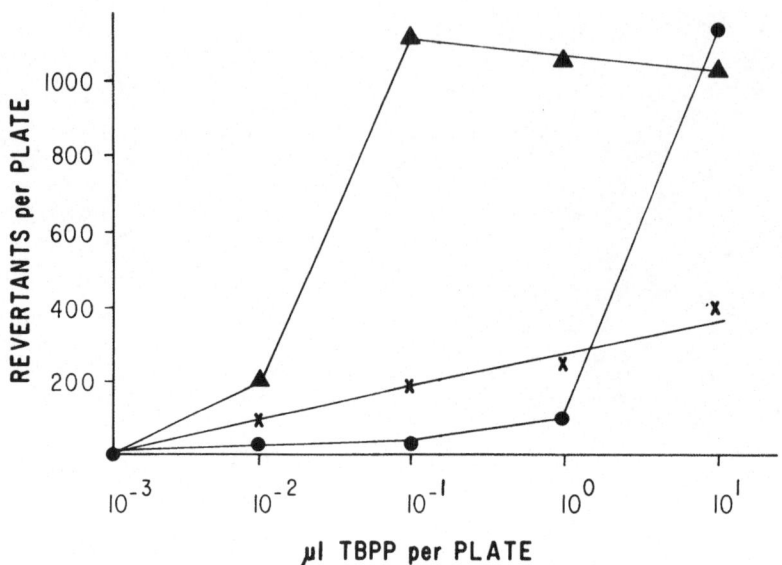

Figure 5. Mutagenicity of the flame retardant tris (2,3-dibromopropyl) phosphate for S. typhimurium TA1535. ● = No microsomes added; (x) = in the presence of rat liver microsomes and co-factors; ▲ = in the presence of microsomes derived from Aroclor 1254-induced rat liver.

it must be remembered that human exposure to TBPP is by contact with wearing apparel that may be saturated (30% of their weight) with this flame retardant. Obviously, therefore, the entry and presumably the metabolic activation of TBPP involves the skin and ingestion is probably a minor factor. We, therefore, studied the ability of skin enzymes to metabolize TBPP to mutagenic intermediates. Moreover, because it has been shown that various environmental agents play important roles in inducing drug- and carcinogen-metabolizing enzymes [e.g., smoking, insecticides, PCB (23-25)], the effect of such inducers when applied to the skin on liver and skin enzymes was investigated.

It was thus shown (Table 4) that microsomal preparations derived from the skin of newborn animals were capable of transforming TBPP to a mutagenic intermediate and that this

Table 4

Ability of Microsomal Preparations of Skin and Liver
to Activate Tris (2,3-dibromopropyl) Phosphate to a Mutagen

Source of Microsomes	Inducer	Site of Induction	Revertants per Plate					
			TA1535 TBPP				TA1538 2-Aminofluorene	
			0	0.1 µl	1 µl	10 µl	0	25 µg
None	----	----	14	28	32	283	13	42
Skin	None	----	14	41	100	1028		51
Skin	Aroclor 1254	Skin Application	15	96	141	987		38
Skin	3-MC	Skin Application	17	24	73	1009		46
Liver	None	----	15	145	231	380		194
Liver	Aroclor 1254	Skin Application	13	358	369	613		569
Liver	3-MC	Skin Application	17	211	334	754		371
Liver (Rat)	Aroclor 1254	Intra-peritoneal Injection	12	392	480	564		672

Abbreviations: TBPP, tris (2,3-dibromopropyl) phosphate; 3-MC, 3 methylcholanthrene.

actively was enhanced if the skin of the animals was previously exposed to Aroclor (PCB) or 3-methylcholanthrene.
Moreover, it was shown that the livers of these animals also
became induced following skin contact with these inducers,
as evidenced by the enhanced ability of the livers to convert
TBPP and 2-aminofluorene to mutagenic intermediates (Table 4).
These experiments also demonstrated a dichotomy by the various
enzyme preparations· in their ability to activate TBPP and 2-
aminofluorene.

These results indicate that in studies involving the
mutagenicity of environmental agents, the site of entry should
be taken into consideration and possible alternatives to the
use of liver enzyme preparations should be considered whenever
appropriate. In addition, it must also be remembered that in
investigating the role of environmental inducers of hepatic
enzymes, some of these inducers can penetrate the skin and
that animals need not receive them by intraperitoneal injection (7).

6.   The Role of the Bacterial Flora in the Activation
of Mutagens. Just as the activation of environmental agents
by mammals can occur at sites other than the liver (e.g.,
the skin), it is also conceivable that such biotransformations can occur extracorporeally as well. The possibility
of these biotransformations is self-evident as each ecological niche has its own flora. Thus, pesticides and polycyclic
hydrocarbon carcinogens can be transformed by soil bacteria
(19,40,55), petroleum products by marine microorganisms (44)
etc. When it is realized that approximately 20×10^6 tons
of manmade sludge is deposited annually in landfills and
lagoons (72) and that large quantities of polynuclear aromatic hydrocarbons are generated from the use of fossil
fuels and forest fires, then the opportunity for biotransformation by omnipresent metabolically versatile aerobic and
anaerobic bacteria cannot be overlooked. This point is
emphasized by the finding that such bacteria can indeed
generate mutagenic intermediates (Table 5).

The role of the microbial flora is not restricted to
extracorporeal niches. In humans, the lower intestinal tract
is extremely rich in varied anaerobic flora numbering as many
as 500 species. These bacteria, which outnumber the aerobes
by a factor of a thousand, are metabolically very versatile
and undoubtedly play a significant role in the etiology of

human colon cancer. Study shows that these bacteria possess unique enzymes capable of activating promutagens (Table 5) and carcinogens (62).

Similarly, the role of plants in converting pesticides to mutagens and, therefore, potential carcinogens has been demonstrated (74).

7.   The Photodynamic Activation of Environmental Agents. In the presence of light, environmental agents may undergo chemical transformations [e.g., dieldrin, aldrin, mevinphos, carbaryl (Sevin), etc. (9,27,50), photochemical urban smog (73), etc.]. Such photogenerated chemicals may either accumulate in the environment or they may be short-lived and undergo further transformation.

Additional study demonstrates that such photodynamically generated chemicals may be mutagenic for the Salmonella tester strains (45,46, Table 6 and Figure 6). The role of visible light in the generation of potential environmental mutagens must, therefore, be explored for the products resulting from each technology.

Some Limitation of the Salmonella Assay

The Salmonella mutagenicity assay system developed by Ames and his associates is probably the most widely used of the procedures available to date. In addition, due to its versatility, the system can be adapted to a number of experimental situations. Certainly its reliability for predicting potential carcinogens is remarkable (58,59). There are, however, a number--although it is small--of known carcinogens that are not detected with the Salmonella assay system but are positive in other microbial assays, e.g., p-rosaniline, auramine O, 1,2-dimethylhydrazine, procarbazine, etc. (83,84, Rosenkranz & Poirier, in preparation). Other substances, such as carcinogenic nitrofurans also are positive in several microbial systems (56,57,100) but were negative in the original Salmonella tester strains (56,57,81,96,100,101). However, the introduction of the plasmid-containing strains (TA100) permitted the detection of their mutagenicity (59,60,96,101). Finally, there are a number of substances of unknown carcinogenicity that are also negative in the standard Salmonella mutagenicity assay but are positive in other microbial assay, e.g., povidone-iodine, tetrabromoethane, and hydroxylaminosulfonic acid (16,80,83,98,99).

Table 5

Activation of 2-Aminofluorene by Cell-Free Extracts
from Anaerobic Bacteria

Source of Extract	Amount of Extract (mg protein per plate)	Condition of Incubation	Mutants per Plate	
			No 2-amino-fluorene	+ 2-amino-fluorene*
Clostridium perfringens	1.0	Anaerobic		288
Clostridium perfringens	2.0	Anaerobic	5	505
Heated C. perfringens	1.0	Anaerobic		21
Heated C. perfringens	2.0	Anaerobic	12	16
Bacteroides fragilis	1.0	Anaerobic		345
Bacteroides fragilis	2.0	Anaerobic	5	549
Heated B. fragilis	1.0	Anaerobic		14
Heated B. fragilis	2.0	Anaerobic	9	17
None	0	Anaerobic	10	18
None	0	Aerobic	4	34
B. fragilis	0.5	Anaerobic	4	533
B. fragilis	0.5	Aerobic	9	68
C. perfringens	0.8	Anaerobic	5	424
C. perfringens	0.8	Aerobic	8	56

*25 μg 2-aminofluorene.

Table 6

Mutagenicity of Methylene Blue in the Presence of Light

MB* (μg per plate)	Filter	Illumination (kJ/m² at 450nm)	Revertants per Plate TA1535	TA1538
0		0	17	9
0	–	9.8	173	
0	+	9.8	33˙	7
30		0	16	12
30	–	9.8	424	10
30	+	9.8	410	
0	–	9.8	166	
10	–	9.8	224	
20	–	9.8	345	
40	–	9.8	588	
100	–	9.8	740	

*MB: Methylene blue. MB was incorporated in the agar overlay. It should be noted that light in the absence of a sensitizer will be mutagenic. This intrinsic mutagenicity can be eliminated by the use of filters that do not allow blue (450nm) light to pass. The photo-induced mutagenicity of MB is dose-dependent.

THE LIQUID SUSPENSION ASSAY

An analysis of the potential mutagenicity of the substances listed above in the Salmonella mutagenicity assay revealed that some of these were non-mutagenic in the standard Salmonella assay because of their strongly bactericidal action which obscured their mutagenic potential. When the mutagenicity assay was carried out in liquid suspension and results expressed as mutants per survivors, the mutagenicity of some of these agents was readily demonstrable (16,43,80, 83,84,98,99).

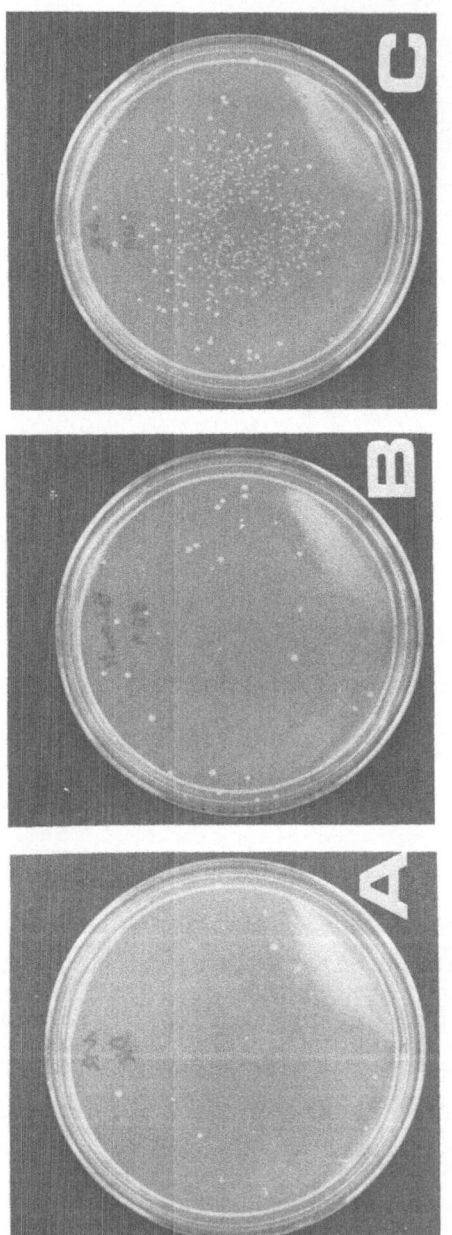

Figure 6. Demonstration of the light-induced mutagenicity of neutral red (NR) for S. typhimurium TA1535. Discs containing either water or 20 μg NR were placed on the surface of agar plates which were incubated in the dark for 14 hours at 37°C to permit diffusion of the NR. Then the discs were removed, and 0.1 ml-portions of the tester strains (10⁸ bacteria) were spread on the surface of the agar. The plates were incubated at 37°C in the dark for 2 hours and then covered with filters designed to remove blue light (450nm) so as to eliminate the "instrinsic" mutagenic activity of blue light. The plates were exposed to white light for 2 hours (4.9 kJ/m² measured at 450nm) and then incubated in the dark for 54 hours. (A) Plate containing NR but kept in dark, (B) Plate devoid of NR but illuminated for 2 hours through a filter, (C) Plate containing NR and illuminated for 2 hours through a filter.

Testing in Suspension

The effect of the test agent on bacterial survival is of great importance when testing for mutagenicity in suspension. Survival is, of course, dependent upon dose, duration of exposure, and temperature of incubation. Also, the choice of the medium in which the bacteria are treated may be of great importance, especially if the test agent requires active transport into the cells. This might not occur in buffer suspension.

When testing agents in suspension culture, the protocol outlined below is followed:

- A determination of the toxic level of the test agent in minimal essential medium is made.

- After a determination of the toxic level, a pilot experiment is carried out using concentrations at either side of 50% toxicity level (total of five concentrations). Cells are treated at 37°C in minimal essential medium with four concentrations of test agent for 60 or 90 minutes. Then dilutions of the cultures are plated on minimal medium supplemented with biotin and a trace of histidine (for determination of number of revertants to histidine-independence, i.e., number of mutants) and on complete medium (for determination of total number of viable cells). Results are expressed as mutants per survivors. Depending upon results the procedure may have to be repeated to find a more appropriate period of exposure (e.g., decrease lethality) or a more suitable concentration of test agent.

- If results obtained above are negative (i.e., lack of mutagenicity) even after changing incubation mixture, then the experiment is repeated using an activation mixture (i.e., S-9 fraction from rat liver plus co-factors). The various variables discussed above are also taken into consideration when the S-9 preparation is used.

- When appropriate conditions for determining mutagenicity have been found (see above), the procedure is repeated using a series of mutagenic dilutions (narrow range) to enable the determination of dose-response curves.

A Strategy for Mutagenicity Testing Using the Salmonella
System

Obviously, there are many variations of the Salmonella
mutagenicity assay system. However, our experience over the
past several years has indicated to us that certain rational
guidelines are possible to enable screening of a large number
of substances without the necessity of carrying out all of
the variations. We have found that the following scheme is
useful:

Step No. 1: All test agents are processed for the Salmonella
spot-test assay wherein the bacteria are incorporated into
the agar overlay and a filter disc containing 200-300 µg of
the test agent is placed on the surface of the agar. Strains
TA1535 and TA1538 are used for this assay.

Step No. 2: All test agents are processed for the E. coli
pol A$^+$/pol A$_1^-$ assay using filter discs impregnated with 200-
300 µg of test agent (see below).

Step No. 3: All test agents are processed in the standard
(quantitative Salmonella assay wherein graded quantities of
the test agent (1/2 log dilutions) and the tester strain are
incorporated into the agar overlay. Initially, the assay is
carried out with strains TA1535, TA1537, and TA1538. Chemicals known to be volatile are processed directly by the pre-
incubation procedure.

Step No. 4: In the event that Step No. 3 yields negative
results, the procedure is repeated using strains TA98 and
TA100. (Again, known volatiles are tested directly by the
pre-incubation modification.)

Step No. 5: In the event that Steps No. 3 and No. 4 do not
yield positive results, first Step No. 3 is repeated using
liver microsomal preparations derived from Aroclor-induced
rats. If the results are still negative, then Step No. 4
is repeated using the microsomal preparations. (As before,
known volatiles are processed directly by the pre-incubation
technique, using microsomal preparations.)

Step No. 6: If none of the procedures using Salmonella as
the indicator strain yields positive results, then the procedure is repeated using the pre-incubation method with and
without microsomes, even if the test agent is known not to
be volatile.

Step No. 7 (Optional): If the E. coli pol A⁺/pol A₁⁻ spot-test assay is positive but none of the Salmonella assays yields a positive result, then a confirmatory test using Salmonella in suspension can be carried out. Generally, only direct-acting bactericidal agents behave in this manner. Therefore, it may not be necessary to repeat the liquid suspension test using microsomal mixtures. In our experience, we have not yet encountered a test agent which was mutagenic for Salmonella in liquid suspension only in the presence of the microsomal activation mixture and negative in the other assays.

Step No. 8 (Optional): In the event that the E. coli pol A⁺/ pol A₁⁻ spot-test gives a "no-test" result (see below), yet one of the Salmonella assays is positive, then a confirmatory test using E. coli pol A⁺ and pol A₁⁻ in suspension (in the presence and absence of a microsomal activation mixture) can be carried out. The usual reason for a "no-test" is lack of diffusion of the test agent from the filter disc.

Step No. 9 (Optional): Depending upon the ecological niche of the test agent or the environmental mixture, other procedures for activation may be attempted, e.g., skin enzymes, bacterial extracts, anaerobiosis, photodynamic activation, etc.

A Note on the Use of Strains TA1535 and TA98

Strain TA100 is a plasmid-carrying derivative of TA1535, and is more sensitive in its response to a number of mutagens. However, strain TA100 has also lost some of its mutagenic specificity as it responds (to a small extent) to frameshift mutagens (Table 7). In our experience, all agents that respond in strain TA1535 also evoke a mutagenic response in strain TA100. However, because we are also interested in ascertaining the mutagenic specificity of the test agents, we have retained strain TA1535 in our panel of tester strains. Replacement of strain TA1535 by TA100 does not permit the unequivocal assignment of a mutagenic response as being due to a base substitution mutation.

DNA-Modifying Activity of Potential Mutagens and Carcinogens

"Normal" cells exposed to noxious agents which alter the cellular DNA will attempt to overcome this effect by excising portions of modified DNA and resynthesizing the correct

Table 7

Mutagenic Specificity of Tester Strains

		Revertants per Plate			
Substance	Amount (µg)	TA1535	TA100	TA1538	TA98
2-Nitrofluorene	0	8	83	3	45
	10	14	251	184	544
4-Hydroxylamino-	0	27	98	11	15
Quinoline-N-oxide	10	69	631	205	321
MNNG*	0	17	129	7	17
	10	1779	1256	12	38
Glycidaldehyde**	0	13	74	8	9
	10	300	589	10	37

 *MNNG: N-methyl-N'-nitro-N-nitrosoguanidine
**Done by the "pre-incubation" procedure

sequence. The enzyme DNA polymerase has been implicated in
this repair process (both in the repair replication step and
the excision step in excision-repair). It is to be expected,
therefore, that cells lacking this repair enzyme will be more
sensitive to the action of agents that react with cellular
DNA. The availability (28) of E. coli mutants (pol A_1^-) lacking this enzyme has permitted verification of this prediction.
Using normal (pol A^+) and DNA polymerase-deficient (pol A_1^-)
strains of E. coli, it was shown that pol A_1^- was much more
sensitive than pol A^+ to a large number of agents known to
alter cellular DNA. These agents included known mutagens
and carcinogens (8,68,83,87). One such example (N-hydroxyurethan) (68) is illustrated in Figure 7. This carcinogenic
(14,64) chemical, thought to be the active intermediate of
urethan, a well-known carcinogen, clearly causes a preferential killing of the pol A_1^- strain (Figure 7). It is very
interesting that N-hydroxyurethan is not mutagenic in the
standard Salmonella mutagenicity assay (unpublished results),
presumably because of its bactericidal action. Based upon
these observations we developed a simple assay procedure (87).
Bacteria (pol A^+ or pol A_1^-) are spread onto the surface of
agar plates, and discs containing the substance to be tested

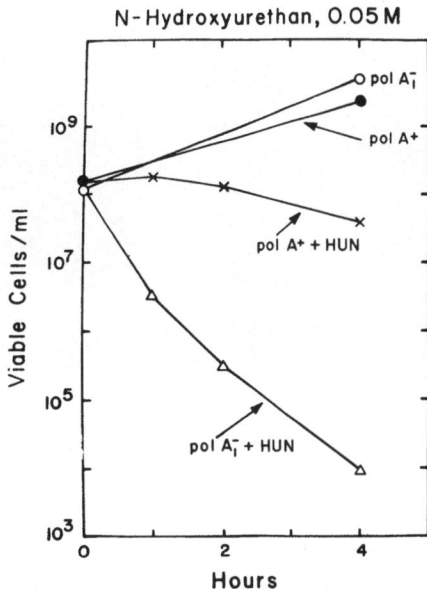

Figure 7. Preferential killing of E. coli pol A_1^- by N-hydroxyurethan. Bacteria were brought to the exponential growth phase whereupon portions of the cultures were supplemented with N-hydroxyurethan (final concentration, 0.05M). The cultures were aerated at 37°C and at intervals portions were withdrawn and processed for the determination of the number of viable cells.

are placed on the plates. After incubation (12 hours) the diameters (or areas) of the zones of inhibition are measured (Figure 8). Agents known to alter cellular DNA were found (87, Table 8) to produce larger zones of inhibition on the pol A_1^- plates than on the corresponding pol A^+ ones. Agents known not to alter the cellular DNA (e.g., cycloserine, chloramphenicol, methicillin, etc., Table 8) gave equal zones of inhibition on both sets of plates. No conclusions could be drawn for substances which inhibited neither strain (i.e., "no-test" result) since this action could be due to inertness of the substances, or inability to penetrate into the cell, or the fact that the test substances normally require metabolic activation but that this is beyond the metabolic capability of the bacterium (see above).

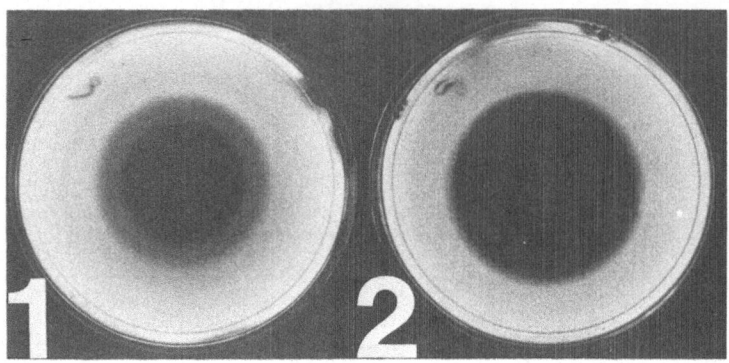

Figure 8.   Effect of methyl methanesulfonate on the growth of
E. coli pol A⁺ (1) and pol A₁⁻ (2).   Each disc contained
10 μl of the test agent.

     Using this procedure, we and others tested a number of
environmental agents, some of which gave positive results.
Some intercalating agents which were negative in the mutagenicity assay (4,47) gave positive tests in the pol A₁⁻
system [e.g., Miracil D (79) p-rosaniline, auramine O (83)].
In addition, the carcinogens 1,2-dimethylhydrazine, safrole,
procarbazine, etc., which are also negative in the Salmonella
mutagenicity assay preferentially inhibit the pol A₁⁻ strain
(see also ref. 31,32).

     It must be stated that the same limitations relating to
metabolic activation which apply to the Salmonella system
apply to the DNA polymerase-deficient E. coli system as well.
Some of these, however, just as with the Salmonella system,
can be overcome by incorporating a cell-free activation
system into the assay procedure (83,87).

     Although this assay works quite well for readily diffus-
ible substances (Table 9, group I), it does not yield interpretable results with poorly diffusible molecules (Table 9,
group II).   In order to overcome this shortcoming, a simplification of the standard liquid suspension procedure was devised (83):   growing bacteria are diluted to a density of
approximately 10,000 per ml, and 1 ml aliquots are exposed
to the test agent for predetermined periods of time whereupon 0.1 ml-portions are plated onto the surface of agar

Table 8

Effects of Various Agents on the Growth of a
DNA Polymerase-Deficient Strain

Agent	Amount		Size of Zone of Inhibition (mm)	
			Parent	DNA Polymerase-Deficient
			(pol A$^+$)	(pol A$_1^-$)
Penicillin	3	units	9	8
Erythromycin	15	µg	9	9
Cycloserine	50	µg	62	62
Chloramphenicol	30	µg	30	30
Streptomycin	10	µg	26	26
Kanamycin	30	µg	18	18
Methyl methanesulfonate	10	µl	44	60
Ethyl methanesulfonate	10	µl	0	20
N-Methyl-N-Nitrosourea	0.5	µmoles	45	85
N-Ethyl-N-Nitrosourea	0.5	µmoles	0	13
N-Methyl-N-Nitrosourethan	0.1	µmoles	2	46
N-Ethyl-N-Nitrosourethan	0.5	µmoles	0	16
Urethan	20	µmoles	0	0
N-Hydroxyurethan	20	µmoles	12	21
Nitrosofluorene	0.5	µmoles	0	15
N-Hydroxylaminofluorene	0.5	µmoles	0	12
MNNG*	250	µg	22	30
Epichlorohydrin	0.13	µmoles	8	32
Natulan	250	µg	16	22
Mitomycin C	1	µg	23	37

*MNNG: N-methyl-N'-nitro-N-Nitrosoguanidine

plates for the determination of the number of viable cells.
In this procedure results are expressed as "Survival Indices,"
(S.I., % Survivors Pol A$_1^-$/% Survivors Pol A$^+$). S.I. values
below 1.0 indicate a preferential killing of the pol A$_1^-$
strain. By these criteria methyl methanesulfonate and N-
acetoxyfluorenylacetamide are preferential inhibitors of the
pol A$_1^-$ strain (Table 10). On the other hand, streptomycin,
although it induces lethality in both indicator strains, does

Table 9

Preferential Inhibition of DNA-Polymerase-Deficient E. coli
(Standard Assay)

Group	Substance	Amount		Diameter of Zone Inhibition (mm) Pol A$^+$	Pol A$_1$$^-$
I	Methyl methanesulfonate	10	µl	43	57
	Ethyl methanesulfonate	10	µl	0	18
	N-Hydroxyurethan	20	µmoles	12	21
	N-Methyl-N-nitrosourea	0.5	µmoles	42	79
II	N-Acetoxy-N2-fluorenylacetamide	250	µg	0	0
	2-Nitronaphthalene	250	µg	0	0

not preferentially kill the pol A$_1$$^-$ strain (S.I. = 1.12).
This procedure is compatible with metabolic activation. Thus,
the procarcinogen 2-aminofluorene, which requires metabolic
activation by hepatic enzyme, does not preferentially inhibit
the pol A$_1$$^-$ strain in the absence of rat liver microsomes,
but does so in the presence of this preparation (Table 10).

A guide to the rational use of each of these assays
has been discussed earlier.

A Note Concerning the Pol A$^+$/Pol A$_1$$^-$ Spot-Test

Studies in our laboratory indicate that the composition
of the liquid and semi-solid medium is of great importance
in these assays (see, for example, 82). Moreover, the results
reflect differential inhibitions dependent upon the diffusion
of the test agent in the agar. It is, therefore, essential
that the volume, moisture content, and age of the agar plates
be controlled carefully. This is best done by pouring exact
amounts of agar (25 ml). Upon solidification of the agar,
the plates are incubated overnight at 37°C to remove water
of condensation and to eliminate contaminated plates. The
plates are then stored at 4°C in plastic bags. Only plates

Table 10

Preferential Inhibition of Pol A_1^- Cells by DNA-Modifying Agents

(Liquid Suspension)

Additions	Amount	Enzyme*	Survivors per Plate				
			Pol A^+	% Survivors	Pol A_1^-	% Survivors	Survival Index**
None	0	–	1873	100	1632	100	1.00
Methyl methane-sulfonate	0.025M	–	292	15.6	44	2.7	0.17
N-Acetoxy-N2-fluorenyl-acetamide	10 µg/ml	–	543	29.0	208	12.8	0.44
2-Amino-fluorene	10 µg/ml	–	1703	90.0	1514	92.8	1.02
2-Amino-fluorene	10 µg/ml	+	910	48.6	579	35.5	0.73
None	0	+	1927	102.9	1670	102.3	1.0
Streptomycin	10 µg/ml	–	270	14.4	263	16.1	1.12

*Enzyme: Microsomal (S-9) preparation supplemented with cofactors. Exposure period: 90 min.

**% Survivors Pol A_1^-/% Survivors Pol A^+

from a single prepared batch can be used at one time. The following controls are routinely included in each batch: chloramphenicol (30 µg): negative control; methyl methane-sulfonate (10 µl): positive control.

Care must be taken to always ascertain the pol A_1^- character of the appropriate strain in view of the fact that pol A^+ revertants appear to have a selective advantage. The medium (HA+T) used in these assays minimizes this advantage; contrariwise nutrient-rich media maximize it.

CONCLUSIONS

The present discussion deals mainly with the properties of the standard <u>Salmonella</u> mutagenicity assay, because this assay has been used most extensively thus far. However, there is still controversy on the significance of "border-line" results, such as when the total number of revertants increased by a factor of 1.5-2. This, in the case of strain TA1535 or TA1538, may mean an increase from a background of 10 to a count of 16. On the other hand, with strain TA100 the increase may be from a background of 100 to a count of 150. The interpretation of such results remains difficult and supporting evidence is best obtained using additional tester strains, performing careful dose-response experiments, and using other assay systems as well. Also, as a result of collaborative studies, unless the procedures are carefully standardized with respect to inoculum size and state of growth (middle exponential growth phase), results obtained in different laboratories may not be comparable.

It would also seem that better procedures for storing the tester strains are required. The continual use of frozen stocks may not be satisfactory. It would appear that one safeguard against spurious results involves two simple procedures:

  a)   The regular testing of the tester strains for their <u>rfa</u> (deep-rough) character, ultraviolet sensitivity, and the resistance of strains TA98 and TA100 to ampicillin (conversely the sensitivity of the other tester strains) to this antibiotic must also be determined.

b)   At regular intervals the dose-response of the tester strains to known mutagens should be evaluated: sodium azide for TA1535 and TA100, 2-nitrofluorene for TA1538 and TA98 and 9-aminoacridine for TA1537.

Undoubtedly, as these microbial mutagenicity procedures are refined further, their usefulness will increase. One of their main advantages is their adaptability to different experimental situations. Full use of the microbial mutagenicity procedures' potentials as pre-screens will not be made unless investigators adapt the procedures to their requirements.

ACKNOWLEDGMENTS

The studies carried out in our laboratory were supported by the National Cancer Institute (NO 1-CP-65855 and NO 1-CP-75949), the National Institute of Environmental Health Sciences (NO 1-ES-6-2124), and the U.S. Environmental Protection Agency (EPA-68-01-4718).

REFERENCES

1. Abelson PH: The tris controversy. Science 197:113, 1977

2. Abramson S, Lewis EB: The detection of mutation in Drosophila melanogaster. In: Chemical Mutagens, Vol. 2 (Hollander A, ed.). New York, Plenum Press, pp 461-487, 1971

3. Ames BN: The detection of chemical mutagens with enteric bacteria. In: Chemical Mutagens, Vol. 1 (Hollander A, ed.). New York, Plenum Press, pp 267-282, 1971

4. Ames BN: A bacterial system for detecting mutagens and carcinogens. In: Mutagenic Effects of Environmental Contaminants (Sutton HE and Harris MI, eds.). New York, Academic Press, pp 57-66, 1972

5. Ames BN, Durston WE, Yamasaki E, Lee FD: Carcinogens are mutagen. A simple test system combining liver homogenates for activation and bacteria for detection. Proc Natl Acad Sci US 70:2281-2285, 1973

6. Ames BN, Lee FE, Durston WF: An improved bacterial test system for the detection and classification of mutagens and carcinogens. Proc Natl Acad Sci US 70:782,786, 1973

7. Ames BN, McCann J, Yamasaki E: Methods for detecting carcinogens and mutagens with the Salmonella/mammalian-microsome mutagenicity test. Mutat Res 31:347-364, 1975

8. Anderson MD, Rosenkranz HS: Greenhouse fungicide-Environmental carcinogen? Henry Ford Hosp 22:35-40, 1974

9. Anonymous: Pesticide photolysates may be carcinogenic. Chem Eng News, Sept 27, 40, 1969

10. Anonymous: Does this chemical cause cancer in man? Lancet 2:629-630, 1971

11. Anonymous: Are 90% of cancers preventable? Lancet 1:685-687, 1977

12. Anonymous: Ban on flame retardant tris appears imminent. Chem Eng News, April 11, p 8, 1977

13. Bartsch H: Predictive value of mutagenicity tests in chemical carcinogenesis. Mutat Res 38:177-190, 1976

14. Berenblum I, Ben-Ishai D, Haran-Ghera N, Lapidot A, Simon E, Trainin N: Skin-initiating action and lung carcinogenesis by derivatives of urethan and related compounds. Biochem Pharmacol 2:168-176, 1959

15. Blum A, Ames BN: Flame retardant additives as possible cancer hazards. Science 195:17-22, 1977

16. Brem H, Stein B, Rosenkranz HS: The mutagenicity and DNA-modifying effect of haloalkanes. Cancer Res 34:2576-2579, 1974

17. Bridges BA: Simple bacterial systems for detecting mutagenic agents. Lab Practice 21:413-419, 1972

18. Bruce WR, Varghese AJ, Furrer R, Land PC: A mutagen in the feces of normal humans in "Origins of Human Cancer" (Hiatt HH, Watson JD, Winston JA, eds.). New York, Cold Spring Harbor, pp 1641-1646, 1977

19. Campacci EF, New PB, Tchan YT: Isolation of amitrole-
     degrading bacteria. Nature 266:164-165, 1977

20. Champe SP, Benzer S: Reversal of mutant phenotypes by
     5-fluorouracil: An approach to nucleotide sequences in
     messenger-RNA. Proc Natl Acad Sci US 48:532-546, 1962

21. Chrisp CE, Fisher GL, Lammert JE: Mutagenicity of
     filtrates from respirable coal fly ash. Science 199:73-
     75, 1977

22. Commoner A, Vithayathil J, Henry JI: Detection of meta-
     bolic carcinogen intermediates in urine of carcinogen-
     fed rats by means of bacterial mutagenesis. Nature
     249:850-852, 1974

23. Conney AH, Burns JJ: Metabolic interaction among
     environmental chemicals and drugs. Science 178:576-586,
     1972

24. Conney AH, Pantuck EJ, Hsiao KC, Garland WA, Anderson KE,
     Alvares AP, Kappas A: Enhanced phenacetin metabolism in
     human subjects fed charcoal broiled beef. Clin Pharmacol
     Therap 20:633-642, 1976

25. Conney AH, Pantuck EJ, Hsiao KC, Kantzman R, Alvares AP,
     Kappas A: Regulation of drug metabolism in man by
     environmental chemicals and diet. Federation Proc 36:
     1647-1652, 1977

26. Connor TH, Stoeckel M, Evrard J, Legator MS: The contri-
     bution of metronidazole and two metabolites to the muta-
     genic activity detected in urine of treated humans and
     mice. Cancer Res 37:629-633, 1977

27. Crosby DG: The nonmetabolic decomposition of pesticides.
     Ann NY Acad Sci 160:82-96, 1969

28. de Lucia P, Cairns J: Isolation of an E. coli strain
     with a mutation affecting DNA polymerase. Nature 224:
     1164-1166, 1969

29. de Serres FJ: The utility of short-term tests for muta-
     genicity. Mutat Res 38:1-2, 1976

30. de Serres FJ: Prospects for a revolution in the methods
     of toxicological evaluation. Mutat Res 38:165-176, 1976

31. de Serres FJ: Mutagenicity of chemical carcinogens.
     Mutat Res 41:43-50, 1976

32. de Serres FJ: Long-range planning for effective in vitro
     tests for carcinogenesis. In: Screening Tests in Chemi-
     cal Carcinogenesis (Montesano R, Bartsch H, Tomates L,
     eds.). Lyon, IARC Scientific Publications (No. 12),
     pp 29-37, 1976

33. Durston WE, Ames BN: A simple method for the detection
     of mutagens in urine: Studies with the carcinogen 2-
     acetyl-aminofluorene. Proc Natl Acad Sci US 71:737-741,
     1974

34. Epler JL, Hardigree AA, Young JA, Rao TK: Feasibility
     of application of mutagenicity testing to aqueous
     environmental effluents. Abstracts, 8th Ann Meeting
     Amer Environ Mutagen Soc p 47, 1977

35. Epler JL, Larimer FW, Nix CE, Ho T, Rao TK: Comparative
     mutagenesis of test materials from the synthetic fuel
     technologies. Abstract, 2nd Intern Conf on Environ
     Mutagens, p 106, 1977

36. Ficsor G, Bordas SM, Wade SM, Muthiani E, Wertz GF,
     Zimmer DM: Mammalian host- and fluid-mediated mutageni-
     city assays of captan and streptozotocin in Salmonella
     typhimurium. Mutat Res 48:1-16, 1977

37. Frantz CN, Malling HV: The quantitative microsomal
     mutagenesis assay method. Mutat Res 31:365-380, 1975

38. Garen A, Siddiqi O: Suppression of mutation in the
     alkaline phosphatase structural cistron of E. coli.
     Proc Natl Acad Sci US 48:1121-1127, 1962

39. Garner RC, Miller EC, Miller JA: Liver microsomal
     metabolism of aflatoxin B_1 to a reactive derivative
     toxic to Salmonella typhimurium TA1530. Cancer Res
     32:2058-2066, 1972

40. Gibson DT, Mahadevan V, Jerina DM, Yagi H, Yeh HJC:
     Oxidation of the carcinogen benzo[a]pyrene and benzo[a]-
     anthracene to dihydrodiols by a bacterium. Science 189:
     295-297, 1975

41. Gorini L, Kataja E: Suppression activated by strepto-
    mycin and related antibiotics in drug sensitive strains.
    Biochem Biophys Res Comm 18:656-663, 1965

42. Green MHL, Muriel WJ: Mutagen testing using trp+ rever-
    sion in Escherichia coli. Mutat Res 38:3-32, 1976

43. Green MHL, Rogers AM, Muriel WJ, Ward AC, McCalla DR:
    Use of a simplified fluctuation test to detect and
    characterize mutagenesis by nitrofurans. Mutat Res 44:
    139-143, 1977

44. Gutnick DL, Rosenberg E: Oil tankers and pollution:
    A microbiological approach. Ann Rev Microbiol 31:379-
    396, 1977

45. Gutter B, Speck WT, Rosenkranz HS: A study of the
    photoinduced mutagenicity of methylene blue. Mutat
    Res 44:177-182, 1977

46. Gutter B, Speck WT, Rosenkranz HS: Light-induced
    mutagenicity of neutral red (3-amino-7-dimethylamino-
    2-methylphenazine hydrochloride). Cancer Res 37:1112-
    1114, 1977

47. Hartman PE, Levine K, Hartman Z, Berger H: Hycanthone:
    a frameshift mutagen. Science 172:1058-1060, 1971

48. Hutton J, Hackney C: Metabolism of cigarette smoke
    condensates by human and rat homogenates to form muta-
    gens detectable by Salmonella typhimurium TA1538.
    Cancer Res 35:2461-2468, 1975

49. Kier D, Yamasaki E, Ames BH: Detection of mutagenic
    activity in cigarette smoke condensates. Proc Natl
    Acad Sci US 71:4159-4163, 1974

50. Kuhr RJ: The formation and importance of carbamate
    insecticide metabolites as terminal residues. In:
    Pesticide Terminal Residues. London, Butterworth and
    Co, pp 199-220, 1971

51. Legator JS: The host-mediated assay, a practical pro-
    cedure for evaluating potential mutagenic agents. In:
    Chemical Mutagenesis in Mammals and Man (Vogel F,
    Rohrborn G, eds.). Heidelberg, Springer-Verlag, pp
    260-270, 1970

52. Legator MS, Pullin TG, Connor TH: The isolation and detection of mutagenic substances in body fluid and tissues of animals and body fluid of human subjects. In: Handbook of Mutagenicity Test Procedures (Kilbey BJ, Legator M, Nichols W, Ramal C, eds.). Amsterdam, Elsevier Scientific Publishing Co, pp 149-159, 1977

53. Loper JC, Schoeny RS, Tardiff RG: Evaluation of organic extracts of drinking water by bacterial mutagenesis. Abstract, 2nd Intern Conf Environ Mutagens, p 35, 1977

54. Malling HV: Dimethylnitrosamine: formation of mutagenic compounds by interaction with mouse liver microsomes. Mutat Res 13:425-429, 1971

55. Matsumura F, Patil KC, Boush GM: Formation of "photodieldrin" by microorganisms. Science 170:1206-1207, 1970

56. McCalla DR, Voustinos D: On the mutagenicity of nitrofurans. Mutat Res 26:3-16, 1975

57. McCalla DR, Voustinos D, Olive PL: Mutagen screening with bacteria: niridazole and nitrofurans. Mutat Res 31:31-37, 1975

58. McCann J, Ames BN: Detection of carcinogens as mutagens in the Salmonella/microsome test assay of 300 chemicals: Discussion. Proc Natl Acad Sci US 73:950-954, 1976

59. McCann J, Choi E, Yamasaki E, Ames BN: Detection of carcinogens as mutagens in the Salmonella/microsome test: Assay of 300 chemicals. Proc Natl Acad Sci US 72:5135-5139, 1975

60. McCann J, Spingarn NE, Koburi J, Ames BN: Detection of carcinogens as mutagens. Bacterial tester strain with R-factor plasmid. Proc Natl Acad Sci US 72:979-983, 1975

61. McCoy EC, Burrows L, Rosenkranz HS: Genetic activity of allyl chloride. Mutat Res, 57:11-15, 1978

62. McCoy EC, Speck WT, Rosenkranz HS: Activation of a procarcinogen to a mutagen by cell-free extracts of anaerobic bacteria. Mutat Res 46:261-264, 1977

63. Meselson M, Russell K: Comparisons of carcinogenic and
     mutagenic potency in "Origins of Human Cancer" (Hiatt
     HH, Watson JD, Winston JA, eds.). New York, Cold Spring
     Harbor, pp 1473-1481, 1977

64. Mirvish SS: The conversion of N-hydroxyurethan to
     urethan in the mouse. Biochim Biophys Acta 93:673-674,
     1964

65. Mizusaki S, Okamoto H, Akiyama A, Fukuhara Y: Relation
     between chemical constituents of tobacco and mutagenic
     activity of cigarette smoke condensate. Mutat Res
     48:319-326, 1977

66. Mohn GR, Ellenberger J: The use of Escherichia coli
     K12/343/113 λ as a multi-purpose indicator strain in
     various mutagenicity testing procedures. In: Handbook
     of Mutagenicity Test Procedures (Kilbey BJ, Legator M,
     Nichols W, Ramel C, eds.). Amsterdam, Elsevier Scien-
     tific Publishing Co, pp 95-118, 1977

67. Moreau P, Bailone A, Devore R: Prophage λ induction in
     Escherichia coli K12 envA uvrB: A highly sensitive test
     for potential carcinogens. Proc Natl Acad Sci USA 73:
     3700-3704, 1976

68. Mullinix KP, Rosenkranz HS: Recovery from N-hydroxy-
     urethan-induced death. J Bacteriol 105:565-572, 1971

69. Nagao M, Suzuki E, Yasuo K, Yakagi T, Seino Y, Sugimura
     T, Okada M: Mutagenicity of N-butyl-N-(4-hydroxybutyl)
     nitrosamine, a bladder carcinogen, and related compounds.
     Cancer Res 37:399-407, 1977

70. Nagao M, Honda M, Seino Y, Yahagi T, Sugimura T: Muta-
     genicities of smoke condensates and the charred surface
     of fish and meat. Cancer Letters 2:221-226, 1977

71. National Cancer Institute: Summary of program, staff
     and data evaluation. Risk Assessment Subgroup Conclu-
     sions on Bioassay Reports, September 24, 1977

72. Nelson N, et al.: Human Health and the Environment –
    Some Research Needs (Report of the Second Task Force for
    Research Planning in Environmental Health Science), DHEW
    Publication No. NIH-77-1277, Washington, DC, US Govern-
    ment Printing Office, 1977

73. Pitts JN, Smith JP, Fitz DR, Grosjean D: Enhancement of
    photochemical smog by N,N'-diethylhydroxylamine in pol-
    luted ambient air. Science 197:255-257, 1977

74. Plewa MJ, Gentile JM: Mutagenicity of atrazine: A
    Maize-microbe bioassay. Mutat Res 38:287-292, 1976

75. Prival M, McCoy EC, Gutter B, Rosenkranz HS: Tris
    (2,3-dibromopropyl) phosphate: Mutagenicity of a widely
    used flame retardant. Science 195:76-78, 1977

76. Purchase IFH, Longstaff F, Ashby J, Styles JA, Anderson
    D, Lefevre DA, Westwood FR: Evaluation of six short term
    tests for detecting organic chemical carcinogens and
    recommendations for their use. Nature 264:624-627, 1976

77. Rao TK, Hardigree AA, Young JA, Epler JL: Correlation
    of mutagenic activity of energy related effluents with
    organic constituents. Abstracts, 8th Ann Meeting, Amer
    Environ Mutagen Soc, pp 47-48, 1977

78. Rosenkranz HS: Aspects of microbiology in cancer
    research. Ann Rev Microbiol 27:383-401, 1973

79. Rosenkranz HS: Miracil D: Inhibition of deoxyribonu-
    cleic acid polymerase-deficient Escherichia coli.
    Antimicrob Ag Chemother 3:530-531, 1973

80. Rosenkranz HS: Hydroxylamine-O-sulfonic acid: In vitro
    and possible in vivo reaction with DNA. Chemico-Biol
    Interactions 7:195-204, 1973

81. Rosenkranz HS: Studies on the mutagenicity of nitro-
    furans in Salmonella typhimurium. Biochem Pharmacol
    26:896-898, 1977

82. Rosenkranz HS, Carr, HS, Morgan C: Unusual growth pro-
    perties of a bacterial strain lacking DNA polymerase.
    Biochem Biophys Res Comm 44:546-549, 1971

83. Rosenkranz HS, Gutter B, Speck WT: Mutagenicity and DNA-modifying activity: A comparison of two microbial assays. Mutat Res 41:61-70, 1976

84. Rosenkranz HS, Speck WT, Gutter B: Microbial assay procedures: Experience with two systems. In: In vitro metabolic activation in mutagenesis testing (de Serres FJ, Fouts JR, Bend JR, Philpot RM, eds.). Amsterdam, North-Holland Publishing, pp 337-363, 1976

85. Rosenkranz S, Carr HS, Rosenkranz HS: 2-Haolethanols: Mutagenicity and reactivity with DNA. Mutat Res 26: 367-370, 1974

86. Simmon VF, Kauhanen K, Mortelmanns K, Tardiff R: Mutagenic activity of chemicals identified in drinking water. Abstr 2nd Intern Conf Environ Mutagens, p 36, 1977

87. Slater EE, Anderson MD, Rosenkranz HS: Rapid detection of mutagens and carcinogens. Cancer Res 31:970-973, 1971

88. Speck WT, Ellner PD, Rosenkranz HS: Mutagenicity testing with Salmonella typhimurium strains. I. Unusual phenotypes of the tester strains. Mutat Res 28:27-30, 1975

89. Speck WT, Rosenkranz HS: Mutagenicity of azathioprine. Cancer Res 36:108-109, 1976

90. Speck WT, Stein AB, Rosenkranz HS: Mutagenicity of metronidazole: Presence of several active metabolities in human urines. J Natl Cancer Inst 56:283-284, 1976

91. Sugimura T, Sato S, Nagao M, Yahagi T, Matsushima T, Seino Y, Takeuchi M, Kawachi T: Overlapping of carcinogens and mutagens. In: Fundamentals in Cancer Prevention (Magee PN, Takayama S, Sugimura T, Matsushima T, eds.). Baltimore, MD, University Park Press, pp 191-213, 1976

92. Talcott R, Wei E: Airborne mutagens bioassayed in Salmonella typhimurium. J Natl Cancer Inst 58:449-451, 1977

93. Teranishi K, Hamada K, Watanabe H: Quantitative relationship between carcinogenicity and mutagenicity of polyaromatic hydrocarbons in Salmonella typhimurium mutants. Mutat Res 31:97-102, 1975

94. Tokiwa H, Morita K, Takeyoshi H, Takahashi K, Ohnishi Y: Detection of mutagenic activity in particulate air pollutants. Mutat Res 48:237-248, 1977

95. US Congress Office of Technology Assessment: Cancer Testing Technology and Saccharin. Washington, DC, Superintendent of Documents, US Government Printing Office (Stock No. 052-003-00471-2), 1977

96. Wang CY, Maraoku, Bryan GT: Mutagenicity of nitrofurans, nitrothiophenes, nitropyrroles, nitroimidazoles, aminothiophenes, and aminothiazoles in Salmonella typhimurium. Cancer Res 35:3611-3617, 1975

97. Wislocki PG, Miller EC, Miller JA, McCoy EC, Rosenkranz HS: Carcinogenic and mutagenic activities of safrole, 1'-hydroxysafrole, and some known or possible metabolites. Cancer Res 37:1883-1891, 1977

98. Wlodkowski TJ, Rosenkranz HS: Mutagenicity of sodium hypochlorite for Salmonella typhimurium. Mutat Res 31: 39-42, 1975

99. Wlodkowski TJ, Speck WT, Rosenkranz HS: Genetic effects of Povidone-Iodine. J Pharm Sci 64:1235-1237, 1975

100. Yahagi T, Nagao M, Hara K, Mutsushima T, Sugimura T, Bryan GT: Relationship between the carcinogenic and mutagenic or DNA-modifying effects of nitrofuran derivatives, including 2-(2-furyl)-3-(5-nitro-2-furyl) acrylamide, a food additive. Cancer Res 34:2266-2273, 1974

101. Yahagi T, Matsushima T, Nagao M, Seino Y, Sugimura T, Bryan GT: Mutagenicities of nitrofuran derivatives on a bacterial tester strain with an R factor plasmid. Mutat Res 40:9-14, 1976

102. Yahagi T, Nagao M, Seino Y, Matsushima T, Sugimura T, Okada M: Mutagenicity of N-nitrosamines on Salmonella. Mutat Res 48:121-130, 1977

103. Yamasaki E, Ames BN: Concentration of mutagens from urine by absorption with the nonpolar resin XAD-2: Cigarette smokers have mutagenic urine. Proc Natl Acad Sci USA 74:3555-3559, 1977

104. Zimmermann FK: Procedures used in the induction of mitotic recombination and mutation in the yeast Saccharomyces cerevisiae. Mutat Res 31:71-86, 1975

MUTAGENESIS OF MAMMALIAN CELLS BY CHEMICAL CARCINOGENS AFTER METABOLIC ACTIVATION

Eliezer Huberman
Biology Division
Oak Ridge National Laboratory
Oak Ridge, Tennessee

Robert Langenbach
Eppley Institute for Research in Cancer
University of Nebraska Medical Center
Omaha, Nebraska

INTRODUCTION

Currently there is an increased interest in developing
short-term bioassays for carcinogens in view of the studies
which implicate a large number of environmental chemicals
in causing cancer. While the mechanism by which chemicals
induce cancer is unknown, one of the simplest explanations
is that carcinogenesis is initiated by a somatic mutation.
Indeed, chemical carcinogens are capable of binding to the
DNA of susceptible mammalian cells (1-6), and can induce
mutations at different genetic loci (25). Some of these
mutations could also involve the genes that control the expression of malignant transformation (7-10). Studies of the
mutagenic activity of carcinogens in mammalian cells should
therefore provide an important technique for detecting cancer-causing agents and possibly for elucidating the mechanism
of carcinogenesis (10-15). However, most compounds encountered in the environment are chemically nonreactive and have
to be enzymatically activated before they can manifest biological activity (16-17). Furthermore, many mammalian cell
lines which are suitable for studies on mutagenesis are not
able to metabolically activate these chemicals (18-22). To
overcome this limitation for using mutable mammalian cells in
culture, such cells have to be used with an exogenous metabolic activating system. Two in vitro approaches have been
developed to metabolize the chemicals to intermediates which
are mutagenic to mammalian cells. Intact viable cells or
tissues and enzyme preparations (microsomes) obtained by
tissue homogenization have been employed to metabolically
activate the chemicals. The recent developments in the use

of these activating systems when coupled with mutable mammalian cells for the detection of carcinogenic substances will be discussed.

CELL LINES AND GENETIC MARKERS

To date the mammalian cell lines used as the mutable cells in the systems for studying chemicals that require metabolic activation have been derived from rodents. Chinese hamster V79 cells, a rat liver epithelial cell line, and a mouse mammary carcinoma cell line have been used. Most studies have been done with Chinese hamster V79 cells and this cell line, when combined with an appropriate metabolic activation system, can be mutated by most of the known chemical carcinogens which have been studied (see below). However, other mutable cell lines may also be employed and special emphasis should be given to developing and using mutable human cells.

Two selective techniques have been used predominantly to detect mutant cells. One of these markers is the development of resistance to 6-thioguanine (TG) or 8-azaguanine (AZ) (15,23). Cells capable of growth in the presence of these agents are believed to have inactive or altered forms of the enzyme hypoxanthine guanine phosphoribosyl transferase. The second commonly used marker is the induction of resistance to the cardiac glycoside, ouabain (23,24). Ouabain is an inhibitor of plasma membrane $Na^+/K^+ATPase$. Cells capable of growth in the presence of this drug are believed to have lost the receptor site for ouabain but still maintain enzyme activity. To insure that cells initially resistant to the drug maintain their phenotype, they should be grown in the absence of the selective agent for about 30 population doublings and then retested for resistance to the drug.

An important consideration when developing a system for the detection of mutant cells is allowance of adequate time for development of resistance to the drug. This period, called "expression time," is the time required, after the DNA altering event has occurred, for cells to grow in the presence of the selective agent. The expression time is a function of the time required for fixation of the mutagenic event, and the rate of turnover of the altered enzyme or protein. The expression time varies with the selective agent. For resistance to ouabain, 48 hr postcarcinogen treatment time is sufficient to obtain an optimal number of mutants (25,26,27). The expression time for TG or AZ resistance appears to be longer and ranges from 7 to 10 days. The density at which the cells are seeded for mutant selection is also important as cross-feeding can occur and resistant colonies can be made sensitive to the drug by this mechanism.

CELL-MEDIATED MUTAGENESIS

The cell- and tissue-mediated mammalian cell mutagenesis systems are listed in Table 1. The cell-mediated mutagenesis approach was developed by Huberman and Sachs (13,25) with embryonic fibroblasts as the metabolizing cells and V79 cells as the target cells. The system was developed with both carcinogenic and noncarcinogenic polycyclic aromatic hydrocarbons. The protocol for the fibroblast-mediated assay is as follows. Chinese hamster V79 cells are seeded at 3×10^5 cells in 4 ml of medium into a 50 mm petri dish containing 2×10^6 lethally irradiated (5000 R) polycyclic hydrocarbon-metabolizing secondary golden hamster embryo cells. The hydrocarbons are added in 1 ml of medium 5 hr later. Forty-eight hr after addition of the chemical, cocultivated cells are dissociated with trypsin-EDTA, counted with a hemocytometer, and reseeded into 50 mm petri dishes at 200 cells per dish for determination of cloning efficiency and at 10^5 cells per dish for determination of mutation frequency. Ouabain is added to a final concentration of 1 mM 48 hr after the reseeding of the cells for mutant selection. The dishes are stained with Giemsa 6-8 days later to determine cloning efficiency, and stained 14-16 days later to determine the frequency of ouabain-resistant colonies. The mutation frequency for resistance to ouabain is calculated per 10^6 survivors, based on cloning efficiency and number of cells seeded for mutant selection.

The mutagenic activities of 11 hydrocarbons with different degrees of carcinogenicity were tested in the fibroblast-mediated mutagenesis system (Table 2). After cocultivation, 4 carcinogenic hydrocarbons [7,12-dimethylbenz(a)anthracene (DMBA), benzo(a)pyrene (BP), 3-methylcholanthrene (MCA), and 7-methylbenz(a)anthracene (7-MBA)] induced ouabain-resistant mutants, whereas 5 noncarcinogenic hydrocarbons [benzo(e)-pyrene, benz(a)anthracene, phenanthrene, pyrene, and chrysene] were not mutagenic. The mutagenic activities of the carcinogenic hydrocarbons were dependent on the number of metabolizing cells present and on the concentration of the carcinogen. Dibenz(a,c)anthracene and dibenz(a,h)anthracene, which have

Table 1

Cell- and Tissue-Mediated Mammalian Cell Mutagenesis System

Metabolizing Cell or Tissue	Target Cell	Chemicals	References
Hamster embryo cells	V79 cells	Polycyclic hydrocarbons	Huberman and Sachs (13,25)
BHK 21 cells	V79 cells	7-M BA and BP	Newbold et al. (30)
Rat embryo cells	V79 cells	7,8-diol BP	Kuroki and Drevon (41)
Rat hepatocytes	Liver epithelial cell line	DMBA, DMN, AAF	San and Williams (34)
Rat hepatocytes	V79 cells	Nitrosamines and aflatoxins	Langenbach et al. (31)
Rat hepatocytes and rat embryonic cells	V79 cells	AF and BP	Langenbach et al. (33)
Human alveolar macrophages	V79 cells	BP and 7,8-diol-BP	Harris et al. (26)
Human bronchus	V79 cells	BP and 7,8-diol-BP	Hsu et al. (27)

Table 2

Induction of Ouabain-Resistant Mutants
in the Fibroblast-Mediated Assay
by Different Carcinogenic Hydrocarbons*

Hydrocarbon	Concentration of Hydrocarbon (µg/ml)	Number of Ouabain-Resistant Mutants per 10^6 Survivors
Control	0	1
Benzo(e)pyrene	1	1
Phenanthrene	1	1
Pyrene	1	1
Benz(a)anthracene	1	2
Chrysene	1	2
Dibenz(a,c)anthracene	1	3
Dibenz(a,h)anthracene	1	4
7-Methylbenz(a)anthracene	1	24
3-Methylcholanthrene	1	108
Benzo(a)pyrene	1	121
7,12-Dimethylbenz(a)anthracene	0.1	66

*The data are based on results from Huberman and Sachs (25)
and from Huberman (42).

been reported to be noncarcinogenic in golden hamsters (28)
showed a weak mutagenic effect. In the presence of aminophylline, which enhanced polycyclic hydrocarbon metabolism
(29), there was a two- to fourfold increase in mutagenicity
with BP and MCA (Table 3). Dibenz(a,c)anthracene, which
had a low degree of mutagenicity without aminophylline,
showed a less than twofold increase in mutagenicity with
aminophylline (Table 3). Dibenz(a,h)anthracene, which

Table 3

Induction of Ouabain-Resistant Mutants in the
Fibroblast-Mediated Assay by Carcinogenic Polycyclic
Hydrocarbons After Treatment with or without Aminophylline*

Hydrocarbon	Number of Ouabain-Resistant Mutants per 10^6 Survivors	
	Without Aminophylline	With Aminophylline
Control	1	1
Pyrene	1	1
Phenanthrene	1	1
Dibenz(a,c)anthracene	3	5
Dibenz(a,h)anthracene	4	46
3-Methylcholanthrene	108	413
Benz(o)pyrene	121	214

*Cell were treated with 1 µg/ml of the polycyclic hydrocarbons and 0.1 mM aminophylline. Data are based on results from Huberman and Sachs (25).

exhibited mutagenic activity similar to that of dibenz(a,c)-anthracene without aminophylline, showed a tenfold increase in mutagenicity with aminophylline. These results indicate that there is a relationship between mutagenesis in the cell-mediated system and the degree of carcinogenicity in vivo of polycyclic hydrocarbons.

Newbold et al. (30) have used the cell-mediated mutagenesis system to study the correlation between carcinogenicity, mutagenicity, and the reaction of BP and 7-MBA with DNA. BHK21 cells were used as the metabolizing cells and conversions of V79 cells from ouabain and AZ sensitivity to resistance were the genetic markers. Both hydrocarbons were mutagenic in the system and the number of mutants per µmole of hydrocarbon-DNA product were of the same order of magnitude for both compounds. Furthermore, the reaction products of the hydrocarbons with the DNA of the V79 cells, after activation by the BHK cells, were indistinguishable from the products observed in vivo under conditions where tumorigenesis occurs. These findings further support the use of cell-mediated muta-genesis as a screen for carcinogenic substances and indicate that basic mechanisms of carcinogenesis can be studied with this system.

While fibroblastic cells are capable of metabolizing carcinogenic hydrocarbons to mutagenic intermediates, they are not capable of activating some other classes of chemical carcinogens, such as those which cause cancer of the liver. Therefore, to expand the spectrum of compounds which can be studied in the cell-mediated assay, we have developed a sys-tem using primary cultures of rat hepatocytes to activate the carcinogens and V79 cells to detect mutagenic intermediates (31). The basic procedure for the hepatocyte-mediated muta-genesis system was similar to the fibroblast-mediated assay. Primary hepatocytes were prepared from 6-8-week-old male Sprague-Dawley rats by enzymatic perfusion of the liver (32). The hepatocytes at about 10^7 cells per 8 ml medium were then seeded into 25-cm² T-flasks which had been seeded 18 hr earlier with 2×10^5 V79 cells. The plating efficiency of the liver cells was about 20 percent, and the maximum number of viable liver cells was attached by 3 hr after seeding. At this time the medium was changed to 8 ml of fresh medium containing the compound to be tested. Forty-eight hr after the addition of the chemical the V79 cells were reseeded for determination of cytotoxicity and the number of mutants.

The data in Table 4 demonstrate that in the presence of hepatocytes, three liver carcinogens, dimethylnitrosamine (DMN), diethylnitrosamine (DEN), and aflatoxin B_1 (AF) were metabolized to intermediates mutagenic to V79 cells. None of these compounds was mutagenic to V79 cells in the absence of hepatocytes. Methyl-tert-butylnitrosamine and aflatoxin G_2, which are noncarcinogenic analogues, were not mutagenic to the V79 cells in the presence or absence of the hepato-cytes. Thus, the compounds tested in the hepatocyte-mediated mutagenesis system, as in the case of polycyclic hydrocarbons, showed a correlation between the degree of mutagenicity in vitro and the degree of carcinogenicity in vivo.

The cell-mediated system has the potential to be used as a means of investigating the tissue- or cell-type speci-ficity of chemical carcinogens. As an initial approach to

Table 4

Induction of Ouabain-Resistant Mutants in the
Hepatocyte-Mediated Assay by Different Liver Carcinogens*

Compound	Concentration (mM)	Number of Ouabain-Resistant Mutants per 10^6 Survivors
Control	0	1
Dimethylnitrosamine	1.4	84
Diethylnitrosamine	4.5	19
Methyl-tert-butylnitrosamine	4.5	1
Aflatoxin B$_1$	3.2×10^{-3}	24
Aflatoxin G$_1$	3.2×10^{-3}	1

*The data are based on results from Langenbach et al. (31).

studying cell-type specificity in vitro we have compared the
abilities of rat embryonic fibroblasts and rat hepatocytes
to activate BP, a potent lung and skin carcinogen, and AF, a
potent liver carcinogen, to intermediates which are mutagenic
to V79 cells (33). Treatment of the V79 cells alone with BP
or AF did not change the mutation frequency from ouabain
susceptibility to resistance, nor was the mutation frequency
altered when the V79 cells were cocultivated with the metabolizing cells. AF in the hepatocyte-mediated assay caused a
forty-one-fold higher mutation frequency than the control at
a concentration of 3 µg/ml (Table 5). AF in the fibroblast-
mediated assay caused only a twofold increase in mutation
frequency. Treatment of the V79 cells with 3 µg/ml of BP in
the fibroblast-mediated assay caused a fiftyfold enhancement
in the mutation frequency (Table 5). In the hepatocyte-
mediated assay BP caused only a twofold increase in the mutation frequency. The mutagenic activity of BP in the fibroblast-mediated assay but not in the hepatocyte-mediated assay,
and the inverse situation with AF are in agreement with the
in vivo activities of these two carcinogens. These results

Table 5

Fibroblast- and Hepatocyte-Mediated Mutagenesis
of V79 Cells by the Carcinogens BP and AF

Metabolizing Cell Type	Compounds	
	BP	AF
	(Number of Ouabain Resistant Mutants per 10^6 Survivors)	
None	1	1
Fibroblast	50	2
Hepatocyte	2	41

The concentration of BP and AF was 3 µg/ml. The data are
based on the results from Langenbach et al. (33).

indicate that a cell-type specificity in chemical carcinogenesis can be investigated by the cell-mediated assay (33).

San and Williams (34) have developed a hepatocyte-
mediated mutagenesis system with primary cultures of hepatocytes for metabolic activation and a rat liver epithelial
cell line as the target cell. Conversion of the epithelial
cell line from AZ sensitivity to AZ resistance was used as
the genetic marker. The carcinogens DMBA, DMN, and AAF were
active in this hepatocyte-mediated system and caused 2.3-,
2.4-, and 1.5-fold enhancement, respectively, in the mutation frequency. This enhancement in mutation frequency with
DMN as the mutagen is lower than when V79 cells are used as
the target cell (compare to data in Table 4). At present
the reason for this difference in response of the target
cells is unknown.

The use of human cells and tissues for carcinogen
activation in mammalian cell mutagenesis systems has been
described by Harris and colleagues (26,27). Both human
pulmonary alveolar macrophages and human bronchus tissue
metabolized BP and its proximate metabolite, the 7,8-diol,
into intermediates which converted V79 cells from ouabain
sensitivity to ouabain resistance. Metabolities of BP were
identified by high pressure liquid chromatography and the

amount of BP binding to bronchial DNA was also measured in
these studies. The mutation frequency was directly related
to the amount of BP bound to bronchial DNA and to the concentration of the hydrocarbon in the medium. This approach is
important because it demonstrates that human cells and tissues can be used in the cell-mediated systems and furthermore
it forms a basis for relating mutagenesis studies with human
cells to the data from in vivo and in vitro rodent cells.
Thus a possible way of extrapolating risk assessment of
environmental chemicals for humans is suggested.

TISSUE HOMOGENATE-MEDIATED MUTAGENESIS

    The tissue homogenate-mediated mammalian cell mutagene-
sis systems are listed in Table 6. Enzyme preparations from
rodent liver tissue have been used for activating chemical
carcinogens to intermediates which mutate mammalian cells.
The enzyme preparations are supplemented with the necessary
cofactors, including NADPH or an NADPH-generating system, and
the incubations carried out with cells in suspension or in
monolayer culture. In some studies the enzyme activities in
the liver homogenate are induced by prior treatment of the
animal. The methodologies for determining cytotoxicity and
mutagenic activity with the tissue homogenate-mediated system are similar to those described above for the cell-mediated
assay.

    Umeda and Saito (34) developed a system for the micro-
some-mediated mutagenesis of FM₃A cells, a C3H mouse mammary
carcinoma cell line, using the carcinogenic nitrosamine DMN.
Microsomes were prepared from mouse livers, and mutations
were detected by the conversion from AZ sensitivity to AZ
resistance. After treatment of the cells in suspension, they
were seeded into agar containing AZ. The carcinogen DMN at
doses of 10 to 100mM caused a sixfold enhancement of the
mutation frequency over background when the microsomes were
prepared from DDD or C3H mouse livers. However, microsomes
prepared from AKR mouse liver appeared less active in converting DMN to mutagenic intermediates. These findings suggest strain differences in the ability of liver microsomes
to activate DMN to mutagenic intermediates. Abbondandolo et
al. (35) have also used mouse liver microsomes to activate
DMN to intermediates which were mutagenic to mammalian cells.
Chinese hamster V79 cells were the target cells and mutagenic
activity was measured by the induction of TG resistant cells.
The 100,000 xg liver enzyme preparation from C3H mice caused

Table 6

Tissue-Homogenate Mediated Mammalian Cell Mutagenesis Systems

Source of Tissue Homogenate	Target Cell	Compound	References
Mouse Liver	FM₃A cells	DMN	Umeda and Saito (34)
Rat Liver	V79 cells	Polycyclic hydrocarbons and aflatoxins	Krahn and Heidelberger (37)
Rat Liver	V79 cells	Nitrosamines	Kuroki et al. (36), Kuroki and Drevon (41)
Mouse Liver	V79 cells	DMN	Abbondondolo et al. (35)

a 110- and 260-fold enhancement in mutation frequency at DMN concentrations of 200 and 500 mM, respectively. A correlation between the N-demethylase activity and the mutagenic activity of the microsomal preparations was observed.

A detailed study of microsome-mediated mutagenesis of mammalian cells with nitrosamines has been reported by Kuroki et al. (36). Microsomes were prepared from control and induced Sprague Dawley rat livers. After incubating the microsomes and nitrosamines with the V79 cells in monolayer culture, mutations were determined by resistance to AZ. Ten carcinogenic and 2 noncarcinogenic nitrosamines were assayed. Of the 10 carcinogenic compounds investigated, only one, N-nitrosomethylphenylamine, was not mutagenic. DMN was the most potent mutagen tested in the system, giving a fifty-five-fold enhancement in the mutatation frequency with microsomes from induced animals. Pretreatment of the rats with phenobarbitone increased approximately twofold the DMN (10-50 mM) induced mutation frequency while MCA pretreatment enhanced the mutation frequency only at higher (50 mM) DMN concentrations. However, aminoacetonitrile pretreatment of the animals reduced the mutation frequency with DMN.

Krahn and Heidelberger (37) developed a rat liver homogenate-mediated mutagenesis system and studied the mutagenic
activity of 2 classes of chemical carcinogens, aflatoxins
and polycyclic aromatic hydrocarbons. Mutagenic activity
was determined by incubating the 9000 xg supernatant and
chemicals with the V79 cells growing in monolayer culture.
Resistance to TG was the genetic marker. AF, DMBA, BP, and
MCA were potent mutagens in the liver homogenate-mediated
system. The mutagenic activity of the aflatoxins and hydrocarbons in these studies paralleled their in vivo carcinogenic activity. The exceptions were the two isomers of
dibenzanthracene which showed an inverse relationship between mutagenic and carcinogenic activity. In general, liver
homogenate-mediated mammalian cell mutagenesis assays are in
agreement with the results obtained with the Ames Salmonella
assay. However, as in the bacterial assay the doses of some
chemicals (DMN for example) are extremely high. Furthermore,
a relationship between the degree of carcinogenesis and mutagenesis cannot be determined with a significant number of
chemicals.

CONCLUSIONS

Further development of mammalian cell mutagenesis systems will provide a valuable method for the detection of
agents which are hazardous to humans. Currently 2 in vitro
mechanisms for the metabolic activation of the chemicals are
being employed: cell- or tissue-mediated activation and
tissue homogenate-mediated activation. Nitrosamines, aflatoxins, and polycyclic aromatic hydrocarbons are classes of
compounds which have been investigated for mutagenic activity
with the 2 activation systems. However, differences between
the 2 types of metabolic activating systems exist. BP was
mutagenic to V79 cells in the liver homogenate-mediated assay
[data of Krahn and Heidelberger (37)] but was not mutagenic
to V79 cells in the hepatocyte-mediated assay (Table 5). As
BP is generally not considered to be a liver carcinogen, the
absence of mutagenic activity with hepatocytes or liver homogenates would be in agreement with the in vivo data. In addition subcellular preparations differ from intact cells in
the profile of metabolites (38,39) and DNA adducts (30,40)
formed after metabolism of carcinogens such as AF, BP, and
DMBA. These results suggest that the use of microsomal
preparations may not truly simulate the in vivo situation.
The cell-mediated system may also be more sensitive than the

microsome-mediated mutagenesis system. In the hepatocyte-mediated assay DMN at 1.4 mM produced as great an enhancement of mutation frequency (Table 4) as 50 mM DMN in the microsome-mediated system [data of Kuroki et al. (36)]. Further studies comparing carcinogen metabolism and activation to mutagenic intermediates by the 2 activation systems are needed to establish the relative merits of each approach.

As indicated above, because the normal balance of metabolism of some environmental chemicals can be altered in cell homogenates, such studies with the cell-mediated assay may be more relevant to the in vivo situation. Studies comparing the cell type specificity of hepatocytes and fibroblasts with the carcinogens AF and BP have already been accomplished (33). In addition to determining mutagenesis in the cell-mediated assay, the metabolic products and the amount of binding of activated intermediates to cellular DNA can be determined simultaneously as has been done with human lung tissue (26,27) and with rodent hepatocytes (33). The accumulation of such data will aid in understanding the causes of carcinogenic activity (or lack of activity) in a given tissue and/or species.

The need of required proximity of the metabolizing cells or enzymes to the target cells in the mutagenesis system has been investigated by Kuroki and Drevon (41). While it is believed that the reactive intermediates of most carcinogens are electrophiles, the stability of these intermediates and thus the distance they can transverse will vary. With DMN and the 7,8-diol of BP as promutagens, separating the cellular or microsomal activating system from the V79 cells by approximately 1 mm prevented mutagenesis. Thus it was concluded that direct or proximate contact between the target cells and the activating system is required for mutagenesis.

The use of human cells or tissues in the cell-mediated approach as conducted by Harris and his collaborators (26,27) demonstrates the potential of making the system relevant for the detection of agents hazardous to humans. Although rodent cells have been used as the target cells in the studies to date, the development of mutable human cell lines will allow the assay to be performed entirely with tissues or cells of human origin.

REFERENCES

1. Miller FC, Miller JA: Mechanisms of chemical carcino-
    genesis: Nature of proximate carcinogens and interac-
    tions with macromolecules. Pharmacol Rev 18, 805-838,
    1966

2. Brookes P, Lawley PD: Evidence for the binding of
    polynuclear aromatic hydrocarbons to the nucleic acids
    of mouse skin: Relation between carcinogenic power of
    hydrocarbons and their binding to deoxyribonucleic
    acid. Nature 202, 781-784, 1964

3. Kuroki T, Heidelberger C: The binding of polycyclic
    aromatic hydrocarbons to the DNA, RNA and proteins of
    transformable cells in culture. Cancer Res 31, 2168-
    2176, 1971

4. Essigmann JM, Croy RG, Nadzan AM, Busby WF, Reinhold VN,
    Buchi G, Wogan GN: Structural identification of the
    major DNA adduct formed by aflatoxin B₁ in $vitro$. Proc
    Natl Acad Sci USA 74, 1870-1874, 1977

5. Jeffrey AM, Weinstein IB, Jenette KW, Grzeskowiak K,
    Nakahishi K, Harvey RG, Authrup M, Harris C: Structures
    of benzo(a)pyrene-nucleic acid adducts formed in human
    and bovine bronchial explants. Nature 269, 348-350,
    1977

6. Huberman E, Sachs L: DNA binding and its relationship
    to carcinogenesis by different polycyclic hydrocarbons.
    Int J Cancer 19, 122-127, 1977

7. Todaro GJ, Huebner RJ: N.A.S. Symposium: New evidence
    as the basis for increased efforts in cancer research.
    Proc Natl Acad Sci USA 69, 1009-1015, 1972

8. Yamamoto T, Rabinowitz Z, Sachs L: Identification of
    the chromosomes that control malignancy. Nature New
    Biol 243, 247-250, 1973

9. Temin HM: On the origin of the genes for neoplasia:
    G.H.A. Clowes memorial lecture. Cancer Res 34, 2835-
    2841, 1974

10. Huberman E, Mager R, Sachs L: Mutagenesis and trans-
    formation of normal cells by chemical carcinogens.
    Nature 264, 360-361, 1976

11. Kao FT, Puck TT: Genetics of somatic mammalian cells
    XII: Mutagenesis by carcinogenic nitroso compounds.
    J Cell Physiol 78, 139-144, 1971

12. Huberman E, Aspiras L, Heidelberger C, Grover PL, Sims
    P: Mutagenicity to mammalian cells of epoxides and
    other derivatives of polycyclic hydrocarbons. Proc
    Natl Acad Sci USA 68, 3195-3199, 1971

13. Huberman E, Sachs L: Cell-mediated mutagenesis with
    chemical carcinogens. Int J Cancer 13, 326-333, 1974

14. Thilly WG, DeLuca JG, Hoppe M, Penman BW: Mutation of
    human lymphoblasts by methynitrosourea. Chem Biol
    Interact 15, 33-50, 1976

15. Chu EHY, Malling HV: Mammalian cell genetics. II.
    Chemical induction of specific locus mutations in
    Chinese hamster cells in vitro. Proc Natl Acad Sci
    USA 61, 1306-1312, 1968

16. Miller JA: Carcinogenesis by chemicals: An overview.
    G.H.A. Clowes Memorial Lecture. Cancer Res 30, 559-576,
    1970

17. Heidelberger C: Chemical carcinogenesis. Ann Rev
    Biochem 44, 78-121, 1970

18. Corbett TH, Heidelberger C, Dove WF: Determination of
    the mutagenic activity to bacteriophage T-4 of carcino-
    genic and noncarcinogenic compounds. Mol Pharmacol 6,
    667-679, 1970

19. Miller EC, Miller JA: The mutagenicity of chemical
    carcinogens: correlations, problems, and interpreta-
    tions. In: Chemical Mutagens: Principles and Methods
    for Their Detection. Vol 1. New York, Plenum Press,
    1971, pp 83-119

20. Gelboin HV, Huberman E, Sachs L: Enzymatic hydroxyla-
    tion of benzo(a)pyrene and its relationship to cytotox-
    icity. Proc Natl Acad Sci USA 64, 1188-1194, 1969

21. Huberman E, Selkirk JK, Heidelberger C: Metabolism
     of polycyclic aromatic hydrocarbons in cell cultures.
     Cancer Res 31, 2161-2167, 1971

22. Huberman E, Sachs L: Metabolism of the carcinogenic
     hydrocarbon benzo(a)pyrene in human fibroblast and epi-
     thelial cells. Int J Cancer 11, 412-418, 1973

23. Arlett CF, Turnbull C, Harcourt SA, Lehmann AR, Collela
     CM: A comparison of the 8-azaguanine and ouabain-
     resistance systems for the selection of induced mutant
     Chinese hamster cells, Mutat Res 33, 261-278, 1975

24. Baker RM, Brunette DM, Mankovitz R, Thompson LH, Whit-
     more GF, Siminovitch L, Till JE: Ouabain-resistant
     mutants of mouse and hamster cells in culture. Cell 1,
     9-21, 1974

25. Huberman E, Sachs L: Mutability of different genetic
     loci in mammalian cells by metabolically activated
     carcinogenic polycyclic hydrocarbons. Proc Natl Acad
     Sci USA 731, 188-192, 1976

26. Harris CC, Hsu IC, Stoner GD, Trump BF, Selkirk JK:
     Human pulmonary alveolar macrophages metabolize benzo-
     (a)pyrene to proximate and ultimate mutagens. Nature
     272, 633-634, 1978

27. Hsu IC, Stoner GD, Atrup H, Trump BF, Selkirk JK,
     Harris CC: Human bronchus-mediated mutagenesis of
     mammalian cells by carcinogenic polynuclear aromatic
     hydrocarbons. Proc Natl Acad Sci USA 75, 2003-2007,
     1978

28. Survey of compounds which have been tested for carcino-
     genic activity, Public Health Service, National Insti-
     tutes of Health, 1972-1973

29. Huberman E, Yamaski H, Sachs L: Independent regulation
     of two types of aryl hydrocarbon [benzo(a)pyrene] hy-
     droxylase in mammalian cells. Int J Cancer 18, 1976

30. Newbold RF, Wigley CB, Thompson MH, Brookes P: Cell-
     mediated mutagenesis in cultured Chinese hamster cells
     by carcinogenic hydrocarbons. Nature and extent of the
     associated hydrocarbon-DNA reaction. Mutat Res 43, 101-
     116, 1977

31. Langenbach R, Freed HJ, Huberman E: Liver cell-mediated mutagenesis of mammalian cells with liver carcinogens. Proc Natl Acad Sci USA 75, 2864-2867, 1978

32. Williams GM, Bermudez E, Scaramuzzino D: Rat Hepatocyte Primary Cell Cultures. III. Improved Dissociation and Attachment Techniques and the Enhancement of Survival by Culture Medium. In Vitro 13, 809-817, 1977

33. Langenbach R, Freed H, Raveh D, Huberman E: Cell specificity in metabolic activation of the carcinogens aflatoxin B₁ and benzo(a)pyrene to mutagens for mammalian cells. Nature, in press

34. Umeda M, Saito M: Mutagenicity of dimethylnitrosamine to mammalian cells as determined by the use of mouse liver microsomes. Mutat Res 30, 249-254, 1975

35. Abbondanolo A, Bonatti S, Corti G, Fiorio R, Loprieno N, Mazzaccaro A: Induction of 6-Thioguanine-resistant mutants in V79 Chinese hamster cells by mouse-liver microsome-activated dimethylnitrosamine. Mutat Res 46, 365-373, 1977

36. Kuroki T, Drevon C, Montesano R: Microsome-mediated mutagenesis in V79 Chinese hamster cells by various nitrosamines. Cancer Res 37, 1044-1050, 1977

37. Krahn DF, Heidelberger C: Liver homogenate-mediated mutagenesis in Chinese hamster V79 cells by polycyclic aromatic hydrocarbons and aflatoxins. Mutat Res 46, 27-44, 1977

38. Selkirk JK: Benzo(a)pyrene carcinogenesis - A biochemical selection mechanism. J Toxicol Environ Health 2, 1245-1258, 1977

39. Decad GM, Hirsch DPH, Byard JL: Maintenance of cytochrome P-450 and metabolism of aflatoxin B₁ in primary hepatocyte cultures. Biochem Biophys Res Commun 78, 279-287, 1977

40. Bigger CAH, Tomaszewski JE, Dipple A: Differences between products of binding of 7,12-dimethylbenz(a)-anthracene to DNA in mouse skin and in a rat liver microsomal system. Biochem Biophys Res Commun 80, 229-235, 1978

41. Kuroki T, Drevon C: Direct of proximate contact between cells and metabolic activation systems is required for mutagenesis. Nature 271, 368-379, 1978

42. Huberman E: Viral antigen induction and mutability of different genetic loci by metabolically activated carcinogenic polycyclic hydrocarbons in culture mammalian cells. In: The Origins of Human Cancer, Vol. 4, CSH Conferences on Cell Proliferation (Hiatt HH, Watsons JD, Winstein JA, eds.). Cold Spring Harbor, New York, Cold Spring Harbor Laboratory Publications, 1977, pp 1521-1535

ONCOGENIC TRANSFORMATION OF MAMMALIAN CELLS BY CHEMICALS AND VIRAL-CHEMICAL INTERACTIONS

Bruce C. Casto
BioLabs, Inc.
Northbrook, Illinois

Short-term tests for the identification of potential chemical carcinogens are urgently needed. Presently, the systems receiving the most attention for prediction of carcinogenic activity are the mutagenesis assays in microbial cells. Eventually, a battery of in vitro tests should be available that are reliable, rapid, inexpensive and generate a low percentage of false positives and no false negatives. In vitro mammalian cell systems, using fibroblast-like cells from hamster, rat, mouse, guinea pig, and human subjects have been used for assays of chemical carcinogens (Table 1). These assays show that the above cell types can be reproducibly transformed by various carcinogens and, especially with hamster cells, the capacity to transform shown to correlate with the in vivo activity of known negative or positive chemicals.

TYPES OF ASSAYS USING MAMMALIAN CELLS

In vitro transformation assays are performed using four basic procedures: mass culture, colony assays, focus formation, and assays in soft agar. With the mass culture technique, cells at a relatively high density are treated continuously for several days or repeatedly at selected intervals. Following treatment, the cells are routinely passaged and observed for alterations in morphology and patterns of growth that are not apparent in similarly passaged untreated control cells (Figure 1). The cultural differences between treated

Table 1

Fibroblast-like Cell Cultures for <u>in</u> <u>vitro</u>
Assays of Chemical Carcinogens

<u>Cell Source or Type</u>	<u>Assay Method</u>	<u>References</u>
Syrian hamster:		
embryo	Mass culture	3,23,33,34,52,54
	Colony	24,62
	Focus	13
BHK21	Soft agar	21
	Colony	76,77
Chinese hamster:		
CH/L	Soft agar	5
	Colony	76,77
CH/O	Colony	76,77
Mouse:		
C3H prostate	Focus	17
C3H embryo (10T½)	Focus	69
Balb/3T3	Colony	27
	Focus	27,49
NIH Swiss embryo	Mass culture	70
	Focus	71
Rat:		
embryo	Mass culture-focus	32,34,36
	Mass culture	39,60,64,73
Guinea pig:		
embryo	Mass culture	29
Human:		
Tumor –		
osteosarcoma	Mass culture	72
neurofibrosarcoma	Mass culture	46
Normal –		
skin biopsy	Mass culture	48
newborn foreskin	Mass culture	58

and control cells have been described as:

- Loss of density-dependent inhibition of replication.

- Conversion of an organized, parallel growth pattern to one showing random orientation.

- Increased glycolysis.

- An increasing ability to grow at reduced levels of serum.

The major disadvantage of the mass culture procedure is the lengthy time interval between treatment and recognition of transformed cells, the necessity for continuous passage of treated and control cultures, and the lack of precise quantitation.

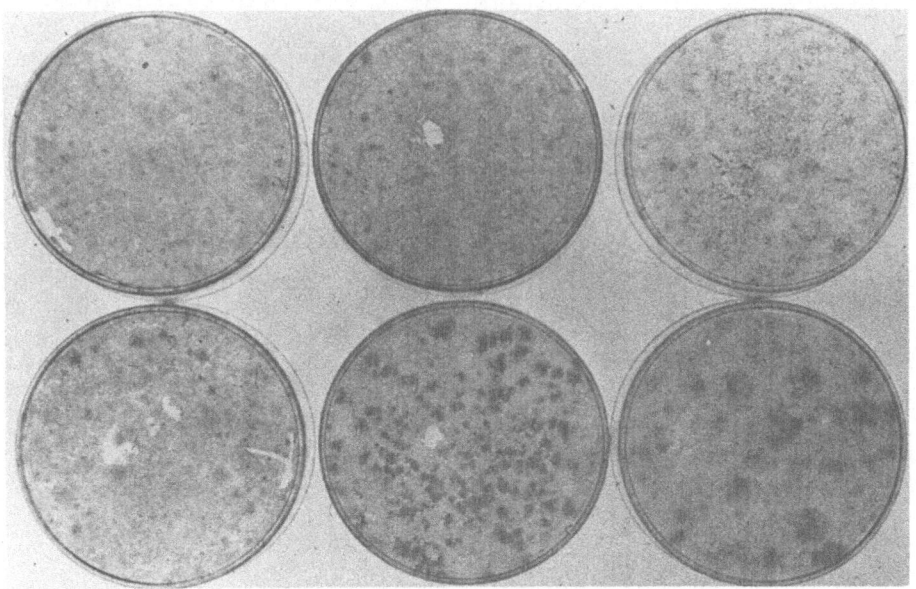

Figure 1. Mass culture assay for chemical carcinogens. Cells were exposed in utero, established in culture, and passaged weekly at a 1:10 split ratio. Top, cells from solvent-treated fetuses at passages 2, 4, and 9 (left to right). Bottom, cells from β-propiolactone treated fetuses at the above passages. x32

For the colony assay, cells are plated for cloning at 100-500 cells per dish and dilutions of test chemical added 24 hr later; the chemical may be removed after 24 hr or remain in the medium for the duration of the experiment. Alternatively, the cells may be treated while at high density and subsequently plated as above immediately following chemical treatment. In many laboratories, the cells for treatment are seeded onto a sparse lawn of x-irradiated cells which provide a "feeder layer" for the assay cells. Colonies of cells are fixed and stained after 8-10 days' incubation and individually examined under a stereomicroscope for properties associated with neoplastic transformation, especially the presence of dense colonies with criss-crossing fibroblast-like cells in the interior and the periphery of the colony (Figure 2). Under proper conditions of medium, pH, temperature, and other factors (serum, cell type, plastic, contamination, etc.), control cells form colonies with parallel arrays of cells, little or no "piling-up," and relatively even margins (Figure 3). The advantages of the colony assay are:

- Survival and transformation assays are done on the same cell populations yielding highly quantitative data.

- The time required between treatment and recognition can be less than 10 days.

- The transformants are easily visualized due to the lack of large numbers of untransformed background cells.

Some of the problems associated with the colony assay are:

- The necessity for the uniform appearance of all untreated colonies.

- The strict cultural conditions that are required to clone some cell strains.

- The technical expertise needed to recognize malignantly transformed colonies from dense areas of untransformed, treated cells.

A third type of assay, developed with continuous mouse cell lines and recently applied to secondary cultures of hamster embryo, involves the development of foci of transformed cells against a background of normal cells (Figure 4). In the focus assay system, cells are plated at densities

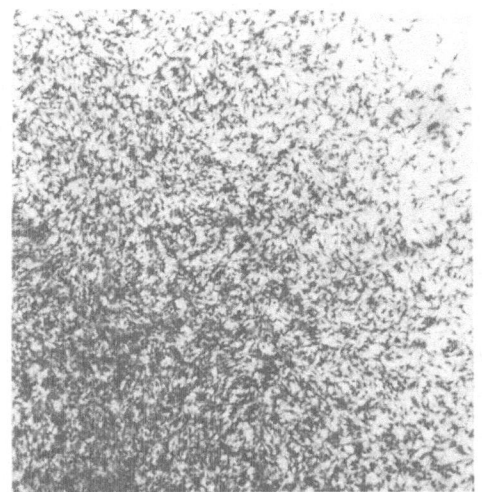 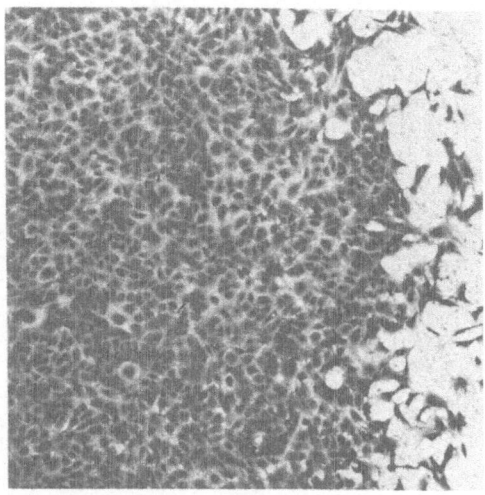

Figure 2. Transformed BALB/
3T3 colony. Cells were
treated for 24 hr with 500
µg/ml of ethyl-methanesul-
fonate, incubated for 10
days, fixed and stained.

Figure 3. Normal colony of
BALB/3T3 cells. Cells were
treated for 24 hr with 0.5%
acetone in medium, incubated
for 10 days, fixed and
stained. x32

ranging from 1,000 (mouse cell lines) to 50,000 (secondary
hamster embryo cells) and chemicals added 24 hr later.
Exposure may be for 6 days or for only 24 hr, after which
the cultures are maintained for periods of 3-4 weeks (hamster) or 6 weeks (mouse). Transformed cell foci appear as
darkly-staining, dense areas of cells that overlay and
invade into the surrounding cell sheet. They are comprised
of piled-up, criss-crossing fibroblast cells (Figure 5), but
may consist of a dense center of nonpolar cells with random-
oriented fibroblastic-like cells at the periphery (Figure 6).
The transformed appearance of the various foci is verified
by examining the cultures with a stereomicroscope. The
advantages of the focus assay in contrast to the colony assay
are:

- Less time is required for examination of the culture dishes since only those dishes showing grossly
  visible, dense areas of cells need be examined.

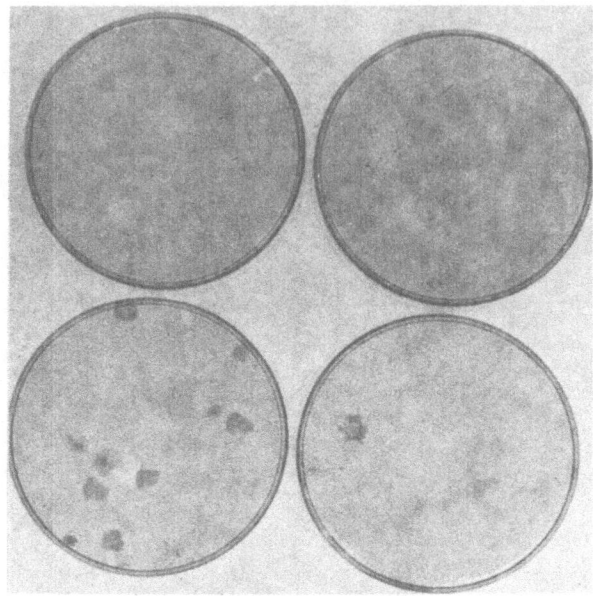

Figure 4. Focus assay for chemical carcinogens. C3H10T½
cells were exposed to 10 µg/ml of 3-methylcholanthrene for
24 hr, incubated for 6 weeks, fixed and stained. Top, solvent treated controls. Bottom, MCA treated cells.

- The morphology of the cells used in the focus assay
  need not be as uniform as those used in the colony
  assay.

- The more stringent cultural conditions necessary
  for colony assays may not be required.

The major disadvantage of the focus assay is the time required for optimum development of the transformed foci (more
than 25 days) necessitating twice-weekly feedings over this
period.

A fourth type of assay makes use of permanent cell
lines of Chinese or Syrian hamsters. In this assay, cells
are treated with chemicals and examined for an increased
ability to clone in soft agar. Chinese hamster lung cells
(CH/L) when treated with certain polycyclic hydrocarbons
also became more tumorigenic for the hamster cheek pouch (5).

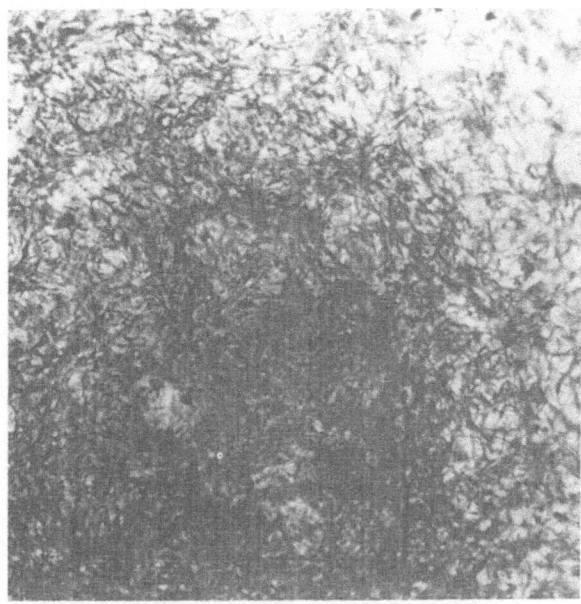

Figure 5. Focus of transformed hamster embryo fibroblasts
demonstrating the random pattern of growth typical for this
assay. Cells were exposed for 6 days to 4 µg/ml of chloro-
dimethyl ether, incubated for an additional 19 days, fixed
and Giemsa stained. x32

BHK21 clone 13 cells treated with selected chemical carcinogens may exhibit a normal morphology at 32°C, but a transformed morphology and capacity to clone in agar at 38.5°C
(21). The agar assay has certain advantages in that a standardized cell line is used, cells may be exposed while in
suspension for brief periods, and the soft agar plating
selects those cells that are malignantly transformed. However, the interpretation of transformation in this assay is
controversial, since the assay cells already have many properties of transformed cells including an increased life span,
aneuploidy, an increased rate of spontaneous transformation,
and are tumorigenic at higher cell doses. The cells also
require relatively strict growth conditions to maintain some
normal cell properties and retard spontaneous transformation.

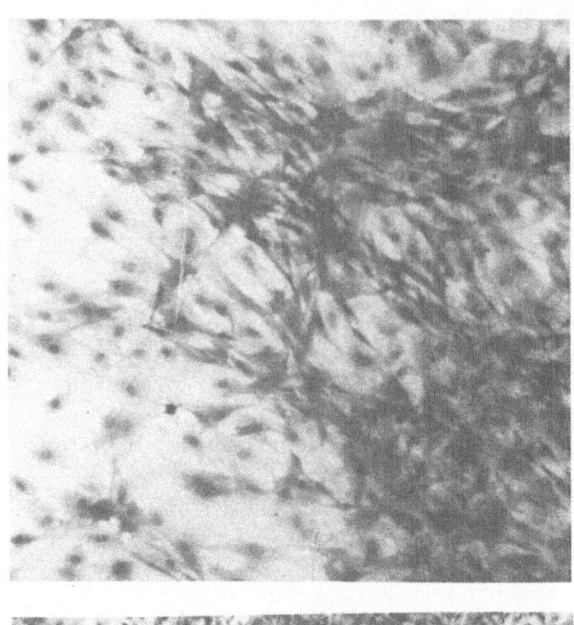
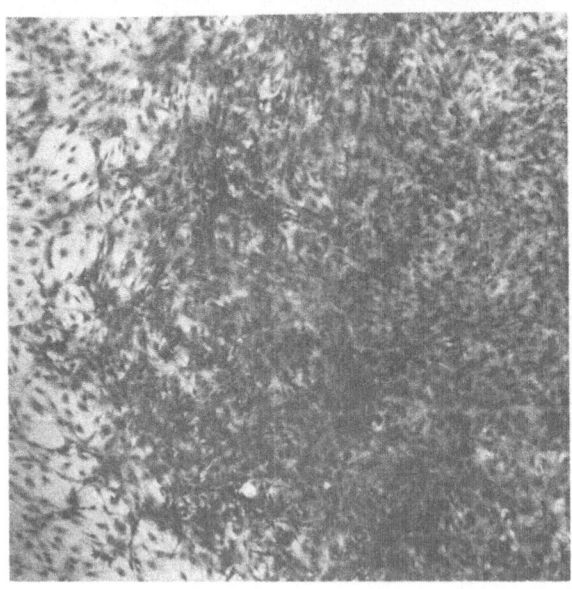

Figure 6. Focus of transformed C3H10T½ cells demonstrating the appearance of cells at the periphery of the focus. Cells were treated for 24 hr with 10 µg/ml of 3-methylcholanthrene, incubated 6 weeks, fixed and Giemsa stained. x32 (left) x125 (right)

TRANSFORMATION SYSTEMS WITH FIBROBLAST-LIKE CELLS

Syrian hamster embryo cells offer many advantages not found in other assay systems for chemical carcinogens. The cells are prepared fresh each time or used from frozen stock prepared from primary cultures and therefore retain the capacity to metabolize chemicals of nearly all classes into ultimate carcinogens. The cells remain diploid and have been well characterized by chromosomal techniques. Transformation usually occurs rapidly, and tests for tumorigenicity can be performed in newborn or weanling hamsters without immunosuppression. Hamster embryo cells underwent transformation using mass culture techniques after treatment with 4NQO (51, 52,54) and polycyclic hydrocarbons (3,23). Application of quantitative cloning techniques (65) by Berwald and Sachs (4) to hamster embryo cells treated with chemical carcinogens resulted in morphologic alteration of a small proportion of the colonies, whereas noncarcinogens were ineffective. Following a series of experiments by DiPaolo et al. (24,25) and others (17,69), it was shown that:

- Chemical transformation does not occur by selection.

- The transformation frequency is directly related to chemical concentration.

- Toxicity and transformation are separate events.

- Transformation is not dependent upon the participation of endogenous oncornaviruses (44,68).

Colony assays have also been applied to continuous lines of hamster cells by Sanders and Burford (76,77), who showed that N-nitrosomethylurea treatment of BHK21/13, CH/O or CH/L cells changed their growth pattern from a normal, parallel orientation of cells in thin layers to colonies of multilayered cells growing with random orientation.

Recently, a focus assay has been described for hamster embryo cells (13) modified from those described by Rhim et al. (71), and Reznikoff et al. (69) with mouse cells. In a series of experiments, the following was shown:

- A number of carcinogens known to cause transformation in the colony assay were also positive in the focus assay.

- Increasing the exposure time from 1 day to 6 days with 3-methylcholanthrene (MCA) increased the transformation frequency at a high dose (2 μg/ml) and resulted in the demonstration of transformation at a lower dose (0.125 μg/ml).

- The number of foci increased relative to the number of cells plated when 5×10^3 to 5×10^4 cells were exposed to MCA.

- The early transformed cells, although differing from control cells with respect to growth patterns, behaved like normal cells when subjected to treatment with certain polysaccharides, low serum concentrations, or soft agar (13).

Focus assays in cell lines from C3H mouse prostate and embryo have been employed by Heidelberger and associates (17, 69) for assay of carcinogens. Three types of foci have been described in the C3H10T½ systems: Type I, composed of normal-appearing cells, tightly packed, but with little tendency to "pile-up"; Type II, a dense, multilayered focus with little criss-crossing of cells; and Type III, a densely stained focus composed of fibroblast-like cells with much criss-crossing especially at the periphery (see Figure 6). Type I foci are not malignantly transformed; however, Types II and III grow as tumor after inoculation into irradiated C3H mice (69).

A third mouse cell system has been used by DiPaolo et al. (27) and Kakunaga (49,40) using both the colony and focus assay methods. Transformation was obtained with the poly-cyclic hydrocarbons, MCA, benzo(a)pyrene (BP), 7,12-dimethyl-benz(a)anthracene (DMBA) and with N-methyl-N'-nitro-N-nitro-soguanidine (MNNG), aflatoxin B_1 (AFB_1) or N-acetoxy-2-fluorenylacetamide (Ac-AAF). The noncarcinogens pyrene and anthracene were ineffective as well as N-nitrosodiethylamine (DENA) which is metabolized poorly if at all by fibroblast cells in vitro. Transformed colonies of two types were observed; one type consisted of dense layers of fibroblast-like cells with extensive criss-crossing of the cells in the interior and the periphery of the colony, a second type was composed of multilayers of non-polar cells. The control colonies retained a flat, epithelial-like appearance with no piling-up of cells (27). Kakunaga (50) has transformed BALB/3T3 cells with 4-nitroquinoline-1-oxide (4NQO) and has shown that at least two cell divisions were required to fix the early interaction leading to the transformation event.

Assays in rat embryo cell cultures have been performed
primarily by mass culture techniques. Gutman et al. (40)
exposed 29th passage monolayers of Wistar rat embryo cells
4 times to N-OH-2-AAF at 3-5 day intervals and observed the
development of foci composed of multilayered criss-crossing
cells. Inoculation of Wistar or Sprague-Dawley rats with
cells from treated cultures gave rise to sarcomas within 6-8
weeks. Olinici and DiPaolo (60) transformed primary or
secondary cultures of Sprague-Dawley rat embryo cells with
DMBA in mass culture with subsequent verification of transformation by the presence of morphologically-altered colonies
at the 15th passage when dishes were seeded with 100 cells.
Rat embryo cells have been shown by others (34,35,64,70,73)
to be relatively resistant to chemical transformation. The
sensitivity to various chemical carcinogens has been increased by infection of passaged rat embryo cells with strains
of mouse leukemia virus. The uninfected cell lines remain
diploid and refractory to transformation by DENA, MCA, DMBA,
BP, extracts of city smog, or cigarette smoke condensates
(33,34,35,36) whereas cell lines infected with Rauscher or
CF-1 murine leukemia show evidence of morphologic transformation within 4-15 subcultures after treatment.

Transformation of guinea pig embryo cells in mass culture by chemical carcinogens has been demonstrated by Evans
and DiPaolo (29). The system differs markedly from those
described earlier for hamster, mouse, and rat in that morphological transformation may not be evident until 4 months
after treatment (>20 subcultures) and often precedes malignant
transformation by several months. The guinea pig system is
not complicated by spontaneous transformation, and untreated
cultures continue to passage, providing control cultures for
comparison with treated cultures. However, the long interval
between treatment and transformation may make the guinea pig
system unsuitable for carcinogen screening assays, but perhaps
useful in following the progression of events leading to neoplastic transformation.

The most obvious assay system for monitoring human carcinogens would be one employing human cells. Early attempts
to transform human cells with chemical carcinogens were not
productive; however, since 1975 there have been several reports of transformation of human cells by chemical carcinogens. Igel et al. (46) reported that of 75 cell strains
derived from tumors or persons with genetic defects, two were
transformed by urethane. Both were derived from neurofibro-
sarcomas but had characteristics of normal cells in culture.
The urethane-treated cultures showed foci of morphologically

altered cells within 3-7 subdivisions after treatment, but
none of the remaining cell strains when treated individually
with two or more chemical carcinogens (MCA, BP, DENA, 4NQO,
DMBA, β-propiolactone, benz(a)anthracene) were shown to
transform. Alternatively, an osteosarcoma cell line used by
Rhim et al. (72) had some attributes of transformed cells
(high saturation density, growth in soft agar, aneuploid)
but was not tumorigenic. Treatment with DMBA caused morphologic and cultural changes in the cells 52-57 days after
treatment, and cells selected from such treated cultures were
now tumorigenic. Normal human cells have been transformed
using mass culture assay systems by Kakunaga (48) and Milo
and DiPaolo (58). Fibroblast cells from a lip biopsy were
transformed after treatment with 4NQO or MNNG. Foci of
altered cells appeared in treated cultures after passage,
and cells selected from these cultures induced tumors in NIH
nude mice following injection of cells. Milo and DiPaolo (58)
have reported the successful transformation of low passage
newborn foreskin fibroblasts by MNNG, 4NQO, Ac-AAF, AFB$_1$, and
propane sultone. Cultural and morphological differences
between control and treated cells were evident after 5-15
population doublings, the lifespan increased from 35 to more
than 60-90 passages, and the cell density increased 4- to
6-fold; treated cells grew at 41°C, formed colonies in soft
agar, and grew in nude mice.

TRANSFORMATION SYSTEMS WITH EPITHELIAL CELLS

A majority of human cancers and, historically, many of
those induced in experimental animals, are of epithelial
origin; therefore, cells derived from epiderm are a logical
choice for use in vitro assays. Most of the epithelial cell
systems are in the developmental stage (Table 2) and have
not been tested with as many classes of carcinogens as those
using hamster, rat, or mouse cells. However, the use of such
systems should be encouraged since the development of malignancy and the metabolism of certain carcinogens may be considerably different in epithelial cells in contrast to
fibroblast-like cells. It has been shown, for example, that
neoplastic transformation of epithelial cells may occur without the attendant morphologic changes found with fibroblast
cells (59,85,86). In the absence of overt morphologic alterations, liver cells treated with chemical carcinogens formed
tumors in isologous hosts and/or cloned in soft agar. Weinstein et al. (83) have found that in the absence of cultural
markers for transformation, growth in soft agar was the single
most reliable criterion for malignant transformation of epithelial cells. The same conclusion was also advanced by Evans

Table 2

Epithelial-like Cell Cultures for in vitro
Assays of Chemical Carcinogens

Cell Source or Type	Assay Method	References
Mouse:		
epidermis	Mass culture	20,28,38,57
salivary gland	Explant	31
prostate	Explant	18
Rat:		
liver	Mass culture	59,83,84,86
submandibular gland	Mass culture	8

and DiPaolo (29) using the guinea pig embryo transformation
system. In addition to growth in soft agar, Montesano et al.
(59) have examined 15 cytologic parameters and production of
plasminogen activator to predict the tumorigenic potential
of transformed liver cells. Two cytologic critera, increased
nuclear/cytoplasmic ratio and cytoplasmic basophilia, were
shown reliable with 94% of the cultures examined, growth in
soft agar 100%, but fibrin lysis failed to show differences
between tumorigenic or nontumorigenic cell cultures derived
from liver.

Epithelial cells from newborn mouse skin have been
treated with DMBA (28,38) and shown to undergo changes in
morphology, acquire an accelerated growth rate, and form
tumors when injected into appropriate hosts. Fusenig et al.
(38) have reported that the transformed cells lose surface
antigens, continue to grow indefinitely, and the tumors from
implanted cells are carcinomas. Colburn et al. (20) have
demonstrated morphologic and cultural transformation of
mouse skin cells with MNNG (rapid growth, increased life
span, loss of keratinization). Miller et al. (57) observed
similar effects in newborn mouse epidermal cells treated with
MCA, MCA-11,12-epoxide in addition to MNNG. Untreated skin
cells were shown to divide and keratinize when first placed
in culture, but the cells phased out after approximately 3
months in vitro. Brown (8) has treated mixed populations
of cells obtained from adult rat submandibular gland with
MCA and observed "piling-up" of epithelial as well as fibroblast cells 11-14 weeks after treatment. Implantation of
the epithelial-like and fibroblast-like transformants yielded

carcinomas and sarcomas respectively. DMBA-transformed
explants of mouse salivary gland produce carinomas when
injected approximately 200 days after treatment (31).

HOST-MEDIATED IN VIVO--IN VITRO SYSTEMS

Strict in vitro screening assays in mammalian cells for
carcinogens may miss a few potent agents due to the failure
of some cell cultures to metabolize the compound into the
ultimate carcinogen. The problem is more apparent with
continuous cell lines in contrast to freshly isolated cells,
but several well-known carcinogens such as DENA, DMNA, urethane or N-2-AAF are metabolized poorly or not at all by
hamster or guinea pig fibroblasts. As a result, transformation assays have been developed that include a period for
metabolism of the test compound in vivo prior to establishing
the cells in culture (Table 3). One such modification has
been reported by DiPaolo et al. and others (26,66,75), who
administered the test chemical to pregnant hamsters, removed
the fetuses 48-72 hr later, and seeded the trypsinized fetal
cells at high cell density. Areas of transformed cells were
observed after 2-8 subcultures (1:10 division of cultures).
Transformed colonies were also observed when passaged cells
from treated fetuses were plated at low cell density (500
cells/dish). A second in vivo--in vitro assay has been described using weanling rats. In these studies (6,7,41), rats
5-6 weeks of age were injected IP with DMNA, and the kidneys
were removed and placed in culture 2 hr to 7 days later. The
cells from kidneys of rats treated with DMNA continued to
passage in vitro, contained areas of morphologically transformed cells, had an increased plating efficiency, and
formed colonies in soft gels. Kidney cells from control
rats could rarely be passaged after the 5th week in culture
(4-5 subpassages).

A third in vivo--in vitro system has been used primarily
to study the early stages in respiratory tract neoplasia (39,
53,56,78). In these studies, pellets of carcinogen in beeswax were placed in the lumen of a tracheal transplant. After
continuous exposure for 2 weeks, explant cultures were established in vitro. Explants from treated tracheas demonstrated
a rapid outgrowth of cells and, in some cases, a multilayered
squamous epithelium was established. In vivo--in vitro methods for studies in liver (84) and bladder (2,18,45) carcinogenesis have been described. With the liver system, after 3
weeks of AAF treatment, cells in culture from excised livers
did not form colonies in agar gels or demonstrate an increased

Table 3

In Vivo--In Vitro Studies with
Chemical Carcinogens

Cell Source of Type	Route of Exposure	Assay Method	References
Syrian hamster embryo	Transplacental	Colony	26,66
		Mass culture	26,66,75
Rat:			
kidney	Intraperitoneal	Mass culture	6,7,41
trachea	Carcinogen pellet in tracheal transplant	Explant	39,53,55, 56
bladder	Intravesicular	Explant	67
liver	Oral	Mass culture	84
Mouse:			
bladder	Oral	Mass culture	2
	Intraperitoneal	Explant	45

survival, although they showed considerable pleomorphism (84).
Mouse bladder treated in vivo with methylazoxymethanol acetate,
when placed in explant culture, had a decreased time for initiation and an increased level of outgrowth in contrast to
bladder explants from untreated mice (45).

PROPERTIES OF CHEMICALLY-TRANSFORMED CELLS

The recognition of neoplastic transformed cells in vitro
necessitates the availability of reliable predictive tests for
assessing potential malignancy in vivo. Several morphologic,
biochemical, and behavioral alterations occur coincidental
with, or subsequent to, chemical transformation of most cells
(Table 4). However, no single criterion can distinguish malignantly transformed cells from control or nontumorigenic,
chemically-treated cells. The one in vitro parameter that
correlates most nearly with the capacity to form tumors is
the ability to replicate in soft agar (29,59,83). However,
in several systems definitive morphologic alterations may
precede this ability by several culture passages (29,43).

Table 4

Properties of Fibroblast-like Cells Transformed
In Vitro by Chemical Carcinogens

Increased saturation density
Morphological alterations
Non-oriented growth patterns
Release from density-dependent inhibition
Production of tumor angiogenesis factor
Loss of surface proteins
Agglutination by concanavalin A or wheat germ lipase
Increase in plasminogen activator
Resistance to certain polysaccharides (heparin, dextran)
Sensitivity to peritoneal exudate cells and lysates
Reduced serum requirement for growth
Loss of anchorage dependence
Capacity to grow in soft agar
Tumor formation in susceptible hosts

High saturation densities need not be directly associated
with neoplastic transformation, since many nontumorigenic
cell lines grow to high cell densities and environmental factors such as pH (16) may influence the growth rate of cells
and consequently the final population density. Density-
dependent inhibition of growth may not be an essential feature
of normal cells, as inhibition of growth or movement can be
released in these cells by alterations in the serum content
of media (1) or by the action of proteolytic enzymes (79).
Responses to concanavalin A or wheat germ agglutinin occur in
transformed cells, but normal cells also respond during cell
division (30) or after enzymatic treatment (47). Hamster
cells transformed by MCA, but not normal cells, were resistant
to the cytotoxicity of concanavalin A in cloning medium; however, this resistance did not correlate with tumorigenicity
(43). Other than growth in soft agar, the capacity of hamster cells to form tumors was closely correlated to their
ability to clone in 1% serum or in the presence of dextran
sulfate or heparin (43).

ENHANCEMENT OF VIRAL TRANSFORMATION AS AN ASSAY FOR
CARCINOGENS OR MUTAGENS

A slightly different system for detecting the mutagenic
or carcinogenic potential of chemicals has been developed by
Casto and co-workers (10,11,12,14,15). This system is based
on the in vivo observations of Rous and Kidd (74) and the in
vitro studies of Stoker (81), Pollack and Todaro (63), Todaro
and Green (82), and Coggin (19). All of the above investigators showed an increase in viral-induced oncogenesis following
pretreatment of cells with chemical or physical carcinogens
(see Casto and DiPaolo for review, ref. 12). Casto (10,11)
and Casto et al. (14,15) have applied these findings to a
large number of chemical carcinogens and mutagens using a
Syrian hamster embryo cell (HEC)--simian adenovirus transformation system. With this technique, HEC after 3-4 days
in culture are treated with chemical (2 or 18 hr), inoculated
with virus, and the cells transferred at 200,000 and 700
cells/dish for transformation and survival assays respectively.
Approximately 160 chemicals have been tested using this system with a 94% correlation between the capacity to enhance
and the known carcinogenic or mutagenic potential of the
chemical. The carcinogenic polycyclic hydrocarbons MCA,
B(a)P, DMBA, DB(a,h)A and DB(a,c)A increase the viral transformation frequency from 1.9 to 22.9-fold depending upon
dose (Table 5). DMBA and B(a)P were found to enhance transformation at concentrations as low as 0.004 and 0.16 μg/ml
respectively, but the noncarcinogenic polycyclic hydrocarbons
pyrene, phenanthrene, and perylene were ineffective (14).
Diverse carcinogens such as aflatoxin B_1, 3,3'-dichlorobenzi-
dene, ethylmethanesulfonate, 4,4-methylenebis-(o-chloroaniline), 4-nitrobiphenyl, β-naphthylamine, propane sultone and
thioacetamide (Tables 6, 7) caused an enhancement of SA7
transformation in addition to several others reported previously (15). The viral enhancement system responds equally
well to the inorganic metal carcinogens and mutagens. Treatment of HEC overnight with the salts of antimony, arsenic,
beryllium, cadmium, chromium, cobalt, copper, iron, lead,
manganese, nickel, platinum, and vanadium increased the frequency of viral transformation (Table 8). Negative results
were obtained with aluminum, barium, calcium, lithium, mag-
nesium, potassium, sodium, strontium, and titanium.

The increase in viral transformation frequency is not
due to selection of transformation-sensitive cells since
absolute increases in viral-transformed foci are observed.
The foci appearing on these dishes are induced only by virus

Table 5

Enhancement of Viral Transformation
by Carcinogenic Polycyclic Hydrocarbons

Chemical[1]	Initial µg/ml	Chemical Dilution				Solvent Control
		1:1	1:2	1:4	1:8	
B(a)P	2	5.0[2]	13.6	3.0	1.9	1.1
MCA	2	8.9	5.8	2.2	0.8	0.9
DB(a,h)A	5	3.7	2.9	2.2	2.1	1.2
DB(a,c)A	10	6.3	4.6	4.1	4.2	0.9
DMBA	0.05	22.9	6.6	2.7	1.9	1.1

[1]B(a)P, benzo(a)pyrene; MCA, 3-methylcholanthrene; DB(a,h)A,
dibenz(a,h)anthracene; DB(a,c)A, dibenz(a,c)anthracene; DMBA,
7,12-dimethylbenz(a)anthracene. Hamster embryo cells were
treated for 18 hr with the various chemicals, inoculated with
virus (SA7), and transferred to new dishes at 200,000 and 700
cells for transformation and survival assays respectively.
Survival assays were fixed and stained after 8 days and transformation assays after 21 days.

[2]Numbers in the table indicate the increase in viral transformation frequency over control values as a result of chemical
treatment.

Table 6

Enhancement of Viral Transformation by 18 hr
Pretreatment of HEC with 3,3'dichlorobenzidine

Dose (μg/ml)	Surviving[1] Fraction	SA7[2] Foci	Enhancement[3] Ratio
200	0.64	2	0.05
100	0.93	14	0.17
50	1.10	242	2.53
25	1.13	271	2.77
12	0.97	204	2.44
A[4]	1.49	101	0.80
C[5]	1.00	86	1.00

[1]Surviving fraction was determined from plates receiving 700 treated or control cells. The number of surviving colonies from treated cells was divided by the number from control cells.

[2]Total SA7 foci from 3.5×10^6 inoculated, treated, or control cells.

[3]Enhancement ratio was calculated by dividing the transformation frequency of treated cells by that obtained in control cells. Transformation frequency for each dilution of chemical was determined by dividing the number of SA7 foci by the surviving fraction.

[4]A = acetone control (0.5%).

[5]C = medium control.

Table 7

Enhancement of Viral Transformation
by Diverse Classes of Carcinogens

Chemical[1]	Initial µg/ml	Chemical Dilution				Solvent Control
		1:1	1:2	1:4	1:8	
AFB	1	13.9[2]	6.7	4.1	4.2	0.9
EMS	200	9.8	2.4	1.9	1.6	0.7
MOCA	20	2.6	2.4	2.1	1.9	0.7
4-NBP	500	2.0	2.5	4.1	1.7	1.4
β-NA	500	3.1	2.6	2.1	2.2	1.2
PS	25	6.0	2.4	2.9	2.4	1.2
TA	500	3.0	3.4	1.7	1.0	1.4

[1]AFB_1, aflatoxin B_1; EMS, ethyl methanesulfonate; MOCA, 4,4'-methylenebis-(o-chloroaniline); 4NBP, 4-nitrobiphenyl; β-NA, β-naphthylamine; PS, propane sultone; TA, thioacetamide. Chemicals were added to hamster embryo cells for 2 hr or 18 hr prior to adding SA7 virus. Cells were transferred to new dishes at 200,000 cells (transformation assays) or 700 cells (survival assays). Survival assays were fixed and stained after 8 days and transformation assays after 21 days.

[2]Numbers in the table indicate the increase in viral transformation frequency over control values as a result of chemical treatment.

Table 8

Enhancement of Viral Transformation by 18 hr
Pretreatment of HEC with Potassium Chromate and Lead Oxide

Dose (μg/ml)	Surviving[1] Fraction	SA7[2] Foci	Enhancement[3] Ratio
K_2CrO_4:			
5.0	0.20	58	9.5
2.5	0.81	75	3.0
1.2	0.89	66	2.4
0.6	1.18	56	1.5
0	1.00	31	1.0
PbO:			
50	1.03	98	5.3
25	1.47	93	3.5
12	1.28	88	3.7
6	1.47	46	1.7
0	1.00	18	1.0

[1]Surviving fraction was determined from plates receiving 700 treated or control cells. The number of surviving colonies from treated cells was divided by the number from control cells.

[2]Total SA7 foci in 10^6 inoculated, treated, or control cells.

[3]Enhancement ratio was calculated by dividing the transformation of treated cells by that obtained in control cells. Transformation frequency for each dilution of chemical was determined by dividing the number of SA7 foci by the surviving fraction.

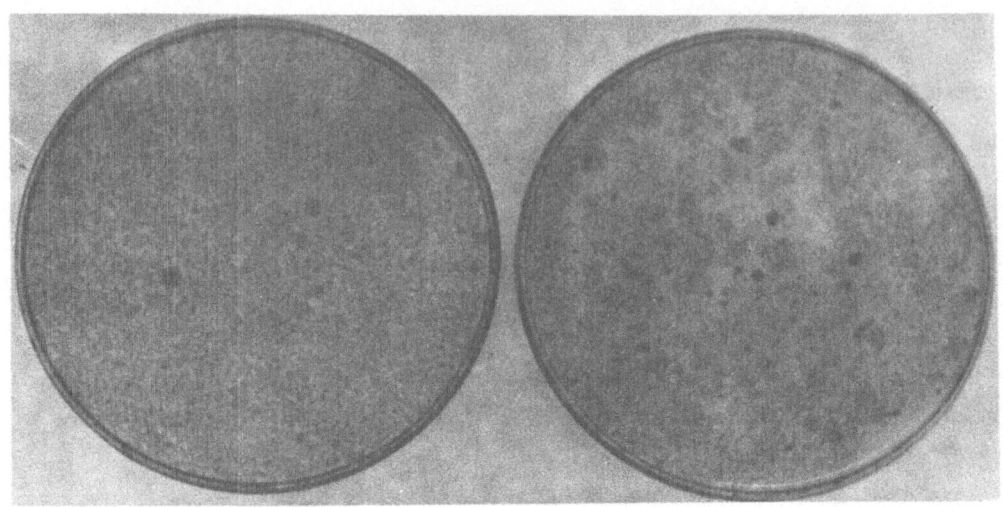

Figure 7. Enhancement of viral transformation assay. Syrian
hamster embryo cells were treated for 24 hr with a chemical
carcinogen, inoculated with a simian adenovirus (SA7), transferred at 200,000 cells/dish, overlaid with 0.3% agar at 6
days, fixed, and stained after 25 days. Left, solvent and SA7
treated cells. Right, carcinogen and SA7 treated cells. x1

since the methods used are inappropriate for expression of
chemically-transformed cells and the viral-induced foci are
distinct morphologically from chemical-induced foci (9,13,
Figure 7). The enhancement of viral transformation in Syrian
hamster embryo cells provides a sensitive, quantitative assay
for carcinogens and mutagens that responds equally well to a
wide variety of chemicals.

DISCUSSION

In vitro transformation studies with chemical carcinogens
in mammlian cells correlate well with in vivo activity. In
many cases, with primary or secondary cell cultures, the progression of events leading to the development of malignant
cells in vitro actually parallels the in vivo state. Using
the colony assay with Syrian hamster embryo cells, Pienta et

al. (62) have tested 87 chemicals with known carcinogenic activity and reported a 90.8% correlation with their current classification; only 8 false negatives and no false positives were found. In the various studies by DiPaolo et al. (24,25), there were also no false positives whereas two of the false negatives (urethane, DENA) were positive following transplacental administration of chemical (26). Overall, DiPaolo and Pienta have tested 30-40 chemicals in common with excellent agreement between test results (22). Chemicals shown to be positive for transformation by the colony assay are also positive when tested by the focus assay in hamster cells (13). Casto (unpublished observations) has tested 44 selected chemicals in the focus assay including 8 noncarcinogens. None of the noncarcinogens were positive, and of the 36 carcinogens tested, 33 induced transformed foci. The 3 negatives included DENA, DMNA, and ethylene thiourea which are presumably not metabolized by hamster fibroblasts in vitro.

With the rat and mouse assay systems, comparative data from different laboratories on a large number of test chemicals are not available. However, consistent results have been obtained with the carcinogenic polycyclic hydrocarbons (23,26, 64,70,71,73). In one study by Freeman et al. (37) encompassing active, weak, or inactive chemicals, 23 of 25 carcinogens were detected in multiple tests, but none of the inactive compounds was considered to transform. There was some variation between repeated tests with certain chemicals; anthracene, fluoranthene, phenanthrene (noncarcinogens) were positive in one test and acetamide, 4-aminobiphenyl, N-OH-N-2-FAA, propane sultone (carcinogens) were negative in one or more tests.

Other than the problem of metabolic activation, negative responses with chemical carcinogens are most probably due to the limited series of test dilutions that can be employed with individual chemicals in a single experiment. The inability to detect known carcinogens may be a reflection of the above in that many chemicals transform over a very narrow dose range (13,37). High concentrations, although not necessarily lethal, are toxic and may interfere with the initiation of transformation; on the other hand, low concentrations may require many replicate, treated cultures and an extended exposure time (13). These problems are magnified when testing complex mixtures, as one component may exert cytotoxic effects at doses lower than the optimal transforming dose of any carcinogen present. Under these conditions, transformation may occur within only a two-fold dilution of the test mixture and the in vitro assay may have to be modified so that more cells are at risk.

The problems of metabolic activation of suspect carcinogens in mixtures have some potential solutions. Primary cultures of hamsters or rats are capable of actively metabolizing
a wide variety of procarcinogens to their active form. For
those assays employing continuous cell lines, cultures of
x-irradiated feeder layers of the above or freshly prepared
liver cell cultures of mouse, rat, or hamster may be employed.
Casto (unpublished observations) has converted some procarcinogens to biologically active materials by using an S-9
hamster liver fraction prepared from animals primed with the
test chemical and simultaneously testing the activated chemicals in the virus enhancement assay.

The _in vitro_ testing of chemicals for carcinogenic
activity in mammalian cell systems has several advantages.

- An impressive correlation has been established
  between _in vivo_ and _in vitro_ activity.

- Quantitative assays are available to assess the
  relative carcinogenic activity of chemical and
  physical agents.

- A large number of compounds can be tested at one
  time under the same conditions at a greatly reduced
  cost.

- In comparison to animal testing, the assays are
  rapid and inexpensive.

- Only small amounts of test material (i.e., fractions
  from complex mixtures) are required for assay.

In addition to the problems of metabolic activation,
proper dose of chemical, length of exposure, and interpretation of test results, there are other technical factors which
influence the results of mammalian cell tests for carcinogens
that need to be recognized.

- The same formulations of media from different sup-
  pliers do not demonstrate the same growth potential
  when cells are plated for clonal assays.

- Fetal bovine sera used in most of the assay systems
  vary according to supplier and often show considerable lot to lot variation from the same supplier.

- Hamster embryonic cells (and possibly others) vary in sensitivity to certain chemicals and in the ease of recognition of chemically-transformed cells depending upon the supplier of pregnant animals.

- Contaminated air or water supplies as well as endogenous contamination of cell cultures (viral, mycoplasma, etc.) can affect the outcome of tests for carcinogens in cell cultures.

REFERENCES

1. Baker JB, Humphreys T: Serum-stimulated release of cell contacts and the initiation of growth in contact-inhibited chick fibroblasts. Proc Natl Acad Sci USA 68:2161-2164, 1971

2. Berky JJ, Zolotor L: Development and characterization of cell lines of normal mouse bladder epithelial cells and 2-acetylaminofluorene-induced urothelial carcinoma cells grown in monolayer tissue culture. In Vitro 13: 63-75, 1977

3. Berwald Y, Sachs L: In vitro cell transformation with chemical carcinogens. Nature, Lond 200:1182-1184, 1963

4. Berwald Y, Sachs L: In vitro transformation of normal cells to tumor cells by carcinogenic hydrocarbons. J Natl Cancer Inst 35:641-661, 1965

5. Borenfreund E, Krim M, Sanders FK, Sternberg SS, Bendich A: Malignant conversation of cells in vitro by carcinogens and viruses. Proc Natl Acad Sci USA 56:672-679, 1966

6. Borland R, Hard GC: Early appearance of "transformed" cells from the kidneys of rats treated with a "single" carcinogenic dose of dimethylintrosamine (DMN) detected by culture in vitro. Europ J Cancer 10:177-184, 1974

7. Borland R, Metcalfe SM, Hard GC: A combined in vivo-in vitro approach to studies of nitrosamine-induced carcinogenesis. In: Screening Tests in Chemical Carcinogenesis (Montesano R, Bartsch H, Tomatis L, eds.). Lyon, IARC Scientific Publications, No. 12,1076, pp 433-444

8. Brown AM: In vitro transformation of submandibular gland epithelial cells and fibroblasts of adult rats by methylcholanthrene. Cancer Res 33:2779-2789, 1973

9. Casto BC: Transformation of hamster embryo cells and tumor induction in newborn hamsters by simian adenovirus SV11. J Virol 3:513-519, 1969

10. Casto BC: Enhancement of adenovirus transformation by treatment of hamster cells with UV, DNA base analogs, and dibenz(a,h)anthracene. Cancer Res 33:402-407, 1973

11. Casto BC: Enhancement of viral oncogenesis by chemical carcinogens. In: Chemical Carcinogenesis. The Biochemistry of Disease Vol. 4 (Ts'o POP, DiPaolo JA, eds.). New York, Marcel Dekker, 1974, pp 607-618

12. Casto BC, DiPaolo JA: Virus, chemicals and cancer. Prog Med Virol 16:1-47, 1973

13. Casto BC, Janosko N, DiPaolo JA: Development of a focus assay model for transformation of hamster cells in vitro by chemical carcinogens. Cancer Res 37:3508-3515, 1977

14. Casto BC, Pieczynski WJ, DiPaolo JA: Enhancement of Adenovirus transformation by pretreatment of hamster cells with carcinogenic polycyclic hydrocarbons. Cancer Res 33:819-824, 1973

15. Casto BC, Pieczysnki WJ, DiPaolo JA: Enhancement of adenovirus transformation by treatment of hamster embryo cells with diverse chemical carcinogens. Cancer Res 34:72-78, 1974

16. Ceccarini C, Eagle H: pH as a determinant of cellular growth and contact inhibition. Proc Natl Acad Sci USA 68:229-233, 1971

17. Chen TT, Heidelberger C: Quantitative studies on the malignant transformation of mouse prostate cells by carcinogenic hydrocarbons in vitro. Int J Cancer 4:166-178, 1969

18. Chopra DP, Wilkoff LF: Induction of hyperplasia and anaplasia by carcinogens in organ cultures of mouse prostate. In Vitro 13:260-267, 1977

19. Coggin Jr, JH: Enhanced virus transformation of hamster embryo cells in vitro. J Virol 3:458-462, 1969

20. Colburn NH, Bates J, Vorderbruegge W, Rossen J: Chemical transformation of mouse epidermal cell cultures. Proc Amer Assoc Cancer Res 17:74, 1976

21. di Mayorca G, Greenblatt M, Trauthen T, Soller A, Giordano R: Malignant transformation of BHK21 clone 13 cells in vitro by nitrosamines--a conditional state. Proc Natl Acad Sci USA 70:46-49, 1973

22. DiPaolo JA: In: Screening Tests in Chemical Carcinogenesis (Montesano R, Bartsch H, Tomatis L, eds.). Lyon, IARC Scientific Publications No. 12, 1976, p 142

23. DiPaolo JA, Donovan PJ: Properties of Syrian hamster cells transformed in the presence of carcinogenic hydrocarbons. Exp Cell Res 48:361-377, 1967

24. DiPaolo JA, Donovan PJ, Nelson R: Quantitative studies of in vitro transformation by chemical carcinogens. J Natl Cancer Inst 42:867-874, 1969

25. DiPaolo JA, Nelson RL, Donovan PJ: In vitro transformation of Syrian hamster embryo cells by diverse chemical carcinogens. Nature 235:278-280, 1972

26. DiPaolo JA, Nelson RL, Donovan PJ, Evans CH: Host-mediated in vivo--in vitro assay for chemical carcinogens. Arch Pathol 95:380-385, 1973

27. DiPaolo JA, Takano K, Popescu NC: Quantitation of chemically induced neoplastic transformation of BALB/3T3 cloned cell lines. Cancer Res 32:2686-2695, 1972

28. Elias PM, Yuspa SH, Gullino MS, Morgan DL, Bates RR, Lutzner MA: In vitro neoplastic transformation of mouse skin cells: morphology and ultrastructure of cells and tumors. J Invest Derm 62:569-581, 1974

29. Evans CH, DiPaolo JA; Neoplastic transformation of guinea pig cells in culture induced by chemical carcinogen. Cancer Res 35:1035-1044, 1975

30. Fox TO, Sheppard JR, Burger MM: Cyclic membrane changes in animal cells: Transformed cells permanently display a surface architecture detected in normal cells only during mitosis. Proc Natl Acad Sci USA 68:244-247, 1971

31.   Franks S: Presented at the First International Congress
      on Cell Biology, Boston, Massachusetts, September 6, 1976

32.   Freeman AE, Igel HJ, Price PJ: Carcinogenesis in vitro
      I. In vitro transformation of rat embryo cells: Cor-
      relations with the known tumorigenic activities of chem-
      icals in rodents. In Vitro 11:107-112, 1975

33.   Freeman AE, Kelloff GJ, Gilden RV, Lane WT, Swain AP,
      Huebner RJ: Activation and isolation of hamster-specific
      C-type RNA viruses from tumors induced by cell cultures
      transformed by chemical carcinogens. Proc Natl Acad Sci
      USA 68:2386-2390, 1971

34.   Freeman AE, Price PJ, Bryan RJ, Gordan RJ, Gilden RV,
      Kelloff GJ, Huebner RJ: Transformation of rat and
      hamster embryo cells by extracts of city smog. Proc
      Natl Acad Sci, Wash 68:445-449, 1971

35.   Freeman AE, Price PJ, Igel HF, Young JC, Maryak JM,
      Huebner RJ: Morphological transformation of rat embryo
      cells induced by diethylnitrosamine and murine leukemia
      viruses. J Natl Cancer Inst 44:65-78, 1970

36.   Freeman AE, Price PJ, Zimmerman EM, Kelloff GJ, Huebner
      RJ: RNA tumor virus genomes as determinants of chemi-
      cally-induced transformation in vitro. In: Unifying
      Concepts of Leukemia, Bibl Haemat, No. 39 (Dutcher RM,
      Chieco-Bianchi L, eds.). Basal, Karger, 1973, pp 617-
      634

37.   Freeman AE, Weisburger EK, Weisburger JH, Wolford RG,
      Maryak JM, Huebner RJ: Transformation of cell cultures
      as an indication of the carcinogenic potential of chemi-
      cals. J Natl Cancer Inst 51:799-808, 1973

38.   Fusenig NE, Samsel W, Thon W, Worst PKM: Malignant
      transformation of epidermal cells in culture by DMBA.
      Inserm 19:219-228, 1973

39.   Griesemer RA, Nettesheim P, Marchok AC: Fate of early
      carcinogen-induced lesions in tracheal epithelium.
      Cancer Res 36:2659-2664, 1976

40.   Gutmann HR, Sekeg LI, Malejka-Giganti D: Malignant
      transformation of rat embryo fibroblasts by fluorenyl-
      hydroxamic acids. Proc Amer Assn Cancer Res 13:32, 1972

41. Hard GC, Borland R, Butler WH: Altered morphology and behaviour of kidney fibroblasts in vitro, following in vivo treatment of rats with a carcinogenic dose of dimethylnitrosamine. Experientia 27:1208-1209, 1971

42. Hatch GG, Balwierz PJ, Casto BC: Evaluation of in vitro tests for malignancy of chemically-transformed hamster cells. In Vitro 13:182, 1977

43. Hatch GG, Balwierz PJ, Casto BC: Evaluation of in vitro tests for malignancy of hamster cells transformed by chemicals. In Vitro (submitted)

44. Hatch GG, Casto BC, McCormick KJ, Trentin JJ: RNA type-C virus antigens in hamster cells transformed by carcinogenic DNA viruses and chemicals. Cancer Res 35:3792-3797, 1975

45. Hodges GM, Muir MD, Spacey G: A scanning electron microscopy study of normal and carcinogen-treated mouse bladder in vivo and in vitro. In: Proceedings of the Workshop on Scanning Electron Microscopy in Pathology. Scanning Electron Microscopy (Part III) (Johari O, Corvin I, eds.). Chicago, IIT Research Institute, 1973, pp 589-596

46. Igel HJ, Freeman AE, Spiewak JE, Kleinfeld KL: Carcinogenesis In Vitro II. Chemical transformation of diploid human cell cultures: A rare event. In Vitro 11:117-129, 1975

47. Inbar M, Sachs L: Interaction of the carbohydrate-binding protein concanavalin A with normal and transformed cells. Proc Natl Acad Sci USA 63:1418-1422, 1969

48. Kakunaga T: Chemical transformation of rodent cells and human cells. In: In Vitro Carcinogenesis. Guide to the Literature, Recent Advances and Laboratory Procedures (Saffiotti U, Autrup H, eds.). Washington, DC, Natl Cancer Inst Carcinogenesis Technical Report Series No. 44, in press, 1978, DHEW Publ No (NIH) 78-844

49. Kakunaga T: A quantitative system for assay of malignant transformation by chemical carcinogens using a clone derived from BALB/3T3. Int J Cancer 12:463-473, 1973

50. Kakunaga T: Requirement for cell replication in the fixation and expression of the transformed state in mouse cells treated with 4-nitroquinoline-1-oxide. Int J Cancer 14:736-742, 1974

51.   Kakunaga T, Kamahora J: Properties of hamster embryonic
      cells transformed by 4-nitroquinoline-1-oxide in vitro
      and their correlations with the malignant properties of
      the cells.   Biken J 11:313-317, 1968

52.   Kamahora J, Kakunaga T: In vitro carcinogenesis of 4-
      nitroquinoline-1-oxide with golden hamster embryonic
      cells.   Proc Jap Acad 42:1079-1081, 1966

53.   Kendrick J, Nettesheim P, Hammons AS: Tumor induction
      in tracheal grafts: A new experimental model for res-
      piratory carcinogenesis studies.   J Natl Cancer Inst 52:
      1317-1326, 1974

54.   Kuroki T, Goto M, Sato H: Malignant transformation on
      hamster embryonic cells by 4-hydroxyaminoquinoline N-
      oxide in tissue culture.   Tohoku J Exptl Med 91:109-118,
      1967

55.   Marchok Ac, Nettesheim P: In vitro growth characteris-
      tics of epithelial cell lines derived from tracheal
      transplants exposed in vivo to 7,12-dimethylbenz(a)-
      anthracene (DMBA).   In Vitro 13:193, 1977

56.   Marchok AC, Rhoton J, Nettesheim P: Differential growth
      and maintenance in cell culture of tracheal epithelium
      exposed in vivo to carcinogen.   Proc Amer Assn Cancer
      Res 17:66, 1976

57.   Miller DR, Viaje A, Bracken WM, Slaga TJ: The effects
      of chemical carcinogens on newborn epidermal cells in
      culture.   In Vitro 13:192, 1977

58.   Milo GE, DiPaolo JA: In vitro transformation of diploid
      human cells with chemical carcinogens.   In Vitro 13:193,
      1977

59.   Montesano R, Drevon C, Kuroki T, Saint Vincent L,
      Handleman S, Sanford KK, DeFeo D, Weinstein IB: Test
      for malignant transformation of rat liver cells in cul-
      ture: Cytology, growth in soft agar, and production of
      plasminogen activator.   J Natl Cancer Inst 59:1651-1658,
      1977

60.   Olinici CD, DiPaolo JA: Chromosone banding patterns of
      rat fibrosarcomas induced by in vitro transformation of
      embryo cells or in vivo injection of rats by 7,12-
      dimethylbenz(a)-anthracene.   J Natl Cancer Inst 52:1627-
      1634, 1974

61. Oshiro Y, DiPaolo JA: Loss of density-dependent regula-
     tion of multiplication of BALB/3T3 cells chemically
     transformed in vitro. J Natl Cancer Inst 44:39-63, 1970

62. Pienta RJ, Poiley JA, Lebherz III WB: Morphological
     transformation of early passage golden Syrian hamster
     embryo cells derived from cryopreserved primary cultures
     as a reliable in vitro bioassay for identifying diverse
     carcinogens. Int J Cancer 19:642-655, 1977

63. Pollack EJ, Todaro GJ: Radiation enhancement of SV40
     transformation in 3T3 and human cells. Nature 219:520-
     521, 1968

64. Price PJ, Freeman AE, Lane WT, Heubner RJ: Morphological
     transformation of rat embryo cells by the combined action
     of 3-methylcholanthrene and Rauscher leukemia virus.
     Nature New Biol (Lond) 230:144-146, 1971

65. Puck TT, Marcus PI, Cieciura SJ: Clonal growth of mam-
     malian cells in vitro. Growth characteristics of col-
     onies from single HeLA cells with and without a "feeder"
     layer. J Exp Med 103:273-277, 1956

66. Quarles JM, Sega MW, Schenley CK, Tennant RW: Rapid
     screening for chemical carcinogens: Transforming activ-
     ity of selected nitroso compounds detected in a trans-
     placental host-mediated culture system. Natl Cancer
     Inst Monogr, Series 51, in press, 1978

67. Reese DH, Friedman RD, Smith JM, Sporn MB: Organ culture
     of normal and carcinogen-treated rat bladder. Cancer Res
     36:2525-2527, 1976

68. Reitz MS, Saxinger WC, Ting RC, Gallo RC, DiPaolo JA:
     Lack of expression of Type C hamster virus after neo-
     plastic transformation of hamster embryo fibroblasts
     by benzo(a)pyrene. Cancer Res 37:3585-3589, 1977

69. Reznikoff CA, Bertram JS, Brankow DW, Heidelberger C:
     Quantitative and qualitative studies of chemical trans-
     formation of cloned C3H mouse embryo cells sensitive
     to postconfluence inhibition of cell division. Cancer
     Res 33:3239-3249, 1973

70. Rhim JS, Cho HY, Joglekar MH, Huebner RJ: Comparison of
     the transforming effect of benzo(a)pyrene in mammalian
     cell lines in vitro. J Natl Cancer Inst 48:949-957, 1972

71. Rhim JS, Creasy B, Huebner RJ: Production of altered cell foci by 3-methylcholanthrene in mouse cells infected with ARK leukemia virus. Proc Natl Acad Sci, USA 68:2212-2216, 1971

72. Rhim JS, Kim CM, Arnstein P, Huebner RJ, Weisburger EK, Nelson-Rees WA: Transformation of human osteosarcoma cells by a chemical carcinogen. J Natl Cancer Inst 55: 1291-1294, 1975

73. Rhim JS, Vass W, Cho HY, Huebner RJ: Malignant transformation induced by 7,12-dimethylbenz(a)anthracene in rat embryo cells infected with Rauscher leukemia virus. Int J Cancer 7:65-74, 1971

74. Rous P, Kidd JG: The carcinogenic effect of a papilloma virus on the tarred skin of rabbits. I. Description of the phenomenon. J Exp Med 67:399-428, 1938

75. Sabharwal PS, Garrett NE, Lidgerding B, Chortyk OT: Induction of transformation by tobacco smoke condensate using host-mediated system. In Vitro 13:183, 1977

76. Sanders FK, Burford BO: Morphological conversion of cells in vitro by N-nitrosomethylurea. Nature 213:1171-1173, 1967

77. Sanders FK, Burford BO: Morphological conversion, hyperconversion and reversion of mammalian cells treated in vitro with N-nitrosomethylurea. Nature 220:448-453, 1968

78. Schreiber H, Schreiber K, Martin DH: Experimental tumor induction in a circumscribed region of the hamster trachea: Correlation of histology and exfoliative cytology. J Natl Cancer Inst 54:187-198, 1975

79. Sefton BM, Rubin H: Release from density dependent growth inhibition by proteolytic enzymes. Nature 227: 843-845, 1970

80. Smith HS, Scher CD, Todaro GJ: Induction of cell division in medium lacking serum growth factor by SV40. Virology 44:359-370, 1971

81. Stoker M: Effect of x-irradiation on susceptibility of cells to transformation by polyoma virus. Nature (Lond) 200:756-758, 1963

82.   Todaro GJ, Green H: Enhancement by thymidine analogs
       of susceptibility of cells to transformation by SV40.
       Virology 24:393-400, 1964

83.   Weinstein IB, Yamaguchi N, Gebert R, Kaighn MF: Use
       of epithelial cell cultures for studies on the mechanism
       of transformation by chemical carcinogens.   In Vitro
       11:130-141, 1975

84.   Williams GM: Functional markers and growth behaviour
       of preneoplastic hepatocytes.   Cancer Res 36:2540-2543,
       1976

85.   Williams GM, Elliott JM, Weisburger JH: Carcinoma
       after malignant conversion in vitro of epithelial-like
       cells from rat liver following exposure to chemical
       carcinogens.   Cancer Res 33:606-612, 1973

86.   Yamaguchi N, Weinstein IB: Temperature sensitive (TS)
       mutants of chemically transformed rat liver epithelial
       cells.   Proc Natl Acad Sci USA 72:214-218, 1975

HIGHER PLANT SYSTEMS AS MONITORS OF ENVIRONMENTAL MUTAGENS

Frederick J. de Serres
National Institute of Environmental
Health Sciences
Research Triangle Park, North Carolina

INTRODUCTION

When we monitor the environment for the mutagenicity of chemical pollutants, we have a diversity of purposes: we are concerned not only for the biological evaluation of chemicals that have already been introduced into the environment, but also for the new chemicals that are being introduced each year. We are not only interested in evaluating effects of exposure to these chemicals on man himself, but also in evaluating the effects on all other organisms.

Our concern in monitoring is really with the biome at large and not simply that part which affects man himself. This vitally important point tends to be overlooked or minimized by many laboratory scientists as well as other scientists and administrators in various government agencies, as though the only important work in environmental research is directed toward risk estimation for the human population, and only such studies are worthy of funding. This attitude that only effects on the human population are of any importance is alarmingly widespread and has stifled research for many years in other allied areas. This attitude has also resulted in the shutdown of potentially useful research facilities, implying that man can exist on this planet alone and that effects of man-made chemicals on organisms are of no great concern. However, from what I have seen of what man can do to the environment, all organisms on this planet could exist quite well, and probably even better, without us!

Effects of man-made chemicals on plant systems, in particular, have often been overlooked or even excluded from consideration, especially when the objective has been to predict mutagenicity in the human population. This is particularly evident in the Committee 17 Report of the Environmental Mutagen Society (6) as well as the DHEW Position Paper on Environmental Mutagenesis (2). New data show quite clearly that plant systems can be used as monitors of air and water pollutants (18). Plant systems can also detect effects of such man-made chemicals as herbicides which may not only be a serious genetic hazard for the plants themselves but also for those animals, including man, who use these plants and plant products as food. From the proceedings from a small workshop that NIEHS organized on January 16-18, 1978, in Marineland, Florida (1), I have extracted the highlights of the general utility of plant systems as monitors of environmental mutagens.

ASSESSMENT OF PLANT SYSTEMS AS MONITORS

Variety of Test Systems

Much of plant literature in the area of chemical mutagenesis, and the literature amounts to hundreds of papers, is concerned with the use of chemicals as a means to induce genetic variability. The new mutants that have come out of this work have been of great practical importance in plant breeding.

Numerous plant systems have been developed, however, that can detect a wide variety of genetic damage. The best reviews are those in Volumes 2 and 4 of Chemical Mutagens by Ehrenberg (3), Kihlman (9), and Nilan and Vig (14). These assays include point mutations and deficiencies in structural genes, mutations in regulatory genes, changes in chromosome number and structure, sister-chromatid exchange, somatic crossing over, and recombination. A wide variety of systems has been developed to study somatic mosaicism on leaves, stamen hairs, and petals, as well as gametic mutations that appear in the second generation following treatment of seeds. For example, the scoring of chlorophyll-deficient mutants can be detected in such plants as barley, Arabidopsis, rice, pea, tomato, and corn.

Other test systems have been developed at the single cell level. The best known and most widely used are those using pollen. Two major genetic characteristics that can

be studied are self-incompatibility or waxiness as a result
of a change in the type of starch accumulated in the pollen
grain. The waxy mutations can be found in many different
plant species, and determining the frequency of waxy pollen
grains (wx) among normal pollen grains (Wx) provides a
simple quantitative assay for mutation induction in those
plants.

Genetic Effects of Pesticides

Genetic effects of pesticides on various plant systems
occur in the literature as early as 1931 (8). Nearly all
known types of cytological aberrations have been reported in
plants following treatment with pesticides. Such effects have
been reviewed recently by W.F. Grant (8,20), and a wide variety of effects has been observed. Pesticides that cause
chromosome aberrations in plant cells also produce chromosome
aberrations in cultured human cells. Frequently, the aberrations are identical. For example, studies have shown that
compounds which have a C-mitotic effect on plant cells also
produce a similar effect in animal cells. This has been
demonstrated for several mercurial compounds (4,5) and
griseofulvin (15) as well as other chemicals.

In general, an assay for chromosome aberrations,
primarily in root tips, is one of the oldest, simplest, most
reliable, and least expensive methods available for testing
the genetic effects of pesticides or any chemical agent.

PLANTS AS MONITORS OF AIR AND WATER POLLUTANTS

At least four different approaches can be used to
monitor air and water pollutants for mutagenicity.

- Collection and concentration of samples with
  assays for mutagenicity on eluted fractions
  with a laboratory-based assay using Salmonella
  or other microorganisms.

- Use of naturally occurring organisms that
  accumulate chemicals from the ecosystem by
  testing extracts of these organisms with
  laboratory-based assay systems.

- Measurement of genetic damage in selected populations of organisms occurring naturally in a given area.

- Introduction of an assay system into a polluted environment for a short period and then determination of the changes in mutation rate.

In the past few years plant systems have been used in all four approaches. Usually lower plants such as yeast, fungi, and bacteria are used in the laboratory for monitoring the mutagenicity of air and water pollutants, but exciting new work has been performed with higher plant systems in the other three areas.

In Situ Assay with Higher Plants

The recent studies of Klekowski (10,12,13) with various fern species illustrates the utility of this approach for monitoring water pollutants. The life cycle of ferns is characterized by an alternation of an independent haploid gametophyte generation and a diploid sporophyte generation. The sporophytes are what we generally recognize as ferns. The gametophytes are small and inconspicuous in most species. Both generations can be examined for the presence of chromosome damage as well as gene mutations that produce lethal or other detrimental mutations.

Homosporus ferns are hermaphroditic. At sexual maturity the haploid gamophyte produces both male and female germ cells (gametangia) through mitotic divisions. Thus, all the gametes formed by a single gametophyte have identical genotypes. Cell fertilization results in a completely homozygous zygote. If the gametophyte contains a recessive mutation affecting only genes expressed in the sporophyte generation, that mutation will be homozygous and expressed. Where this mutation is a recessive sporophyte lethal, the zygote will abort. Other kinds of mutations can result in sporophytes with aberrant phenotypes.

Klewoski has capitalized on these unique features of the fern life cycle by collecting spores from mature ferns in the field and determining the percentages that can form abnormal (haploid) gametophytes. These spores are readily cultured in the laboratory, and large numbers of spores can be germinated on a petri plate and gametophytes can be cultured individually in small Erlenmeyer flasks. Furthermore,

since the spores formed at the apex of the frond result from
more recent cell divisions than spores at the base and because
many ferns reproduce asexually as huge clones, the occurrence
of genetic damage can be traced in both time and space. For
example, Klekowski has found high frequencies of genetic
damage in ferns growing with rhizomes emersed in polluted
streams whereas low or normal frequencies of genetic damage
can be found in the same clone located several yards away from
the bank of that stream. The most widely studied species in
his work is Osmunda regalis, the royal fern, which grows well
in western Massachusetts.

Three papers (op. cit.) have appeared on the use of this
system in the past two years, and Klekowski is preparing a
review which will appear in Volume 5 of Chemical Mutagens (11).
There are numerous well characterized higher plant systems
that could be exploited in a similar fashion and there is high
potential in this area for additional exploratory work.

Another approach for monitoring air pollutants is to
plant particularly suitable species in areas of high pollution and then to compare the genetic load in these plants
versus that found in the same plants grown in an unpolluted
area. This method was used by Ehrenberg (19) with barley
measuring chlorophyll mutations (which result from mutations
in about 150 genes) to map environmental contamination from
ethylene oxide in industrial areas of Stockholm.

The use of the Tradescantia stamen hair system in a
mobile laboratory (18) is another assay available to measure the mutagenicity of various types of air pollutants
in heavily-industralized areas of various American cities.
The basic philosophy of this work is to take an assay system
known to be highly sensitive to both radiation and chemical
mutagens in the laboratory out into the field. In this
exploratory work, Tradescantia is being exposed to a variety
of air pollutants, both organic and particulate, to evaluate
the efficacy of the assay system. In addition, sites in
various counties around the country with high incidence of
cancer have been selected to evaluate the mutagenicity of air
pollutants.

The success of this work makes it desirable to develop
a battery of indicator organisms in this mobile laboratory
to insure a comprehensive evaluation as well as a duplication
of positive or negative data in different assay systems.
This is another area that needs additional exploratory work.

PLANTS AS DETECTORS OF MUTAGENIC ACTIVITY OF HERBICIDES AND
THEIR METABOLITES

Recent studies with corn have shown that the s-Triazine
herbicides (atrazine, simazine, and cyanazine) are not only
mutagenic to the corn plant itself but also that plant
extracts grown on soils treated with these herbicides are
mutagenic to other laboratory organisms. This was a major
discovery--that crop plants can activate nonmutagenic herbicides to a form that is not only mutagenic to the plant
itself but also that plant extracts contain a metabolite
mutagenic to other organisms. The first papers on this new
work appeared in Mutation Research in 1975 and 1976 (7,17).

In these studies, the corn plant is grown in soil
treated with a given herbicide and the tissues of the plant,
at various ages, are homogenized and plant extracts are
made. Similar extracts are made from control plants, and
the mutagenicity of extracts from both control and treated
plants are assayed for mutagenicity using short-term tests
for mutagenicity in combination with in vitro metabolic
activation. The chemicals themselves were not generally
found to be mutagens when tested directly with the microbial
assay systems.

The genetic endpoint of studies to determine the effect
of these same herbicides on the corn plants themselves is
reversion of the waxy locus as measured in the pollen grains.

Chromosomal Damage with s-Triazine Herbicides

As early as 1966 (20) there were data in the literature
which showed that the s-Triazine herbicides could cause
chromosomal damage in both mitosis and meiosis in such diverse plants as barley, Vicia, Tradescantia, and sorghum.
But this demonstration of mutagenic activity was essentially
ignored, and the use of these herbicides has increased dramatically over the past ten years, especially in the
production of corn. More recent data show that all three
herbicides can be biologically activated into agents that
produce point mutations. Whereas the original chemicals
are not mutagens in Salmonella or in yeast, extracts from
plants grown in soil treated with atrazine induced reversions
in Salmonella as well as gene conversion in yeast. All
three chemicals have been found to produce dominant lethal
mutations in Drosophila, and both atrazine and simazine will

produce recessive lethals when injected peritoneally. When
administered by larval feeding, atrazine induces both sex-
linked recessive lethals as well as increased rates of X or
Y chromosome loss.

Effects of these s-Triazine herbicides which demonstrated mutagenic effects on the corn plant itself (16)
came from studies of reversion of the recessive gene waxy.
This recessive mutation results in the starch of the endosperm containing only amylopectin in waxy kernels but both
amylopectin and amylose in the starch of wild-type kernels.
Endosperm of normal corn kernels stains dark blue-black
when reacted with iodine, whereas endosperm of waxy pollen
grains stains reddish-brown. Because large populations of
pollen can be obtained from corn plants, plants homozygous
for the recessive allele can be studied for reversion back
to wild type. Revertants will stain blue-black whereas
homozygous recessive pollen stains reddish-brown. At least
three- to fivefold increases in the reversion frequency of
waxy (which reverts spontaneously at about 3-5 x 10^{-5}) were
found with atrazine, simazine and cyanazine; two-and-a-half-
to threefold increases were also found with heptachlor and
chlordane. Significant increases (three- to fourfold) were
also found with various combinations of these herbicides
(since they are often used in combination in the field).

SUMMARY AND CONCLUSIONS

There is a wide range of applications for plant systems
in evaluating the mutagenic activity of environmental
chemicals. It is quite clear that there are data on many
chemicals already in the literature and that these data
are not irrelevant but are a valid demonstration of mutagenic
activity. Comparison of mutation frequencies in plants in
vivo and in animals in vivo has shown that the quantitative
response can be quite similar and is related to DNA content
per genome. A wide array of genetic damage can be studied,
all of which is important to man.

Plant systems can be characterized as providing low
cost assays, large numbers of cells or organisms that can
be analyzed, short life cycles, and well-studied genetic
systems.

In conclusion, if it is feasible to extrapolate from
Salmonella to man, then it certainly should be feasible to
extrapolate from higher plants to higher animals, perhaps

with even greater confidence since they are both eukaryotic organisms.

REFERENCES

1.   de Serres FJ, Shelby MD, eds.: Higher plants as
      monitors for environmental mutagens. Environmental
      Health Perspectives, in press, 1978

2.   Drake JW, Abrahamson S, Crow JF, Hollaender A, Leder-
      bert S, Legator MS, Neel JV, Shaw MW, Sutton HE, von
      Borstel RC, Zimmering S, de Serres FJ, Flamm WG:
      Environmental mutagenic hazards, Science 187:503-514,
      1975

3.   Ehrenberg L: Higher plants. In: Chemical Mutagens:
      Principles and Methods for Their Detection (Hollaender
      A, ed.) Vol 2, Plenum Press, New York, 1971, pp 356-365

4.   Fiskesjö G: Some results from Allium tests with
      organic halogenides. Hereditas 62:314, 1969

5.   Fiskesjö G: The effect of two organic mercury com-
      pounds on human leukocytes in vitro. Hereditas 64:142,
      1970

6.   Flamm WG, Valcovic LR, Pertel P, Roderick TH, Ray V,
      de Serres FJ, D'Aquanno W, Fishbein L, Green S, Malling
      HV, Mayer V, Prival M, Wolff G, Zeiger E: Approaches
      to determining the mutagenic properties of chemicals:
      Risk to future generations. J Environ Path Tox 1:301-
      352

7.   Gentile JM, Plewa MJ: A bioassay for screening host-
      mediated proximal mutagens in agriculture. Mutat
      Res 31:317, 1975

8.   Grant WF: Chromosome aberrations in plants as a
      monitoring system. Environmental Health Perspectives,
      in press, 1978

9.   Kihlman BA: Root tips for studying the effects of
      chemicals on chromosomes. In: Chemical Mutagens:
      Principles and Methods for Their Detection (Hollaender
      A, ed.) Vol 2, Plenum Press, New York, 1971, pp 489-514

10. Klekowski EJ, Jr.: Mutational load in a fern popu-
    lation growing in a polluted environment. Amer J Bot
    63:1024-1030, 1976

11. Klekowski EJ, Jr: Detection of mutational damage in
    fern populations: An in situ bioassay for mutagens in
    aquatic ecosystems. In: Chemical Mutagens: Principles
    and Methods for Their Detection (Hollaender A, de Serres
    FJ, eds.) Vol 5, Plenum Press, New York, 1978, pp 77-99

12. Klekowski EJ, Jr, Berger BB: Chromosome mutations in
    a fern population growing in a polluted environment:
    A bioassay for mutagens in aquatic environments.
    Amer J Bot 63:239-246, 1976

13. Klekowski EJ, Jr, Davis EL: Genetic damage to a
    fern population growing in a polluted environment:
    Segregation and description of gametophyte mutants.
    Canad J Bot 55:542-548, 1977

14. Nilan RA, Vig BK: Plant test systems for detection of
    chemical mutagens. In: Chemical Mutagens: Principles
    and Methods for Their Detection (Hollaender A, ed.) Vol.
    4, Plenum Press, New York, 1973, pp 143-170

15. Paget GE, Walpole AL: Some cytological effects of
    griseofulvin. Nature (London) 182:1320, 1958

16. Plewa MJ: Activation of chemicals into mutagens by
    green plants: A preliminary discussion. Environ-
    mental Health Perspectives, in press, 1978

17. Plewa MJ, Gentile JM: Plant activation of herbicides
    into mutagens -- the mutagenicity of field applied
    atrazine on maize germ cells. Mutat Res 38:390, 1976

18. Schairer LA, Van't Hof J, Hayes CT, Burton RM, de Serres
    FJ: Exploratory monitoring of air pollutants for muta-
    genic activity with the Tradescantia stamen hair system.
    Environmental Health Perspectives, in press, 1978

19. Sulouska K, Lindegren D, Eriksson G, Ehrenberg L: The
    mutagenic effects of low concentrations of ethylene
    oxide in air. Hereditas 62:264, 1969

20. Wuu KD, Grant WF: Chromosomal aberrations induced by
    pesticides in meiotic cells of barley. Cytologia
    32:31-41, 1967

THE ROLE OF *DROSOPHILA* IN CHEMICAL MUTAGENESIS TESTING

Carroll E. Nix and Bobbie Brewen
Biology Division
Oak Ridge National Laboratory
Oak Ridge, Tennessee

INTRODUCTION

An important question facing our society is the impact
of numerous chemical insults on the health of man and his
environment. Faced with a staggering array of chemicals and
enormous testing costs, we can test only a few chemicals
for possible carcinogenic effects. Recent results with the
Salmonella/mammalian microsome mutagenesis assay developed
by Ames (2), demonstrating a striking correlation between
carcinogenicity and mutagenicity of many chemical compounds,
offer the possibility that mutagenesis assay systems can
provide a quick identification of potential carcinogens.
Results from microbial assays can serve as a guideline for
further mutagenesis testing as well as identify those compounds requiring more extensive analysis in mammalian
systems.

Unquestionably, man is more closely related to other
mammals than to bacteria, and information regarding pharma-
cokinetics can only be obtained from mammals. Detection of
point mutations and small deletions in mammals, however,
requires considerable costs, time, and labor; thus, the number of chemicals available for investigation is restricted.
Other mammalian assay systems that rely solely on chromosome
breakage do not suffer from these disadvantages, but their
utility as diagnostic tests are questionable in light of recent results obtained with Drosophila. Vogel (10) has
shown that many chemicals are very effective in producing
point mutations and small deletions but do not produce chromosome breakage effects at all, while others produce chromo-

some breakage, but only at concentrations much higher than
that required to produce point mutations. Such compounds
would appear safe in any assay that measured only chromosome
breakage.

    Reliance on the results from a single mutagenic assay
system is rather risky. It seems preferable, in our opinion, to use a battery of tests (the tier approach) that would
include the rapid microbial assays as well as mammalian systems. Also, the use of Drosophila as a bridge between the
microbial and mammalian assays has many desirable features.

ADVANTAGES OF DROSOPHILA AS A TEST ORGANISM

    As a mutagenesis test organism, Drosophila is not as
economical nor as rapid a screen as the microbial assays, but
few higher organisms offer the economy and short generation
time that can be achieved with Drosophila. Drosophila mutagenic assays can be used in pre-screening tests, but perhaps
the most useful approach is to use the assays as a confirmation of results obtained in microbial assays and extension of
the analysis to include genetic end-points that are unattain-
able in the microbial systems.

    Due to the availability of a wealth of tester strains,
the assessment of a variety of induced genetic changes is
readily obtainable with Drosophila. Genetic end-points easily scored cover a wide spectrum, including point mutations
and small deletions, translocations, chromosome loss, non-
disjunction and genetic recombination. Thus mutagenic assays
with Drosophila can detect genetic damage due to both point
mutation and chromosome breakage.

    In many mutagenesis screening programs,the method of ex-
posure is often an important parameter. In those cases, the
advantage of using Drosophila again becomes apparent as the
chemical compound may be administered via feeding, injection,
inhalation, or direct treatment of sperm. Feeding and injection are the most commonly used methods, but inhalation of a
gas or aerosol is also very effective. A serious disadvantage
of the aerosol method is that a considerable volume of the
chemical agent is required. This disadvantage is overcome by
the injection technique in which only microliter quantities
are needed. For a more thorough discussion of the advantages
and disadvantages, see the review by Lee (8). Whatever
method is chosen, one should keep in mind that a negative
result may be due to the particular method of exposure (17).

In these cases, an alternate route of administration should be used.

Chemical mutagens often show a cellular specificity (3), and failure to detect mutagenic activity may result from a stage-specific response. In Drosophila the mutational response to a chemical insult in different germ cell stages may be studied by the brood pattern analysis (a technique whereby the mutation frequency of successive mating is obtained). Though somewhat more time-consuming, the additional information gained can give a more detailed picture of the mutagenic activity of a chemical. The method of brood pattern analysis developed for use in radiation genetics works equally well with chemicals with the exception that one has to consider the lingering effect of chemicals that remain in the body, resulting in exposure of germ cells over a longer period of time.

Another feature that adds to the utility of Drosophila is the presence of a mixed function oxidase system that is similar to that of the mammalian liver in its ability to activate indirect mutagens. In recent years, considerable attention has been focused on the metabolism of certain drugs and pesticides by insects. The crucial step in such metabolism is an oxidative attack by mixed function oxidases that can be isolated as a microsomal fraction (1,5). Drosophila are insects and the evidence that they also possess microsomal activities similar to the mammalian liver is indirect. The evidence is based largely on the fact that some forty to fifty compounds that require metabolic activation are, when tested in Drosophila, effective in inducing recessive lethals (11). These compounds fall into several different groups with widely differing structures. From these studies one can conclude that the Drosophila enzyme systems are similar to the mammalian systems in versatility and lack of substrate specificity. Recently, Baar et al. (4) have presented evidence that isolated Drosophila microsomes possess cytochrome P450 and aryl hydrocarbon hydroxylase activity. Nix et al. (9) have shown that Drosophila microsomes are capable of activating numerous promutagens when they are substituted for rat liver microsomes in the Salmonella histidine reversion assay.

THE USE OF DROSOPHILA IN THE MUTAGENIC ANALYSIS OF COMPLEX MIXTURES

Potential health effects of existing, as well as new, fuel technologies have become of increasing concern. Epler

et al. (6,7) have used the Salmonella/mammalian microsome
test system to assay environmental effluents and crude products from the synthetic fuels technology. Complex mixtures
were fractionated, and each fraction was tested for possible
mutagenic activity. Such procedures identified several fractions as mutagenic and as candidates for further biological
testing. Experiments described here represent an attempt to
extend others' observations of an eukaryotic organism and
to identify other genetic effects. In addition, we describe
the isolation of a crude Drosophila microsome fraction and
the use of such fractions in the Salmonella/mutagenicity
test system.

Muller-5 [In (1) $sc^{S1L}sc^{8R}$+S, $sc^{s1}sc^{8}w^{a}$B] males and
females and Oregon-R wild-type males were collected as
needed from the Oak Ridge stock collection. The Salmonella
strain used was TA98 (hisD3052, uvrB, rfa, frameshift plus R
factor), obtained from Dr. Bruce Ames, Berkeley, California.

Synthetic fuel fractions were dissolved in DMSO and then
diluted with a sterile sucrose solution to a final concentration of 2% sucrose and 4% DMSO. A glass fiber filter paper
was placed into an empty glass vial and then saturated with
175 µl of the appropriate test solution. Wild-type (Oregon-
R) males, 1-2 days old, were starved for 5 h, placed in the
vials containing the test solution (25 males per vial), removed after 24-48 h and mated to virgin Muller-5 females.
In the brood pattern analysis treated males were mated for
five successive 3-day broods. F_1 females were mated and
progeny scored for the presence of X-linked recessive lethals.

In the Salmonella/microsome mutagenicity tests the
standard procedures given by Ames et al. (2) were employed,
except that Drosophila microsomes were substituted for rat
liver microsomes. Concentrations of buffer and cofactors
were as previously described by Ames.

For the isolation of Drosophila microsomes, wild-type
(Oregon-R) flies were grown on standard media that contained
no live yeast. Adults were collected 7-10 days after emergence, administered ether, and placed on ice. Two volumes
(wt/vol) of ice-cold potassium phosphate buffer (ph 7.5)
were added and flies were homogenized by gently pounding in
a mortar until a smooth brei was formed (approximately 120-
150 stokes with the pestle). The homogenate was filtered
through four layers of cheesecloth and the filtrate was
spun at 750 g. The resulting supernatant was spun two times

at 10,000 g, and after the final spin the supernatant was immediately tested in the Salmonella system; the remainder was frozen at -70°C.

Our primary concern in the assay of the mutagenic effects of the synthetic fuels was to confirm the results in a higher organism and then, if possible, to extend the analysis to include other genetic effects. For this purpose, we selected the X-linked recessive lethal assay as it has been shown to be the most sensitive in Drosophila. Vogel (10) has carried out a comparative study of the frequency of induction of recessive lethals, dominant lethals, and chromosome loss by various concentrations of different mutagens. For all mutagens studied the recessive lethal assay was the most sensitive. We find a similar result for a series of cyclic nitrosoamines as shown in Table 1. In addition, we find a very close correlation between mutagenicity as measured by the X-linked recessive lethal assay and carcinogenicity in rats.

Since the crude synthetic fuel is toxic to Drosophila, only selected fractions could be tested. The results of a brood pattern analysis are shown in Table 2. Fractions 7 and 9 are ineffective in inducing X-linked recessive lethals in broods 1-3, although fraction 9 seems to be slightly mutagenic for spermatogonial cells. Using fraction E, the acetone-soluble portion of a more highly purified subfraction of the combined basic fractions from the Stedman fractionation scheme (6 and 7), we find a significant increase in the frequency of lethals in broods 1 and 2 but not in brood 3. This suggests that fraction E is an effective mutagen for mature sperm and spermatids but not meiotic cells. With this in mind, we then fed fractions 7, 9, and 14 at several different concentrations and monitored the production of X-linked recessive lethals in mature sperm and spermatids. Inspection of Table 3 reveals that fraction 7 is ineffective at all concentrations tested. Fractions 9 and 14, at the two lower concentrations tested, increase the frequency of lethals twofold over the spontaneous level but this is not statistically significant. In order to show a significant doubling with a critical region of 0.05 one would need to test 12,000-15,000 chromosomes.

From these results, we can conclude that the basic fractions (7 and 9), which are mutagenic in the "Ames" assay, induce at most only a twofold increase in the frequency of X-linked recessive lethals in Drosophila melanogaster.

Table 1

Induction of X-linked Recessive Lethals and Sex Chromosome
Loss in Drosophila by a Series of Cyclic Nitrosoamines

Mutagenicity in Drosophila

Compound	X-linked Recessive Lethals	Chromosome Loss	Carcino- genicity Rats*
Nitrosopiperi- dine (NP)	+	−	+
2,6-Dimethyl NP	−	−	−
2-Methyl NP	+	−	+
4-Methyl NP	+	−	+
3,4-Dichloro NP	+	−	+
Nitrosopipe- colic acid	−	NT	−
Dinitrosopiper- azine	+	NT	+
2,3,5,6-Tetra- methyldinitro- sopiperazine	−	NT	−
Nitrosomorpho- line	+	NT	+

*The carcinogenicity data was kindly provided by Dr. W.
Lijinsky.
NT = Not tested.

Further purification of these fractions results in a subfraction which shows a slight mutagenic activity in Drosophila; it induces an increase of three to four times over
the spontaneous level. Thus, we confirm the mutagenic
activity of fractionated products of synthetic fuels in a
eukaryotic organism, but the activity is very low compared
to that obtained in the microbial assay.

One of the advantages of Drosophila as a mutagenesis
test organism is the presence of a metabolic activation system. By substituting isolated Drosophila microsomes for rat
liver microsomes in the Salmonella/histidine reversion assay,
one can correlate mutagenic activity of a chemical compound
in vivo with the ability of isolated microsomes to activate
the chemicals. Results of experiments in which we tested
the ability of Drosophila microsomes to activate fractions

Table 2

Brood Pattern Analysis of X-linked Recessive Lethals
Induced in Drosophila melanogaster By Synthetic Fuels

Fraction	Conc. Fed (μg/ml)	Brood	Chromo- somes Tested	Lethals	Percent Lethals
Control	--	1	1334	3	0.22
		2	1839	4	0.22
		3	1318	1	0.08
		4	803	2	0.25
7. BI_a*	994	1	1071	0	0.00
		2	1039	3	0.29
		3	984	3	0.30
9. B_E*	1059	1	1083	4	0.37
		2	1197	1	0.08
		3	1230	2	0.16
		4	1295	7	0.54
E**	500	1	1661	11	0.66
		2	1686	13	0.77
		3	1780	3	0.17

*Basic fractions isolated from a crude synthetic fuel
product by the Stedman fractionation procedure.

**Acetone subfraction of Stedman basic fraction which is
further fractionated by LH20 (Epler et al., these proceedings).

7, 9, and E are shown in Figure 1 and Table 4. Instead of
Aroclor-induced rat liver fractions, 400 μl of Drosophila
10,000 g supernatant was used; all other procedures were as
described by Epler et al. (6,7).

     In light of the in vivo activity, these results are
rather surprising. Figure 1 shows the number of revertants/
plate plotted versus concentration. For all three fractions

Table 3

Induction of X-linked Recessive Lethals in Mature
Sperm and Spermatids of Drosophila melanogaster
By Subfractions of Synthetic Fuels

Fraction*	Conc. Fed (μg/ml)	Chromosomes Tested	Lethals	Percent Lethals
Control	--	1753	4	0.23
6. SA$_w$	15.02	1214	2	0.16
7. BI$_a$	994	1097	0	0.0
	397	1069	3	0.28
	199	1036	1	0.10
9. B$_E$	1059	1346	0	0.0
	423	702	3	0.42
	212	861	4	0.46
14. Neutral	870	1012	2	0.20
	435	1185	6	0.51
	218	1065	5	0.47

*For explanation of fractions see Epler et al., these
proceedings.

we obtained a linear dose-response curve over the concentrations tested. The slope of each induction curve was determined, and these results along with those obtained using
Aroclor-induced rat liver are shown in Table 4. It is of
interest that Drosophila microsomes are just as effective
as Aroclor-induced rat liver microsomes in the activation of
all three fractions and is even more effective with fractions
9 and E. We have tested several pure compounds in the
Salmonella/Drosophila microsome assay. Of these, 2-acetyl-
aminofluorene and aftatoxin B showed the highest mutagenic
activity, 144,000 and 180,000 revertants/mg respectively;
thus, Fraction E gives almost a tenfold increase in mutagenic
activity over any compound we have thus far tested. In these
instances, results with uninduced Drosophila microsomes compare very well with those of induced rat liver.

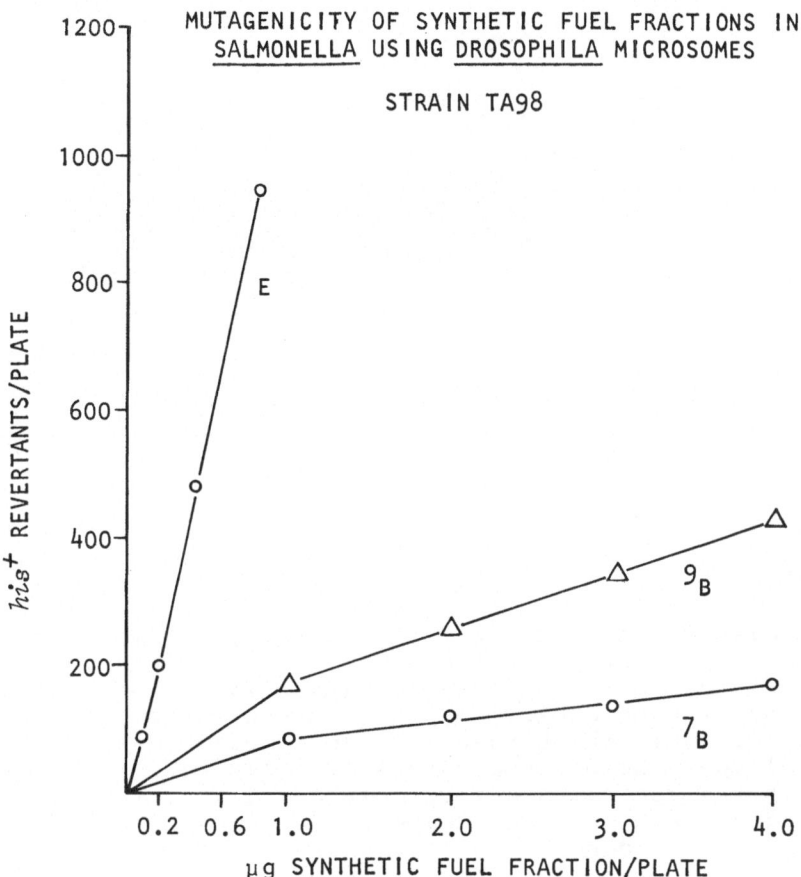

Figure 1. Effect of increasing concentration of synthetic fuel fractions on his reversion in <u>Salmonella</u> strain TA98. All reagents were as described by Epler et al. (6,7) except that 400 microliters of <u>Drosophila</u> S9 were substituted for rat liver S9.
o , 7_B ; △ , 9_B ; O, E

Comparisons based on the number of histidine revertants per milligram of S9 protein are even more striking in favor of <u>Drosophila</u> as typical <u>Drosophila</u> microsome preparations contain about one-fourth the protein of induced rat liver microsomes.

Table 4

Comparison of Mutagenic Activity
of Synthetic Oils Activated by Drosophila
and Aroclor-Induced Rat Liver Microsomes

Fraction*	Specific Activity (rev/mg)	
	Rat Liver	Drosophila
7. BI_a*	45,000	30,000
9. B_E	28,900	85,000
E (acetone)	222,000	1,300,000

*See footnotes to Table 2.

The discrepancy between the in vivo and in vitro mutagenic activity of fractionated complex mixtures is interesting, but at this point we have no explanation. One must keep
in mind that the metabolism of foreign compounds involves
enzymatic pathways which result in toxification as well as
detoxification. The balance between the two will often determine whether a generated active metabolite will remain in
the cell and exert genetic damage or be detoxified before
any damage can be done.

REFERENCES

1. Agosin M, Perry AS: Microsomal mixed-function oxidases.
    In: The Physiology of Insecta, 2nd ed., Vol. V (Rock-
    stein M, ed.). New York, Academic Press, 1974, pp 538-
    596

2. Ames BN, McCann J, Yamasaki E: Methods for detecting
    carcinogens and mutagens with the Salmonella/mammalian-
    microsome mutagenicity test. Mutat Res 31:347-364, 1975

3. Auerbach C: The chemical production of mutations.
    Science 158:1141, 1967

4. Baars AJ, Zijlstra JA, Vogel E, Breimer DD: The occur-
    rence of cytochrome P-350 and aryl hydrocarbon hydroxy-
    lase activity in Drosophila melanogaster microsomes,
    and the importance of this metabolizing capacity for
    the screening of carcinogenic and mutagenic properties
    of foreign compounds. Mutat Res 44:257-268, 1977

5. Casida JE: Insect microsomes and insecticide chemical
    oxidations. In: Microsomes and Drug Oxidations (Gil-
    lette JR, Conney AH, Cosmides GJ, Estabrook RW, Fouts
    JR, Mannering GJ, eds.). New York, Academic Press,
    1969, pp 517-530

6. Epler JL, Larimer FW, Rao TK, Nix CE, Ho T: Energy-
    related pollutants in the environment: The use of
    short-term tests for mutagenicity in the isolation
    and identification of biohazards. Environ Health Per-
    spect, in press

7. Epler JL, Young JA, Hardigree AA, Rao TK, Guerin MR,
    Rubin IB, Ho C-h, Clark BR: Coupled analytical and bio-
    logical analyses of test materials from the synthetic
    fuel technologies: Mutagenicity of crude oils from the
    Salmonella/microsomal activation systems. Mutat Res,
    in press

8. Lee WR: Chemical mutagenesis. In: The Genetics and
    Biology of Drosophila (Ashburner M, Novitski E, eds.).
    New York, Academic Press, Vol 1c, 1976, pp 1299-1341

9. Nix CE, Brewen B, Epler JL: Microsomal activation of
    selected polycyclic aromatic hydrocarbons and aromatic
    amines in Drosophila melanogaster. Mutat Res, in press

10. Vogel E, Leigh B: Concentration-effect studies with
    MMS, TEB, 2,4,6-TriCl-PDMT, and DEN on the induction of
    dominant and recessive lethals: Chromosome loss and
    translocation in Drosophila sperm. Mutat Res 39:383-
    396, 1975

11. Vogel E, Sobels FH: The function of Drosophila in
    genetic toxicology testing. In: Chemical Mutagens:
    Principles and Methods for Their Detection (Hollaender
    A, ed.). New York, Plenum Press, Vol 4, 1976, pp 93-142

12. Vogel E, Luers H: A comparison of adult feeding to in-
    jection in D. melanogaster. Drosophila Inform Serv 51:
    113-114, 1974

THE CELLULAR TOXICITY OF COMPLEX ENVIRONMENTAL MIXTURES

Michael D. Waters and Joellen L. Huisingh
Environmental Toxicology Division
U.S. Environmental Protection Agency
Research Triangle Park, North Carolina

Neil E. Garrett
Environmental Sciences Group
Northrop Services, Inc.
Research Triangle Park, North Carolina

INTRODUCTION

The principal aim of environmental toxicology is to
define the means whereby an agent exerts its biochemical,
physiological, and pathological effects. Such effects can
occur at multiple levels of biological organization from
individual molecules to intact animals. It appears that
toxic responses proceed as a function of dose from the
molecular level to the responding tissue or organ. A complete understanding of the toxic response requires investigation at each level of biological organization. However,
the primary unit of integrated biological organization and
response is the intact cell, and most toxic manifestations
depend ultimately upon changes that occur at the cellular
level.

All living cells require the maintenance of certain
critical metabolic and biosynthetic processes such as the
transcription and translation of DNA, the synthesis of RNA,
and the production of structural and enzymatic protein.
The integrity and the activity of these processes can be
used as a basis for estimating the relative cellular toxicity
of environmental agents. Most, if not all, of these functions are retained by mammalian cells in culture. Thus, the
use of cell and tissue culture techniques has been advocated
as a means to study the direct effects of toxicants in the
absence of the complex neural or humoral influences of the
intact animal (1,5,6,16,19,34,36,38,39,42,43).

To the extent that toxic agents alter basic metabolic
and biosynthetic processes, cell or tissue culture studies
are useful as preliminary screens for potential toxicity of
environmental agents. Certain cell types receive direct
exposure to environmental toxicants and may be especially
valuable in this regard. More often than not, however,
environmental toxicants exert their effects through complex
and indirect mechanisms. In these cases, the neural and
humoral influences of the intact animal may play a decisive
role in the mechanism of toxicity. Because of this and
because in vitro cell systems cannot mimic the multiplicity
of responses obtainable in the intact animal, these systems
are perhaps best used as supplementary tests together with
in vivo studies (2,15,27,28,44). This supplemental use of
in vitro cell systems can take at least two forms: (1)
cytotoxicity screening and (2) mechanistic studies.

The objective of this presentation is to provide some
insight into the current status of short term tests for
cellular toxicity screening and to suggest how these systems may be used to provide information on mechanisms of
cellular toxicity. Cytotoxicity screening is based upon
the detection and quantitation of critical morphological,
biochemical, and cytochemical alterations which signal
functional impairment or impending cell death. The validity
of this approach depends upon a clear understanding of how
toxic substances exert their effects and how these effects
are translated into discernible endpoints of cellular injury.
In other words, effective cytotoxicity screening depends upon
an appreciation of mechanisms of cellular toxicity.

CRITERIA OF CELLULAR TOXICITY

The criteria of cellular toxicity most frequently
employed include morphological changes detectable by light
and electron microscopy, alterations in cell growth and
division, biochemical alterations, and cytochemical change.

Morphological Changes

The correlation of structural, biochemical, and cytochemical alterations observed in vivo with similar changes
in vitro has led to a better understanding of the sequence
of morphological events occurring in cell injury and death.

Trump et al. have examined the morphological changes occurring in cell injury and death and have proposed a model to explain their observations (46). After the application of a sublethal or lethal injury, at least three major courses of events may be followed: (1) If the injurious agent is removed prior to cell death, recovery may be complete such that no morphological alterations remain. (2) Recovery may be incomplete and may proceed to an altered steady state characterized by the presence of numerous secondary lysosomes filled with digestive debris. (3) If injury proceeds to the point that recovery cannot occur even if the injurious stimulus is removed, the cell becomes necrotic. Membrane systems become fragmented, mitochondria become badly swollen, and lysosomes begin to leak. It is the sequence of events leading ultimately to cell death that one attempts to monitor using in vitro cytotoxicity test methods.

Cellular morphology following exposure to chemical toxicants has been studied using tissue culture techniques for many years. Rapid screening approaches based on the use of morphological indicators of cytotoxicity have developed rather slowly. However, recent work in this area by Walton and Buckley (47) shows considerable promise. This group has developed a computer model for cytotoxicity estimation based upon the careful definition of morphological alterations. This model accommodates the various information which can be gleaned from observations of morphological changes. The utility of this approach in the evaluation of complex effluents remains to be tested, but the technique potentially facilitates the gathering of information which may be useful in suggesting plausible mechanisms of cellular toxicity.

Cell Growth and Division

Cell growth and division have been used as indicators of cytotoxic effects for many years. Most of the older literature on the use of continuous cell cultures in pollution research has relied on these endpoints. As an example, Rounds et al. have used growth and cytogenetic alterations in the Chang strain of human conjunctival cells as an indicator of physiologically active components of automobile exhaust (40) and ambient air (41). In the former study, cell monolayers from which serum-containing medium was temporarily removed were flushed with auto exhaust collected from a car operating at simulated cruising speeds. Hydrocarbon and other constituents of the same type of auto

exhaust were extracted by bubbling the gas through chloroform
and redissolving the extracted species in an aqueous nutrient
medium after evaporation of the solvent. Dilutions of this
stock solution were then added to the cell cultures, and the
effects on growth and chromosomal figures at metaphase were
observed.

Exposure of the conjunctival cells to both the total
gaseous exhaust and the chloroform extract produced an
increase in chromosomal clumping and bridging at metaphase,
an observed decrease in total mitotic number, and a stimulation of growth rate in treated cultures as compared to controls. These latter two results suggest that the increased
growth rate resulted in a shorter overall duration of the
mitotic phase of the cell cycle. At any given time, then,
fewer total metaphases could be observed.

This study indicated the sensitivity of the in vitro
cellular system to stimulatory activity of exhaust components
since nearly a 50 percent stimulation in growth rate was
obtained from exhaust components contained in 2×10^{-5} ml of
auto exhaust. Interestingly, even those flasks flushed with
automobile exhaust containing carbon monoxide at high concentrations (not to mention NO, NO_2, SO_2, aldehydes, etc.)
appeared upon microscopic examination to be as grossly "normal" and healthy as the control cultures.

The tedious nature of morphological investigations and
those involving manual quantitation of cell numbers and
associated cytogenetic changes have encouraged the development and use of biochemical methods to quantify cell growth
in cytotoxicity screening. Christian et al. (11,12) employed
such biochemical techniques to measure cell growth in studies
on the effect of aqueous extracts of coal on mouse fibroblasts
(Clone L-929). After exposure of these cells for up to 6 days
to aqueous or serum extracts of the test samples, cytotoxicity
was determined by depression in cell growth as measured by the
protein or DNA content of the cultures. These studies showed
that aqueous extracts of coal samples obtained from a mine in
Pennsylvania inhibited cell growth to a greater extent than
did the extracts from a mine in Utah. The incidence of coal
workers pneumoconiosis is considerably greater in Pennsylvania
than in Utah (25), suggesting that cytotoxicity screening may
be useful in predicting relative in vivo toxicity. The study
points to the need for more in vitro and in vivo comparative
studies to facilitate more definitive interpretation of cellular toxicity studies.

Christian et al. (10) have also applied the L-929 system in the analysis of aqueous solutions of elements found in drinking water and to drinking water itself (13). The elements tested were selected because they are considered toxic at low concentrations and because they are unlikely to be affected by conventional water treatment procedures employed in municipal water plants. Barium, arsenic, lead, silver, and cyanide were added to the cell culture media as water soluble salts. The cells were grown in test tubes with metal-containing media for a period of 5 days. Each day, tubes were removed for protein assay and for cell counting. Within an order of magnitude above the level specified in drinking water standards, there was significant inhibition of cell growth as determined by cell number estimations and protein assays. Within an order of magnitude below the permissible concentrations, cell growth was virtually normal. At the permissible concentrations, partial inhibition or no inhibition of cell growth was observed as was the case for in vivo studies upon which the standards were based.

Thus, the measurement of cellular growth and division is one of the most widely used and accepted criteria of cellular toxicity, as is the measurement of animal growth (weight gain) to assess in vivo toxicity. This criterion has become more useful and precise by the application of biochemical techniques to quantify cell growth and division. Further applications of biochemical methodology have begun to enhance our understanding of more subtle cytotoxic effects.

Biochemical Alterations and Cytochemical Changes

The measurement of biochemical alterations after exposure of whole cells in vitro to toxic agents provides the potential for increasing the sensitivity of cellular toxicity test systems. Biochemical alterations occurring after exposure to a toxin should precede in time and occur at lower concentrations than the gross manifestations of cellular injury and death. As in the case of morphological changes, these biochemical alterations may be entirely reversible or may involve irreversible changes leading to an altered steady state or ultimately to cell death. Since different toxins may selectively affect different biochemical pathways, the measurement of a spectrum of biochemical parameters after exposure of cells in vitro to an agent may provide insight into the mechanism of cellular toxicity. In turn, this information may be used to enhance the sensitivity of cytotoxicity assays.

A single biochemical parameter has in some cases been
employed to evaluate the cellular toxicity of a particular
toxic agent whose mechanism of toxicity is understood. In
evaluating chemicals whose potential mechanism of toxic
action is unknown, a battery of biochemical endpoints is
frequently employed. These parameters may be chosen to
monitor selected sites of cellular metabolism such as macromolecular synthesis, enzyme function, repair mechanisms,
specific metabolite concentrations within the cell, membrane
transport, etc.

The experimental studies that will be described were
undertaken in an effort to integrate a battery of biochemical
measurements and to couple them with the detection of cytochemical changes using fast flow cytophotometry (32). The
latter technique permits quantitation of cell number, size,
viability, and various cytochemical measurements on a per
cell basis.

Applying this approach to cytotoxicity screening using
the diploid human lung fibroblast, strain WI-38, we have
examined a number of biochemical and cytophotometric toxicity
endpoints in an effort to compare their relative sensitivities
and to determine whether they may be used together to provide
some understanding of the cytotoxic mechanism. The materials
and methods used in these investigations have been described
in previous publications (49,53). WI-38 fibroblasts are perhaps the most completely characterized diploid human cells
presently available.

Figure 1 illustrates typical growth curves for strain
WI-38 in passages 22 and 30. One observes an initial lag
phase followed by a period of rapid cell growth. The experiments to be described were performed within a 22-hour interval in this rapid growth phase on the hypothesis that this
interval should be very sensitive to the influence of environmental toxicants.

Our initial investigations (48) with strain WI-38
involved salts of selected metals that may occur in polluted
air, including vanadium, cadmium, platinum, nickel, manganese,
and chromium. Platinum, at that time, was considered a possible emission product of the catalytic converter. Figure 2
illustrates the fitted concentration response curves for cell
viability following a 20-hour exposure to the various metallic ions at concentrations indicated on the X-axis. Viability, by exclusion of trypan blue, and cell number in a trypsinized cell suspension was determined simultaneously by use
of fast flow cytophotometry (32).

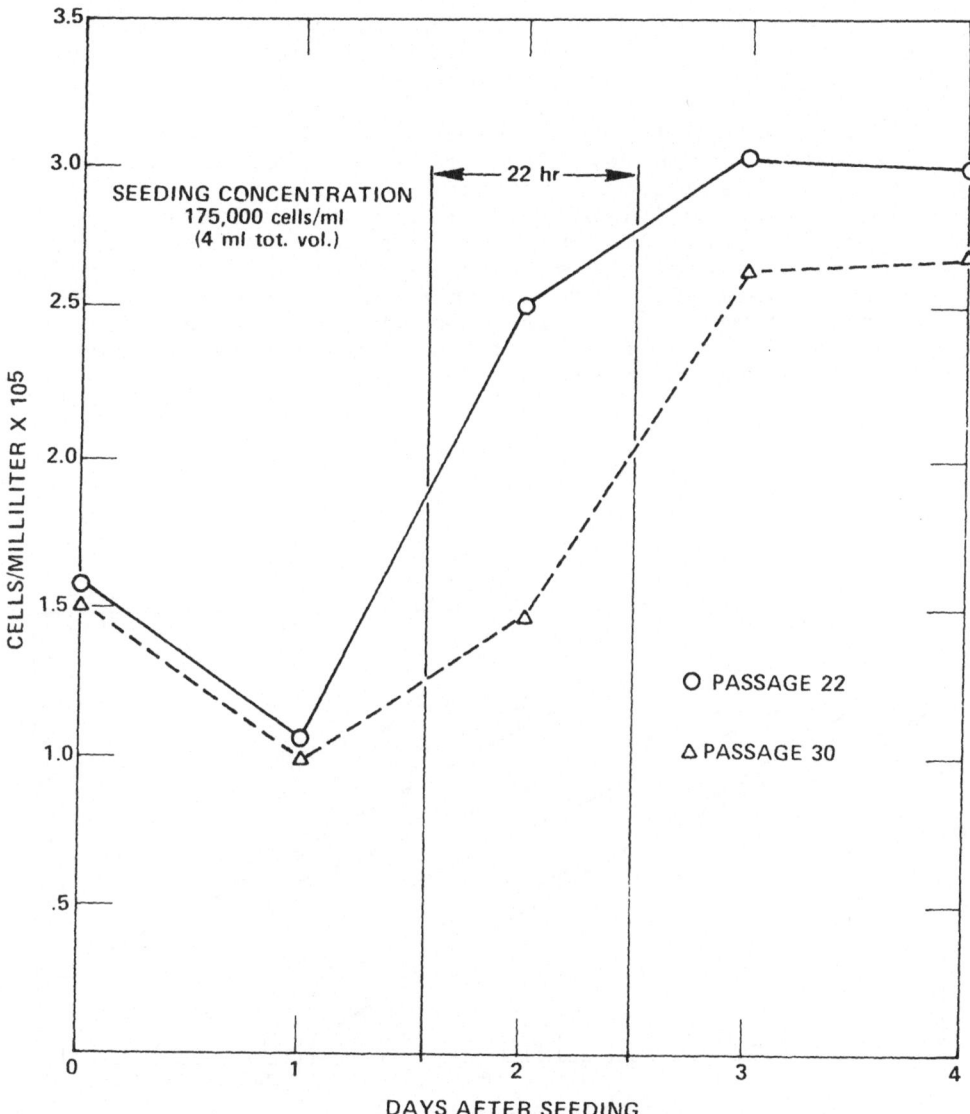

Figure 1. Growth curves for strain WI-38 human lung fibro-blasts (o) in passage 22 and (△) in passage 30. The seeding concentration was 1.75 x 10^5 cells/ml. Note 22-hour period within rapid growth phase in which experiments described were performed. Each point represents the mean of two cultures.

Total ATP was determined by the luciferin-luciferase method to provide additional evidence of cellular toxicity. Table 1 summarizes the results obtained in terms of EC_{50} values. The EC_{50} value is defined as the estimated concentration (mM) of each metal that results in a 50 percent response after a 20-hour exposure. The viability index is the percentage of viable cells as compared to control (cf. Table 1). By all measurements (viability, cell number, viability index, and ATP/million cells), vanadium, cadmium, and platinum were the more cytotoxic of the six metals studied, i.e., they displayed the lowest EC_{50} values. Our efforts then turned to measurement of macromolecular biosynthesis in an attempt to detect cytotoxic responses to these three metals at lower concentrations in the media.

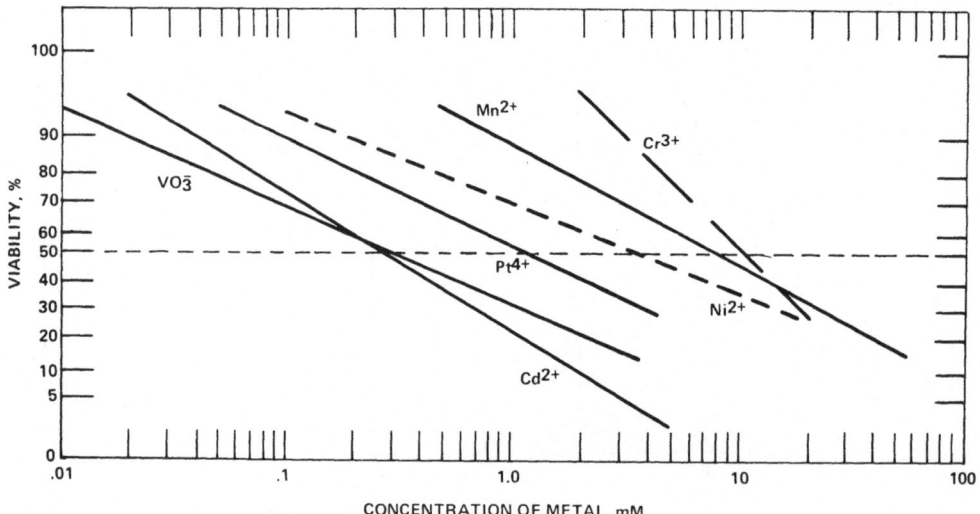

Figure 2. Effect of metallic ions on viability of WI-38 human lung fibroblasts after 20 hours. All compounds were chlorides except for ammonium vanadate. Note arc sine scale on abscissa.

Table 1

EC_{50} Values[1] for Viability, Cell Number, Viability Index[2], and Total ATP in Human Lung Fibroblasts (Strain WI-38) After 20 Hours

	Viability, %	Cell Number % of Control	Viability Index, % of Control	$ATP/10^6$ Cells, % of Control
VO_3^-	0.275	2.11	0.100	0.039
Cd^{2+}	0.270	1.75	0.230	0.110
Pt^{4+}	0.790	1.03	0.426	0.105
Ni^{2+}	3.42	4.35	1.50	0.108
Mn^{2+}	7.84	15.0	1.77	0.951
Cr^{3+}	10.53	Undeterminable	2.7 (est.)	7.24

[1]Concentration (mM) results in 50 percent response after 20 hours.

[2]Viability index = viability (%) x $\dfrac{\text{cell no. expt'l}}{\text{cell no. control}}$

Reduced uptake and incorporation of radiolabeled precursors for DNA, RNA, and protein synthesis can provide a sensitive indication of cytotoxicity in dividing cells. However, to determine whether an inhibitory effect is preferential for one of these major pathways, one must exclude the possibility that a compound causes unbalanced cell growth.

As shown in Figure 3, over the concentration ranges employed, there was no significant difference between culture DNA content (solid lines) and total culture protein (dashed lines) for cultures exposed to cadmium, vanadium, or platinum salts for a period of 22 hours.

Figure 4 illustrates the inhibitory effect of the three metals on the incorporation of radiolabeled thymidine, uridine, and leucine into perchloric acid-precipitable material between 20 and 22 hours after a 20-hour exposure to the respective metals. In the case of cadmium and vanadium there was a parallel concentration-dependent decrease in incorporation of all three precursors. However, in the case of platinum, 50 percent inhibition of thymidine incorporation occurred at a five- to ninefold lower concentration than required to affect uridine or leucine incorporation.

Figure 5 summarizes the data on platinum tetrachloride and illustrates how a series of biochemical and cytological tests, performed simultaneously, can provide information on relative sensitivity of endpoints and may suggest a possible sequence of events leading to cell death as a result of platinum exposure. Indicated in the cross-hatched areas below the curves representing DNA and total protein content are the concentration ranges in which EC_{50} values were observed for the parameters studied. In terms of concentration, the first biochemical lesion observed was the inhibition of thymidine incorporation into acid-precipitable material. Fifty percent inhibition was observed between .007 and .014 mM. At a slightly higher concentration, total cellular uptake of thymidine was also inhibited. These effects occurred at concentrations considerably lower than those required to cause noticeable changes in total protein or DNA content.

The incorporation and uptake of uridine and leucine was inhibited by 50 percent at still higher concentrations of platinum (from .025 to .066 mM). These effects coincided with noticeable decreases in total culture protein and DNA content.

At even higher concentrations of platinum (.078 to .105 mM), total ATP per million cells was depressed by 50 percent as compared to controls. Finally, decreases in viability by dye exclusion were seen between 0.46 and 1.0 mM. The data thus suggests a sequence of concentration-dependent events reflecting the cellular toxicity of platinum. Although the details of the entire sequence obviously cannot be reconstructed from such fragmentary information, the data clearly illustrate the utility of such an approach in suggesting a mechanism of cytotoxicity involving inhibition of DNA synthesis as a possible primary event. These parameters relating to major biosynthetic pathways together with information

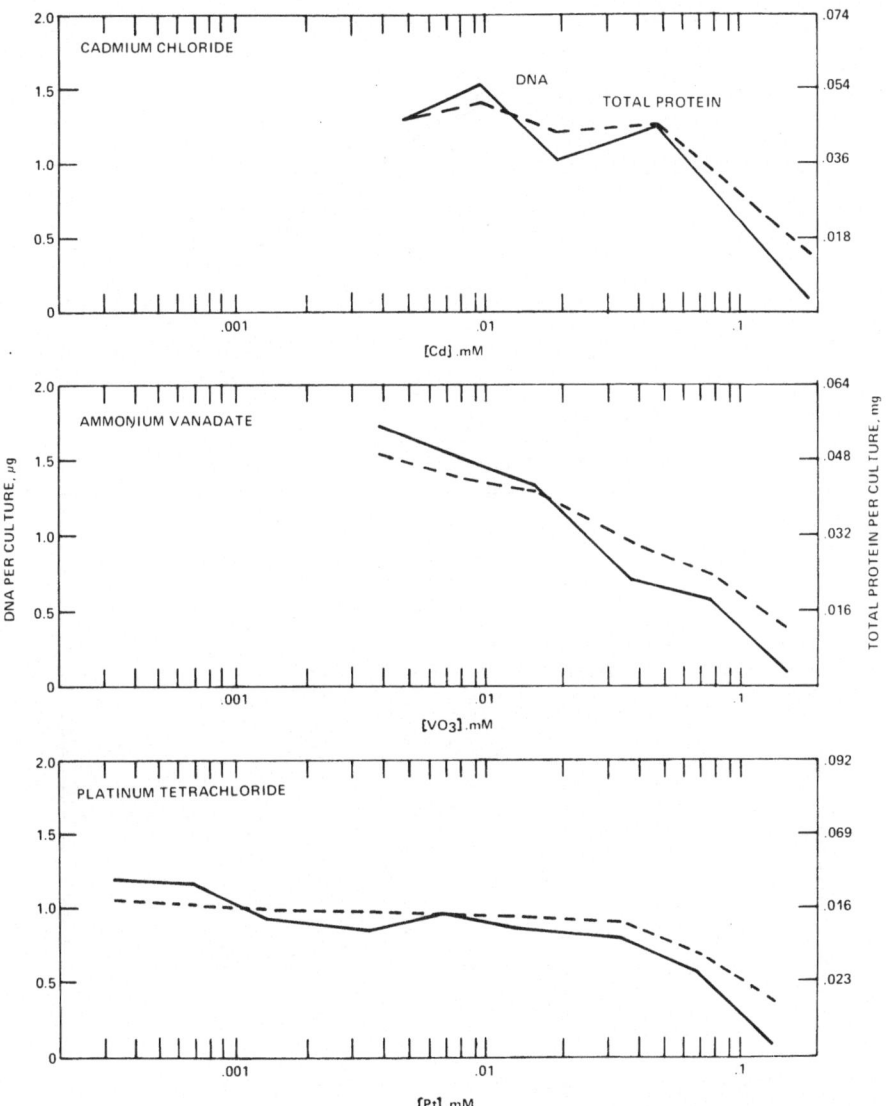

Figure 3. Effect of cadmium chloride, ammonium vanadate, and platinum tetrachloride on (—) DNA and (---) protein in cultures of strain WI-38 human lung fibroblasts after 22 hours. Each point represents the mean of two to six cultures.

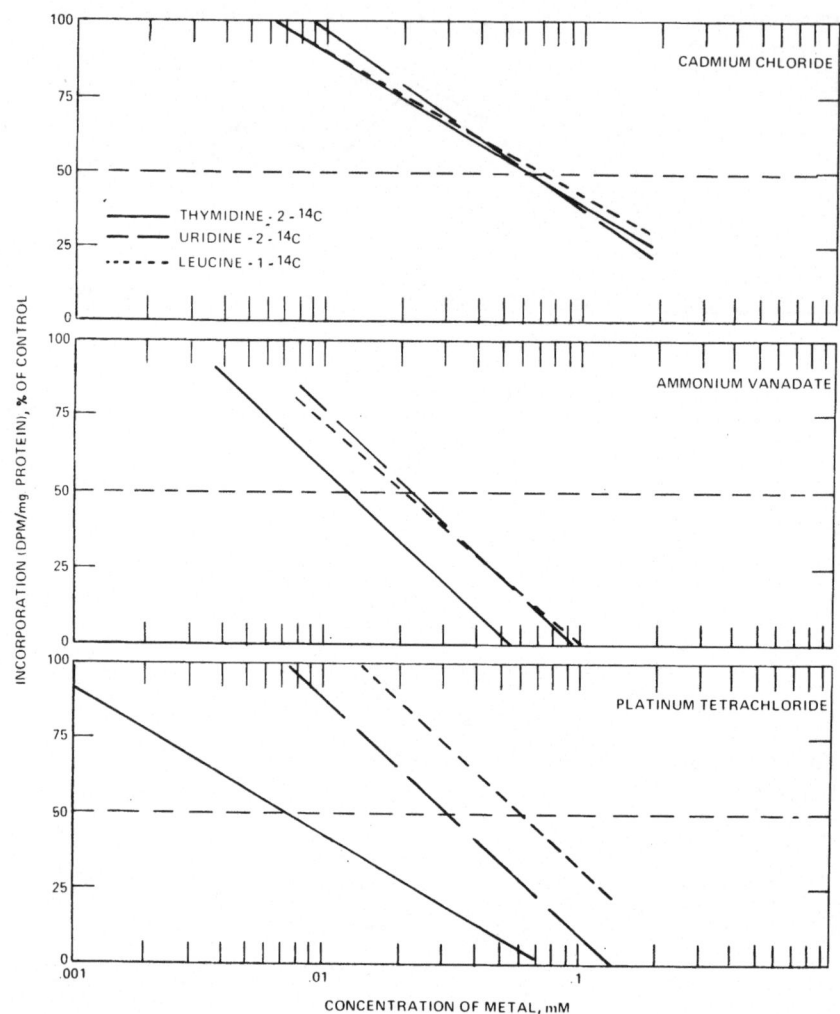

Figure 4. Inhibitory effect of cadmium chloride, ammonium vanadate, and platinum tetrachloride on incorporation into PCA precipitable material of (——) thymidine-2-^{14}C, (— —) uridine-2-^{14}C, and (---) leucine-1-^{14}C in strain WI-38 human lung fibroblasts after 22 hours. Radiolabeled precursors were present only during the last 2 hours of the 22 hour total exposure period. Each line is fitted to data from 12 to 14 cultures.

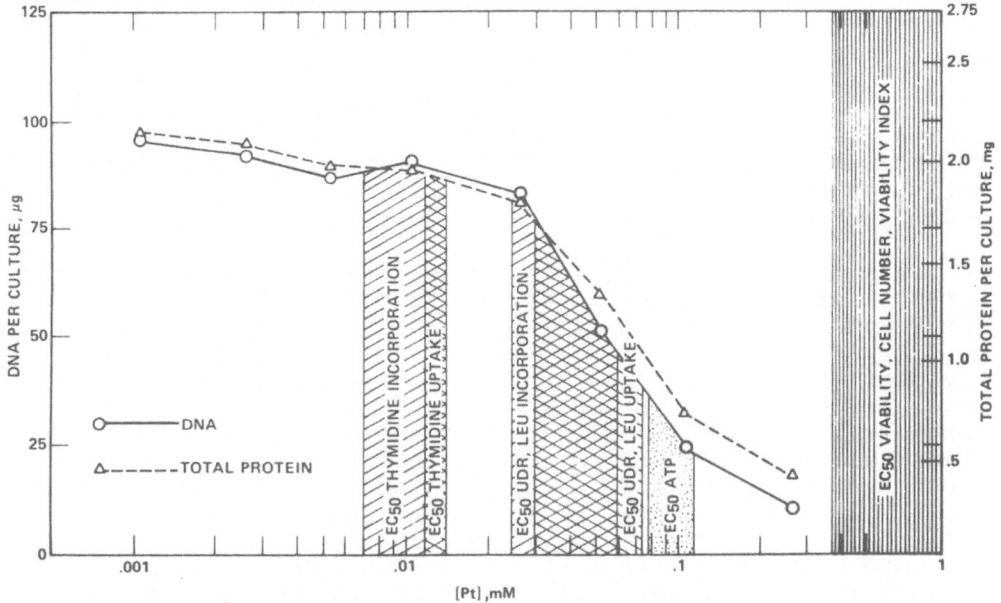

Figure 5. Effect of platinum tetrachloride on (o) DNA and
(Δ) protein in cultures of strain WI-38 human lung fibroblasts after 22 hours. Each point represents the mean of
two to six cultures. Also indicated below curves are concentration ranges in which 50 percent responses were
observed in incorporation and total uptake of thymidine-
2-^{14}C, uridine-2-^{14}C, and leucine-1-^{14}C; adenosine triphosphate per 10[6] cells; viability; cell number; and viability
index.

on the kinetics of uptake, and the intracellular concentrations of the test compound provide an effective and fairly
rigorous means of estimating relative cellular toxicity.

USE OF TARGET CELLS IN CYTOTOXICITY STUDIES

The principal disadvantage of the work cited thus far
is that the continuous cell lines employed, while useful in
detecting general toxic effects on growth and biosynthesis,
do not represent the metabolic and cellular specificity
known to exist in differentiated cell populations. Many

cell types established in continuous culture undergo a rapid
dedifferentiation and lose their metabolic activation and
detoxification capability.

Recent methodologic advances have made it feasible to
examine the response of metabolically-active "target cells"
to chemical toxicants in vitro. Two examples of such target
cells, phagocytic alveolar macrophages and liver parenchymal
cells, are discussed here.

The rabbit alveolar macrophage is a cell type which is
directly exposed to environmental chemicals both in soluble
forms and as components of particulate materials. Other
cell types such as the tracheobronchial epithelial cells,
alveolar epithelial cells, etc., also represent target cells
which are directly exposed to chemical agents. Critical
morphological, functional, and biochemical alterations in
these cell types resulting from direct exposure to chemical
substances at reasonable concentrations can be indicative of
potential toxic effects in the intact animal.

Other target cells such as those derived from liver,
kidney, mammary glands, etc. receive indirect exposure to
chemicals or to their metabolites and are likewise capable
of indicating morphological, functional, or biochemical
changes. In either case, the validity of in vitro toxicity
screening systems is in large part dependent on maintenance
of metabolic fidelity. This aspect of cytotoxicity testing
has been neglected in earlier studies and must be given considerably greater attention in the future.

Alveolar Macrophages

The phagocytic alveolar macrophage plays a central role
in the defense of the lung, especially against inhaled particulate materials. Techniques for recovering these cells from
the lungs of animals were introduced in 1960-61 by LaBelle
and Brieger (24) and by Myrvik et al. (33). Prior to this
time, peritoneal exudates were used as a source of macrophages
for in vitro studies (30,31). Some of the earlier work on
the effects of environmental agents on the alveolar macrophage
was done by Coffin, Gardner, and coworkers (14,17,23) and
by Weissbecker and his associates (54). Our group was
the first to carry out extensive comparative studies on the
influence of environmental metals using the rabbit alveolar
macrophage (RAM) system.

The work of Lee (26) and Natusch (35) has focused on the tendency of certain metals to condense and become concentrated on the surfaces of the smaller particles emitted by high temperature combustion processes. The absorption efficiency for most trace elements in the alveolar region of the lung is reported to be 50 to 80 percent (35). This means that cells lining or residing in the alveoli are exposed directly to soluble as well as insoluble forms of metallic air contaminants.

Our initial experiments with the RAM system employed salts of several metals found in highest concentrations among respirable particulates in ambient air (35). These included cadmium, vanadium, nickel, manganese, and chromium. Data for platinum was also obtained for purposes of comparison. As shown in Table 2, after a 20-hour exposure to the soluble metallic compounds, dye exclusion tests indicated that decreases in cell viability to 50 percent occurred at similar concentrations as observed previously using strain WI-38 human lung fibroblasts. This fact is interesting in view of the species differences and the fact that the rabbit macrophage does not divide in culture while the WI-38 human fibroblast does. These findings suggest that a dividing cell may not always provide a more sensitive subject in a cytotoxicity test system.

Table 2

Concentrations of Metallic Ions Causing Reduction in Viability to 50 Percent in Rabbit Alveolar Macrophages and Human Lung Fibroblasts (Strain WI-38) After 20 Hours

Metallic Ion	Concentration of Metal, mM	
	Rabbit Alveolar Macrophages	Human Lung Fibroblasts
Cd^{2+}	0.099	0.270
VO_3^-	0.234	0.275
Pt^{4+}	0.400	0.790
Ni^{2+}	4.17	3.42
Mn^{2+}	5.29	7.84
Cr^{3+}	5.48	10.5

In investigating further the events related to loss of function and death of the macrophage, we chose as endpoints depression in total ATP and phagocytic activity, changes in hydrolytic enzyme specific activities, loss in cell viability, and cell lysis. Shown in Table 3 are EC_{50} values obtained for viability, cell number, viability index, and acid phosphatase specific activity (52).

The determination of cell numbers indicated that all metals listed in Table 3 except cadmium were cytolytic, that is, they caused decreases in cell numbers over concentration ranges similar to those that had resulted in lowered cell viability. We have determined since that Hg, Cu, and Zn exhibit a nonlytic behavior like cadmium (50). The last column in Table 3 shows the EC_{50} values for acid phosphatase specific activity. This predominantly lysosomal enzyme was depressed at concentrations similar to those that resulted in lowered cell viability and provided corroboration of the viability estimate by dye exclusion. This correspondence suggests that the depression in specific activity or release of lysosomal hydrolases occurs at concentrations of metals sufficient to bring about cell death.

Table 3

Concentration of Metallic Ions Causing Reduction
in Viability, Cell Number, Viability Index,
and Acid Phosphatase Specific Activity to 50 Percent
in Rabbit Alveolar Macrophages After 20 Hours

Metallic Ion	Concentration of Metal, mM			
	Viability	Cell Number	Viability Index	Acid Phosphatase Specific Activity
Cd^{2+}	0.099	*	0.082	0.205
VO_3^-	0.234	0.221	0.101	0.094
Ni^{2+}	4.17	12.8	3.78	3.80
Mn^{2+}	5.29	17.2	4.67	5.31
Cr^{3+}	5.48	8.57	5.06	4.44

*No difference from control, $p < 0.05$.

As mentioned previously, macrophages are active phago-cytes. It is well known that the availability of the high energy intermediate ATP is a limiting factor in the process of phagocytic engulfment (37). A demonstrated correlation between depression in ATP content and depression in phago-cytic activity would suggest that ATP levels may be useful as an indicator of phagocytic capability in cells exposed to particulates. Figure 6 illustrates the results of time course experiments involving measurement of viability, phago-cytic activity, and ATP/million cells on suspension cultures exposed to chlorides of mercury, cadmium, nickel, and zinc. Phagocytic activity and ATP were determined on aliquots of the cell suspension to which were added 1 μm polystyrene latex spheres for a period of one hour. Phagocytosis was monitored microscopically by collection of cells on nucleo-pore filters (5 μm pore size) and dissolution of extracellu-lar spheres with xylene (51). Statistical analysis, control-ling for the effects of time, revealed excellent correlation coefficients (0.94 or better) between ATP/million cells and phagocytic activity in the case of each metal.

It is important to note that ATP and phagocytic activity are considerably more sensitive cytotoxicity endpoints in the macrophage system than viability by dye exclusion. This dif-ference was particularly noticeable in dose response studies with zinc and nickel where 18 to 52 times the concentration of metal was required to reduce viability to 50 percent after 20 hours as was required to cause such an effect on ATP per cell or phagocytic activity (51). The rapidity with which this measurement could be made argued strongly for its inclusion in a battery of biochemical tests for potential impairment of macrophage function.

Having completed these studies with soluble metallic salts, we examined a series of metallic oxides to determine whether similar responses would be demonstrable. Our earlier work with vanadium oxides (49) had shown that toxicity was a function of the amount of soluble vanadium released to the culture medium independent of the original oxidation states. As indicated in Table 4 and as confirmed by atomic absorp-tion spectrophotometry, soluble cadmium was also released by cadmium oxide. As the concentration of the particulate was increased, the viability and viability index decreased. How-ever, as with soluble cadmium, cell number did not exhibit a dose response. Also as shown in Table 4, a number of other oxides displayed minimal solubility in the test system and were judged relatively nontoxic even though they were actively

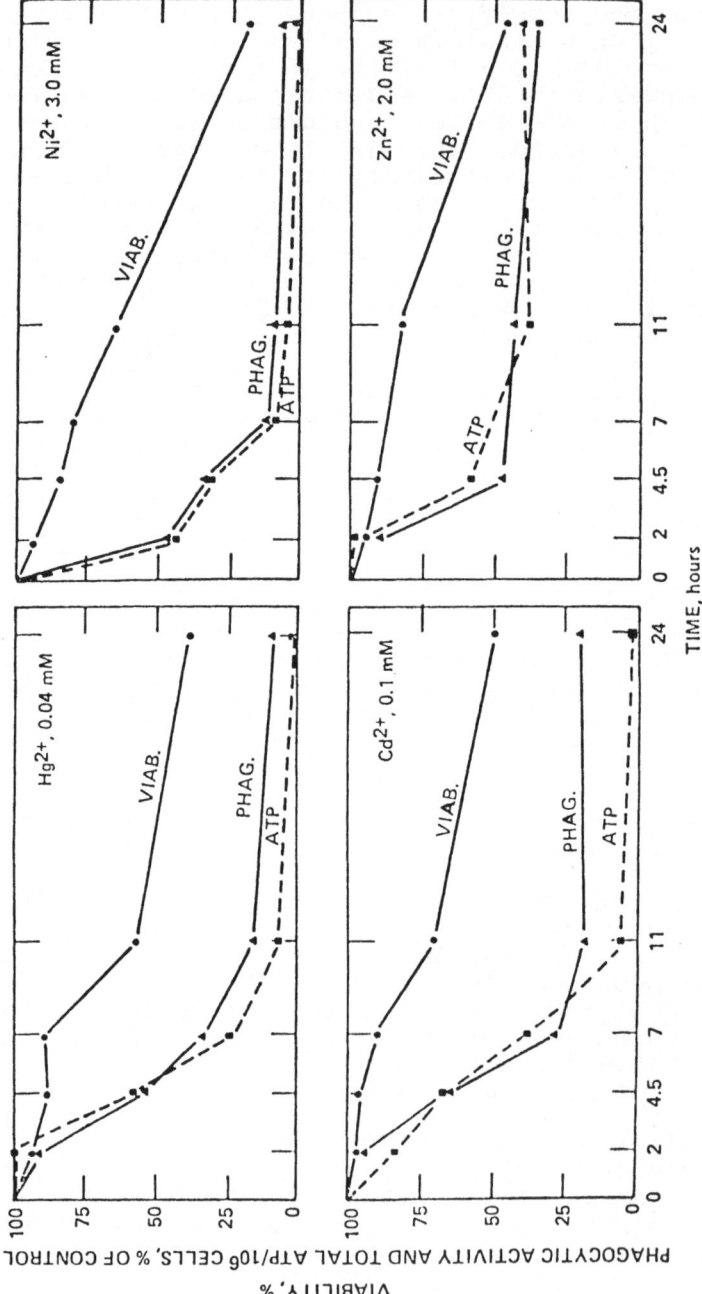

Figure 6. Time course effect of divalent metallic chlorides on rabbit alveolar macrophage (RAM) (●) viability, (▲) phagocytic activity, and (■) adenosine triphosphate per 10⁶ RAM. (n = 3)

Table 4

Effect of Metallic Oxides* on Viability, Numbers
and Viability Index of Rabbit Alveolar Macrophages
After 20 Hours

Compound	Wt. Metal Per Unit Volume	Viability (± SEM)	Cell Number (± SEM)	Viability Index
	µg/ml	%	% of Control	%
CdO	100	63.7 + 5.4	63.9 + 7.4	40.7
	200	46.8 + 2.0	71.4 + 5.0	33.4
	500	15.1 + 2.4	70.3 + 8.1	10.6
NiO	500	91.2 + 1.4	91.2 + 5.4	83.2
MnO_2	500	92.3 + 0.77	80.6 + 4.7	74.4
Cr_2O_3	500	91.7 + 2.6	79.0 + 6.3	72.4
PtO_2	500	95.8 + 0.96	93.7 + 7.8	89.8
PdO	500	96.0 + 0.35	87.1 + 5.7	83.6
RuO_2	500	97.5 + 0.56	98.8 + 3.7	96.3

*Particulate samples washed prior to addition to cultures.

phagocytized by the alveolar macrophages. Figures 7 and 8
include surface maps of two of these metal oxide particles
which had been engulfed by macrophages. These maps were
obtained by use of energy dispersive X-ray analysis (45). The
upper part of Figure 7 is a scanning transmission electron
micrograph of a nickel oxide particle contained within a
macrophage. The lower part of the slide is the corresponding
nickel X-ray map. Extensive mapping showed that the metal
was detectable only in association with the particles. Since
glutaraldehyde fixation was employed, it is possible that
intracellular translocation did occur. However, these
results appeared to confirm the biological observation that
the nickel was not released in sufficient quantity to produce
a detectable cytotoxic response.

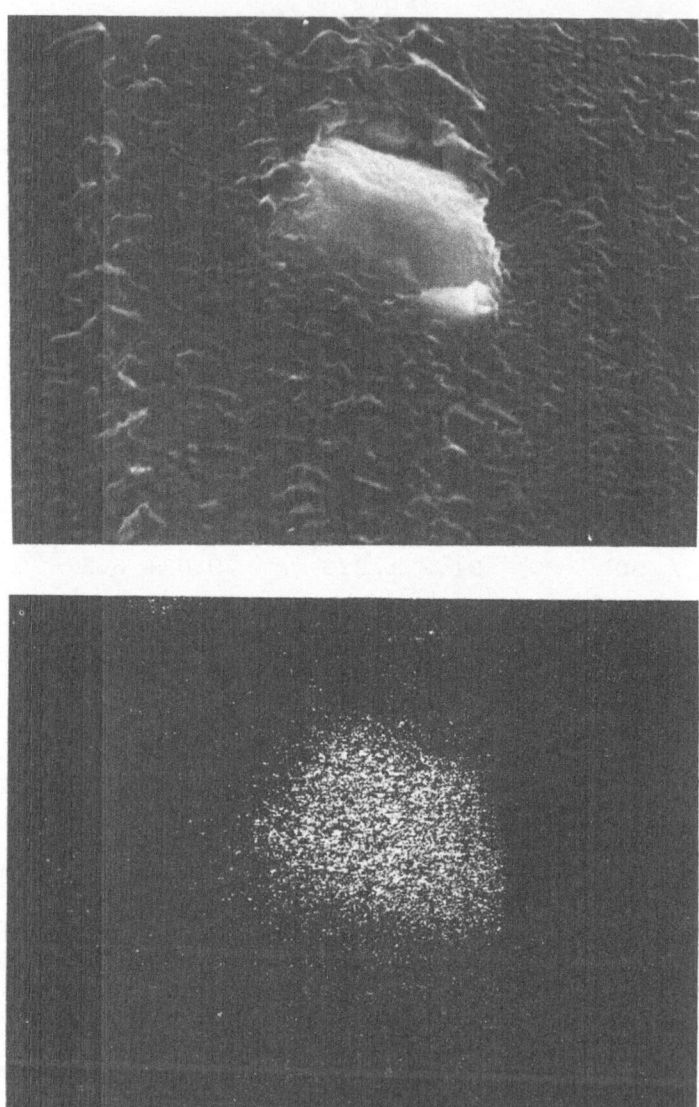

Figure 7. Top: Scanning transmission electron micrograph (STEM) of RAM which has phagocytized particles of nickel oxide. Width 11 μm. Bottom: X-ray map of nickel spectrum from the area shown in the STEM above.

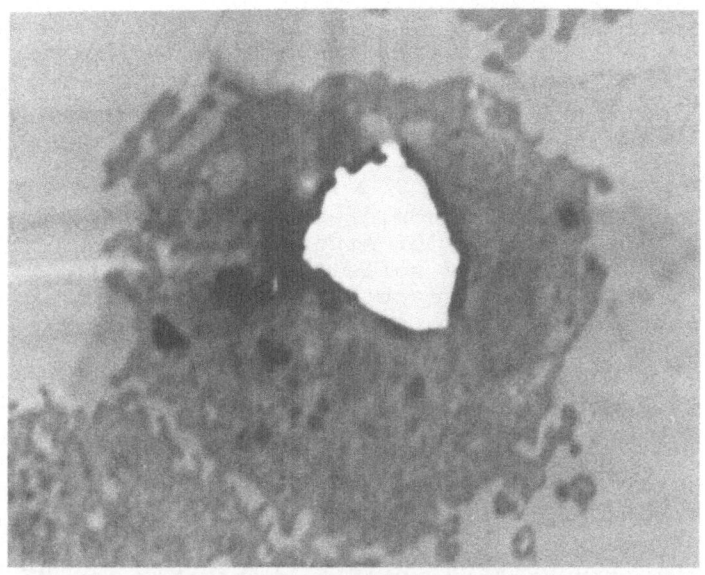

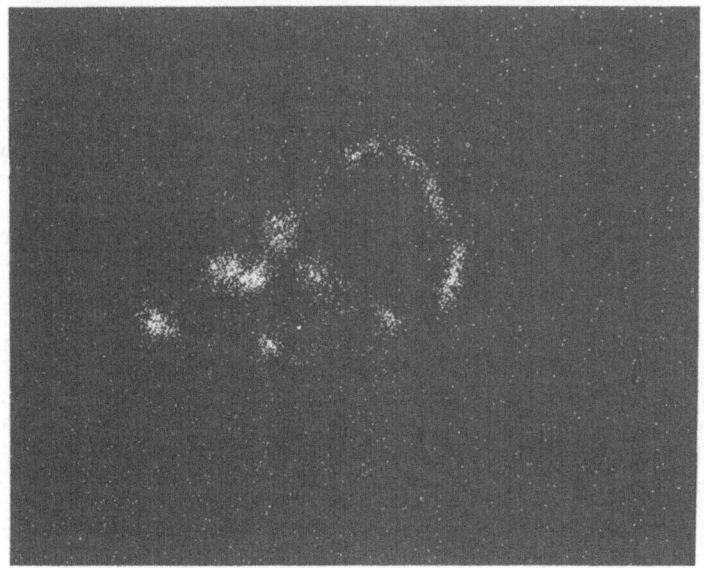

Figure 8. Top: Transmission electron micrograph (TEM) of RAM which has phagocytized particles of manganese dioxide. Bottom: X-ray map of manganese spectrum from the area shown in the TEM above.

Similar results were obtained with manganese dioxide as
illustrated in Figure 8. The transmission electron micrograph is in the upper half of the slide and the manganese
map is in the lower half. No translocation of manganese was
observed indicating that very little of the soluble cation
was released.

The behavior of the nonleachable metallic oxides was
also observed in a series of studies with actual particulate
samples. In cytotoxicity screening studies with sized particulate materials from seventeen different high temperature
combustion processes we determined that approximately 70 percent of the particles displayed cytotoxicity for the rabbit
macrophage without having any soluble or leachable toxic compound (29). As illustrated in Figure 9, the coke oven heater
sample and the sludge incinerator sample represented such
particles. It was not possible to remove any toxic component
from the particles by pre-incubating the particles in culture
medium for 20 hours followed by addition of this "supernatant"
medium to the cells. Thus, it appeared that toxicity was due
to an interaction between surface components of the particle
and the intracellular milieu. This finding suggested that
surface components and, hence, surface area may be important
in the cytotoxic action of particulate materials.

Certain of the industrial particulate samples contained
leachable toxic components, e.g., the oil fired power plant
and copper smelter which are again shown in Figure 9. These
samples will be considered in a subsequent section.

The following series of investigations were carried out
to determine the influence of particle size, surface area,
and metal surface coating on cytotoxicity (3,4). These
studies considered in greater detail the question of the
biological activity of particle surfaces. In an effort to
model the situation represented by nonleachable toxic samples, fly ash particles collected from an electrostatic
precipitator were size-fractionated, coated with lead, nickel,
or manganese, and treated to oxidize the metals (55). For
each coated fly ash, the percentage of metal on the fly ash
was approximately the same regardless of particle size, and
no leaching could be detected by AA spectrophotometry (3).

Figure 10 shows the effect of fly ash concentration and
size on viability of rabbit macrophages after a 21 hour exposure. Within each particle size range, viability decreased
as the concentration of particles was increased. Also at a
given concentration, macrophage viability decreased with the

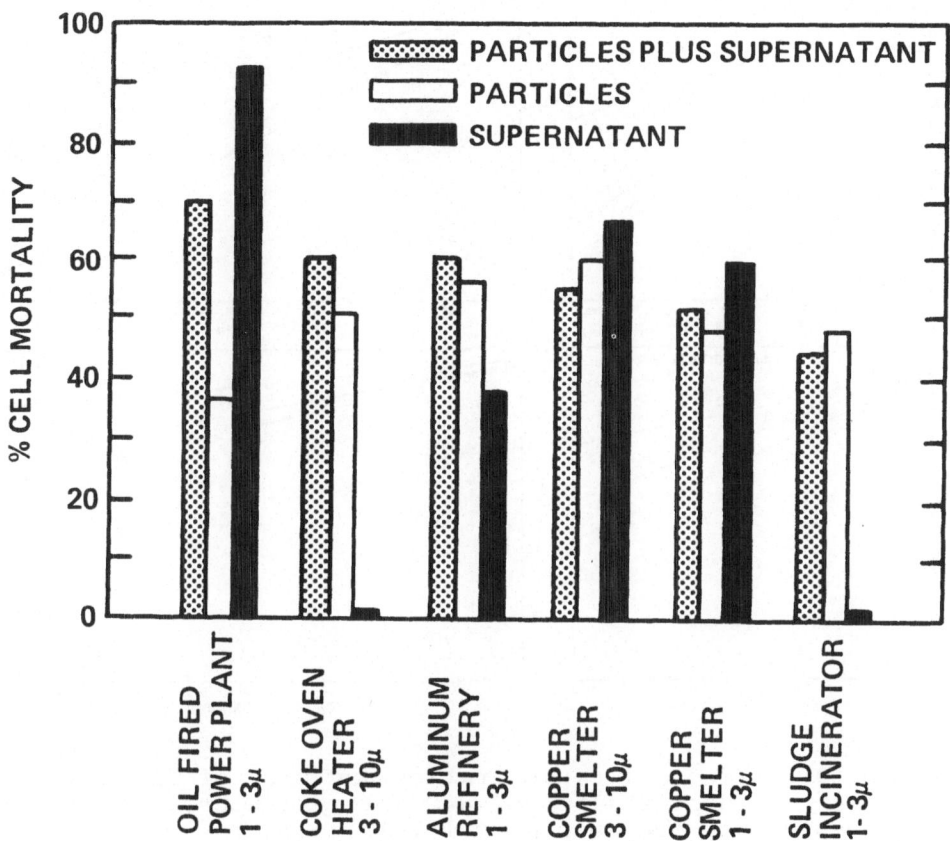

Figure 9. Comparative toxicity of industrial particulate samples tested at 1000 μg/ml in the RAM system as particles preincubated for 20 hours plus supernatant medium, particles without preincubation supernatant medium, preincubation supernatant. Each mode of testing was performed in triplicate and mean values are shown.

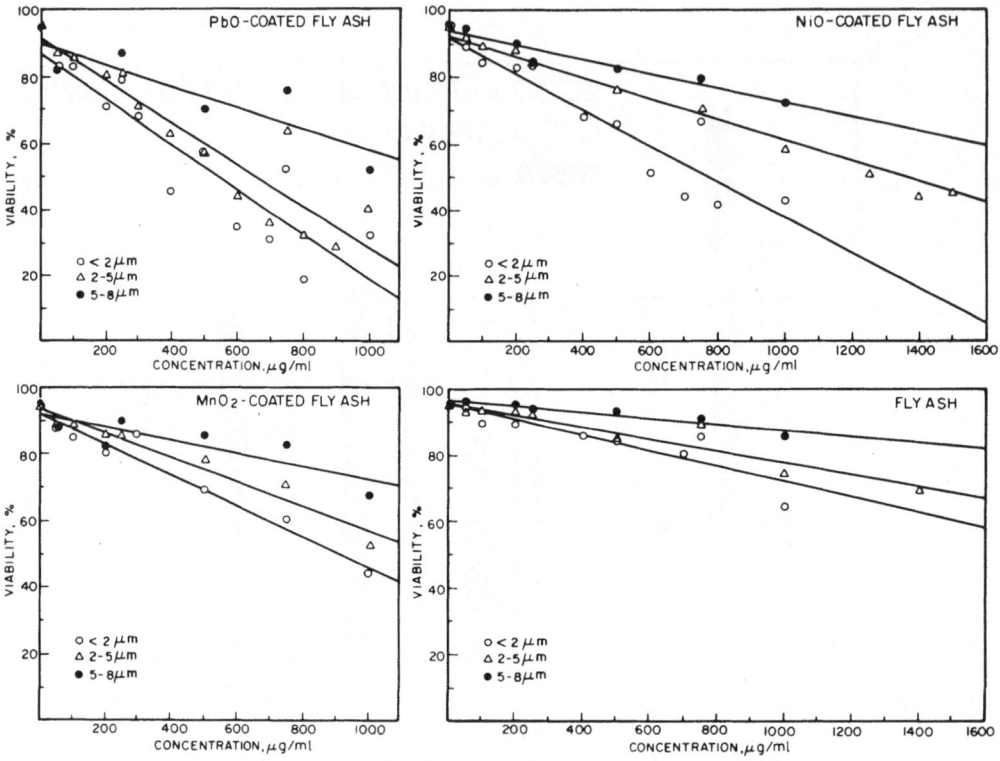

Figure 10. Effect of concentration and size of fly ash particles on the viability of RAM after 21 hours. Each point represents the mean of multiple cultures.

decrease in particle size of the fly ash. Although uncoated fly ash particles reduced macrophage viability to a much lesser extent, the size effect remained clearly evident. Thus, it could be tentatively concluded that the smaller particles, having a greater amount of metal per unit surface area and greater total surface area per unit weight, were more toxic than were the larger particles of the same type having essentially the same percentage of metal by weight.

Table 5 summarizes the data obtained on the concentrations of particles required to reduce macrophage viability to 75 percent in a 21 hour exposure. Units are μg of particles/10^6 macrophages/ml of medium. By comparison with previous data on soluble metal salts, all of the fly ash particles were relatively nontoxic in the dye exclusion test.

Table 5

Estimated Concentration of Particles Required to Reduce Macrophage Viability to 75 Percent After a 21-Hour Exposure

Fly Ash	Particle Concentration, µg 10^{-6} AM ml^{-1}		
	Size Range, 2 µm	Size Range, 2 to 5 µm	Size Range, 6 to 8 µm
Coated With			
PbO	180	270	470
NiO	320	550	870
MnO$_3$	375	510	860
Uncoated	870	1160	*

*Above the tested concentration range.

Several enzyme activities were measured to confirm the influence of the metal coated fly ash on cell viability. As shown in Figure 11, with the lead-coated material, lactic dehydrogenase, a soluble enzyme, was depressed to about the same extent as viability measured by exclusion of trypan blue. LDH is commonly measured as an alternative to the dye exclusion test. Acid phosphatase and β-glucuronidase were depressed to a lesser extent than LDH, and β-glucuronidase specific activity was not significantly altered.

Although the toxicity of metal-coated and uncoated fly ash particles was detectable, greater sensitivity was desirable to permit more careful statistical analysis of the dose-response data. The relative sensitivity of the endpoints: viability, viability index, ATP (expressed both on a per cell and per total protein basis), cell number and total protein was determined using the NiO-coated fly ash. ATP was determined to be the most sensitive endpoint, followed by total protein and viability or viability index and cell number.

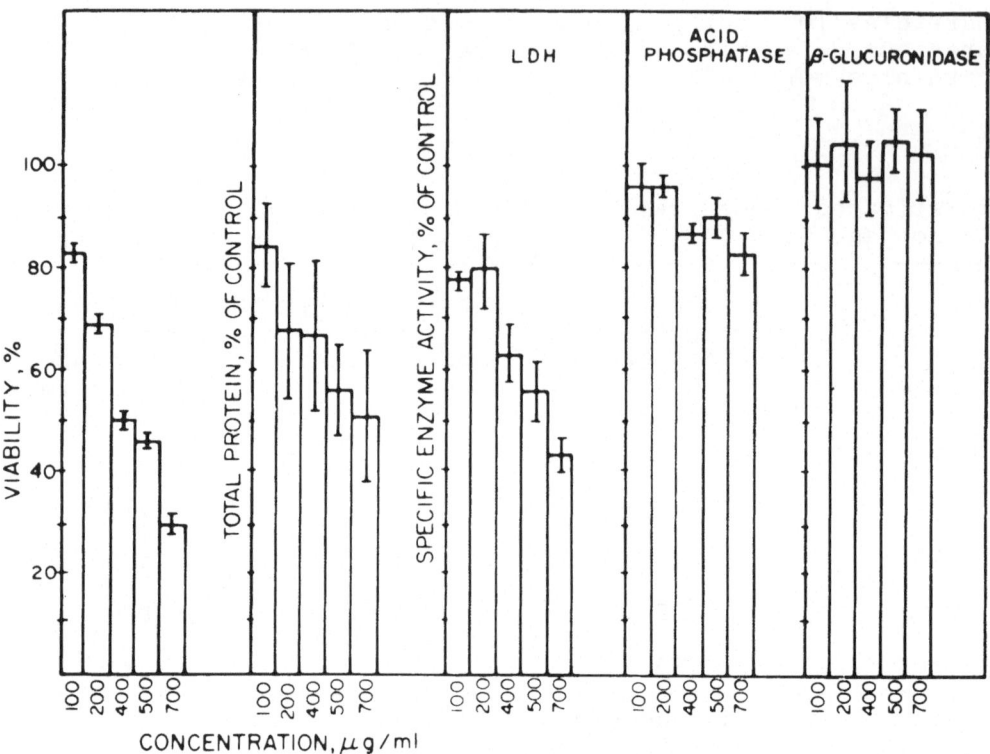

Figure 11. Effect of exposure to PbO-coated fly ash particles of 0-2 μm on RAM viability, total cellular protein content, and specific activity of LDH, acid phosphatase, and
β-glucuronidase. Means ± standard error of three experiments
are shown.

    The same experiment was repeated with several actual
industrial particulate samples. The toxicity of the nickel
coated fly ash particles was not unlike that of the particulate sample shown in Figure 12. This sample was a respirable (0-3 μm) particulate from a conventional coal-fired
power plant. Depressions in viability, viability index,
ATP, and total protein provided evidence of the toxicity of
the sample and ATP was again the most sensitive indicator
of cell damage. These experiments confirmed the relative
sensitivities of the endpoints selected in previous studies
with model particulates.

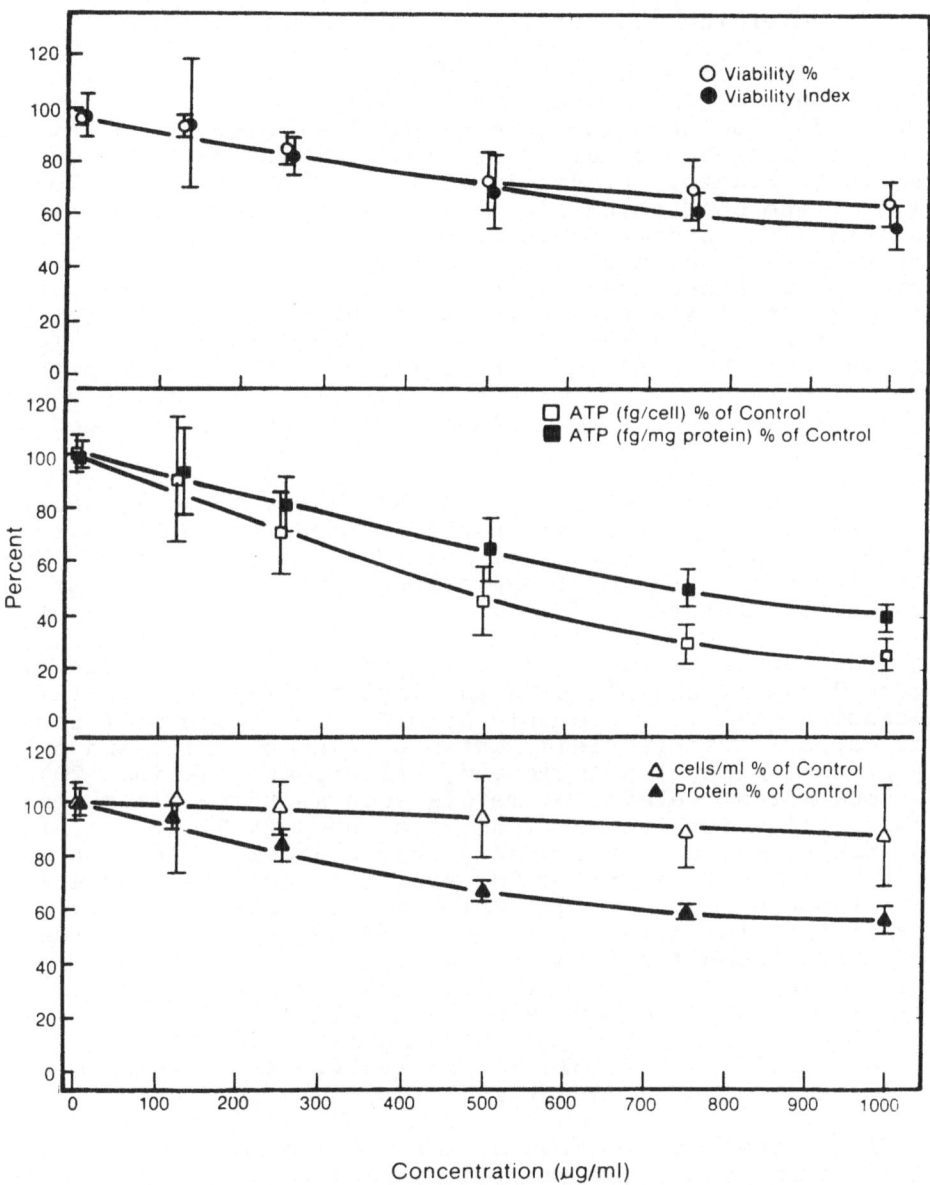

Figure 12. Effect of coal combustion particulate sample (0-3 μm) on the RAM. Mean of two experiments ± standard deviation (n = 6).

To determine whether the overall sensitivity of the
system to particulates could be increased, the effect of
reducing the serum concentration in the culture medium on
the response of the macrophage was investigated. The procedure of reducing serum content has been used to advantage
in making a number of cytotoxicity test systems more sensitive to toxicants. Figure 13 shows that the apparent toxicity of the coal combustion particulate was increased if
the serum concentration was reduced from 10 percent to zero
percent. No effects were seen when the same experiment was
performed with uncoated fly ash. Thus, the sensitivity of
the system to toxic particles can be increased by reducing
the serum concentration of the tissue culture medium. The
data suggest, however, that a particle exhibiting no detectable toxicity cannot be made appreciably toxic in the RAM
system even if the serum content of the media is reduced to
zero.

As discussed previously, the RAM assay was originally
conceived as a system that should respond to leachable toxic
components on the surfaces of respirable particulates. The
possible interaction of these components was a matter of
considerable interest.

The copper smelter sample mentioned in conjunction with
Figure 9 was an example of a particulate likely to contain
leachable metals. The sample itself contained over 1 percent copper, arsenic, lead, antimony, and bismuth; and over
0.1 percent zinc, tin, selenium, silver, and cadmium (20).
The concomitant release of metals such as these offers many
possibilities for interactions. As shown in Figure 14, the
supernatant extract or leachate from 7 μg/ml of the copper
smelter particulate killed 80 percent of cells during a 20
hour exposure (20). To determine whether any of several
known antagonistic interactions would occur in vitro, the
cells were incubated with the same supernatant extract or
leachate in the presence of nontoxic concentrations of
cadmium, mercury, copper, zinc, and selenium. In the presence of 0.15 mM Zn, the percent viability and viability index
increased 1.7- and twofold respectively. Zinc is known to
protect against both copper and cadmium in the intact animal.

In further exploration of this phenomenon using soluble
salts, antagonistic interactions were demonstrated in the
macrophage system between mercury and selenium, cadmium
and selenium, copper and vanadium, and cadmium and zinc
(20). All of these interactions are known or suggested from
whole animal data. Shown in Figure 15 is the interaction

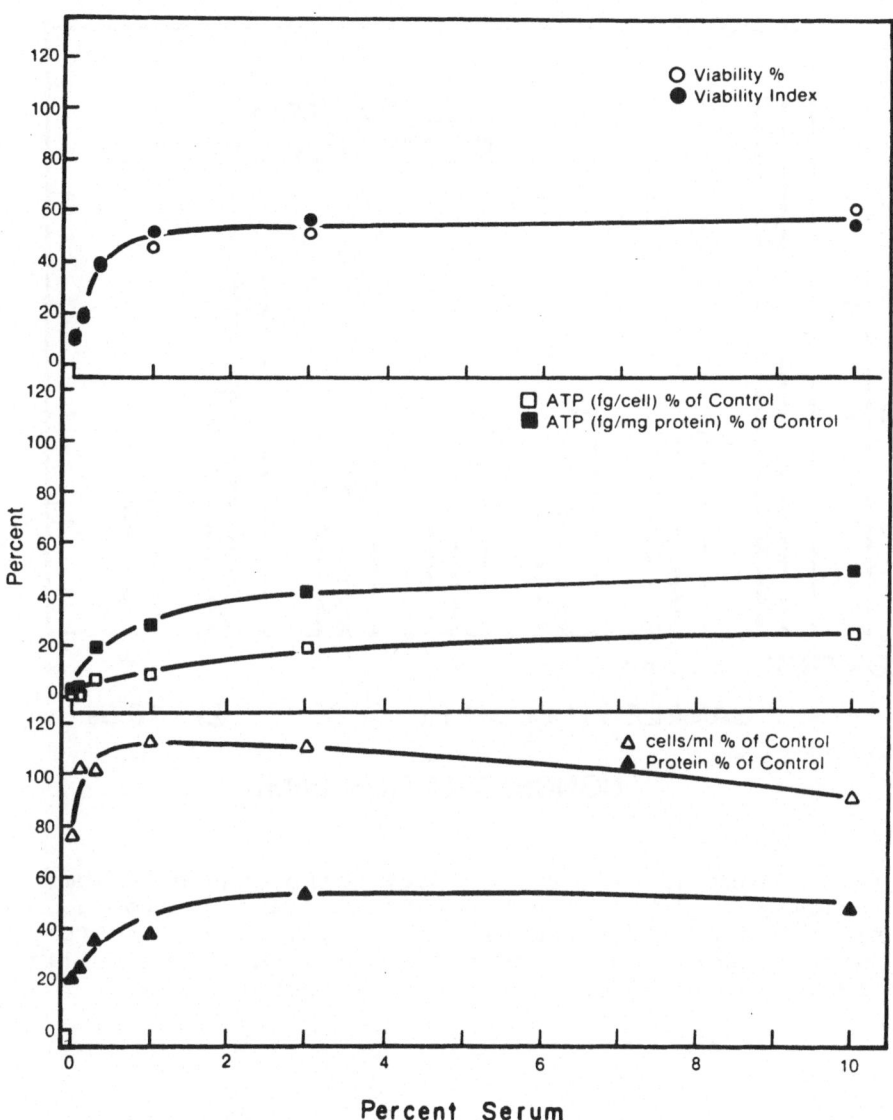

Figure 13. Effect of reduction in fetal bovine serum concentration on the toxicity of coal combustion particulate sample (0-3 μm) on the RAM. Sample tested at 1000 μg/ml of culture medium. Each point represents the mean value of 3 replicate cultures.

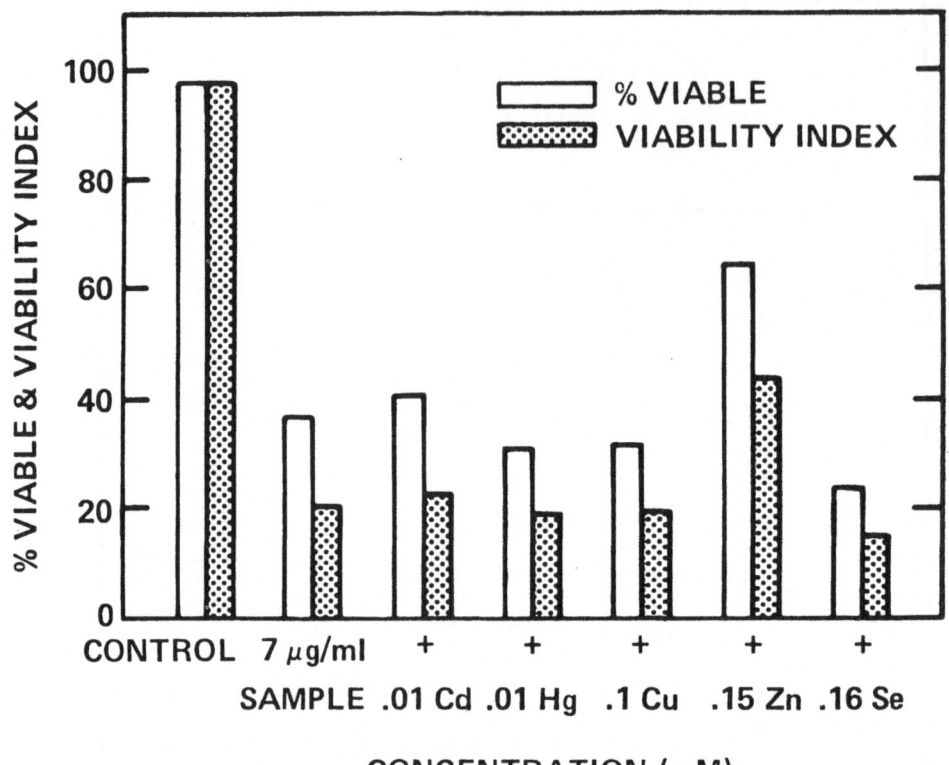

Figure 14. Effect of cadmium, mercury, zinc, and selenium on the toxicity to RAM of the supernatant fraction from a copper smelter particulate sample (3-10 μm). The standard errors per data point for percent viability and for viability index were less than 10 percent of the respective means.

between cadmium and zinc. Exposure of RAM to CdSO₄ (0.08 mM) plus increasing concentrations of ZnSO₄ (from 0.08 to 0.16 mM) showed increasing protection by zinc against cadmium toxicity. All three parameters measured (viability, viability index, and ATP concentrations) increased up to two-, three-, and 4.4-fold, respectively, with increasing concentration of ZnSO₄ up to 0.16 mM. Thus, the optimum Cd:Zn ratio appeared to be

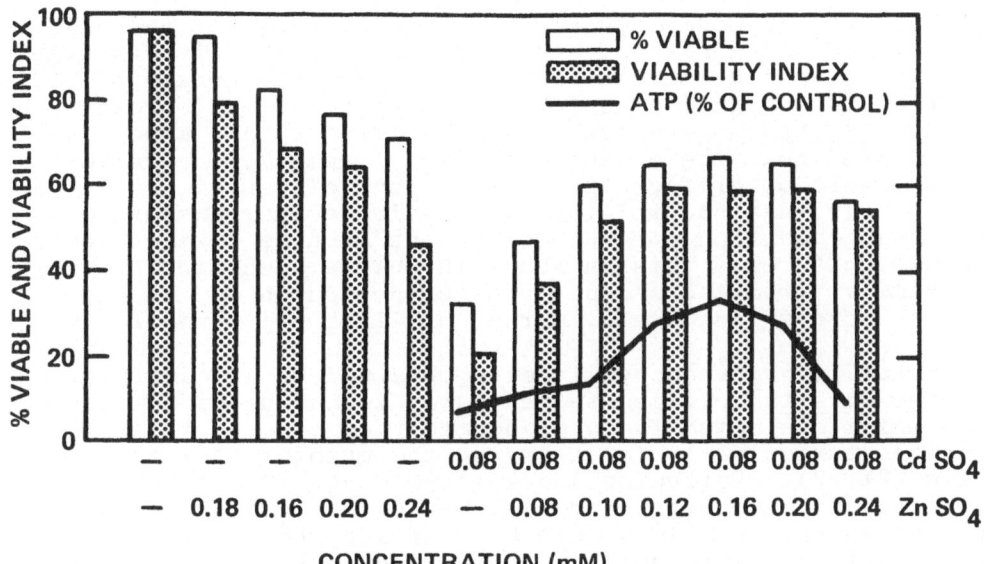

Figure 15. Interaction between $CdSO_4$ and $ZnSO_4$ in RAM after 20 hours. The standard errors for viability and viability index were less than 5 percent of the respective means. The standard errors for ATP content were less than 10 percent of the means.

1:2 and the lack of protection above 0.16 mM $ZnSO_4$ is probably due to increasing toxicity of zinc as indicated by control cultures treated with $ZnSO_4$ alone. Data such as these provide evidence that the RAM system is capable of reflecting toxicant interactions and indicate the potential utility of other isolated cell systems for inhibition and interaction studies.

Liver Parenchymal Cells

      Liver parenchymal cells play a central role in metabolism of both essential and toxic compounds. Virtually every hepatic activity investigated to date is measurable in isolated liver parenchymal cells, although the level of activity may vary with reference to the liver in vivo, especially as a function of time in culture. Some of these activities remain near in vivo levels for several days, including albumin synthesis and secretion, gluconeogenesis from 3-carbon

precursors, glycogen synthesis, responsiveness of the cells
to insulin and glucagon, and inducibility of p-nitroanisole
O-demethylase, a microsomal mixed-function oxygenase (7,9).
Bile salt conjugation, uptake and storage of sulfobromo-
phthalein, and heme catabolism to bilirubin, all appear to be
maintained near the expected in vivo level (7,9). One of
the markers of differentiated functions of hepatic cells in
culture is the inducibility of tyrosine aminotransferase
following addition of glucocorticoids such as dexamethasone
to the cultures. This increase in enzyme specific activity
is similar from the second through the fourth day in culture
(7) and provides a useful marker of liver cell function.

        Rat liver parenchymal cells obtained by a modification
of the in situ collagenase perfusion technique of Bonney (8)
are being evaluated in our laboratory as a metabolically
active "target cell" for studying the cytotoxicity of chemicals (21,22). Perfusion and culture conditions have been
selected to give high viability and cell yields. In this
technique the animal is anesthetized and the liver is per-
fused with buffered collagenase to reduce the liver to a
monocellular suspension. Parenchymal cells are readily
separated from debris and from other cell types by low-speed
centrifugation. After several hours of incubation, the cells
attach, contact adjacent cells, and exhibit a cuboidal morphology. The viability of these cells by the trypan blue
dye exclusion method averages 95 percent after attachment.
Enough primary hepatocytes are obtained from a single rat
to permit a number of cytotoxicity evaluations.

        Exposure of liver cells to soluble metallic salts and
selected organic solvents results in a reduction in cellular
viability, ATP content, and a reduction in the activity and
inducibility of tyrosine aminotransferase. As illustrated
in Table 6, the relative toxicity of a series of metal salts
in this liver culture system as determined by the concentration which reduced the trypan blue viability by 50 percent
(EC_{50}) was generally comparable to that previously described
for RAM. Selenium shows the most significant difference in
toxicity in the two systems; the RAM system is much less
sensitive to the toxic effects of this metal.

        The future value of the rat liver system for evaluating
the toxicity of organic compounds will depend to a great
extent on the metabolic capability of these cells in vitro
as compared to the liver in vivo (21). Studies are currently
in progress to evaluate a series of biochemical and cytological endpoints reflecting the cytotoxicity of organic and
inorganic chemicals in this system (22).

Table 6

Concentrations of Metallic Ions Causing Reduction
in Viability to 50 Percent (EC_{50})
in Primary Rat Hepatocytes After 20 Hours

Ion*	EC_{50} mM	Ion*	EC_{50} mM
Cd^{+2}	0.010	Hg^{+2}	0.100
VO_3^{-1}	0.050	CrO_4^{-2}	1.000
AsO_2^{-1}	0.055	Zn^{+2}	1.700
SeO_2^{-2}	0.060	Cr^{+3}	2.000

*Sodium or chloride salts were employed.

COMPARATIVE CYTOTOXICITY STUDIES

The relative cytotoxic response of different cell systems to environmental samples is of considerable interest
and importance. In order to study and compare the cytotoxic
response of several different cell types, the following
experiments were performed. The toxicity of four liquid
textile mill effluents in vitro was examined using the rabbit alveolar macrophage (RAM), the diploid WI-38 human lung
fibroblast (WI-38), and the aneuploid Chinese hamster ovary
cell (CHO) (18). Cultures of macrophages or WI-38 cells
were incubated in the presence of aliquots of the textile
waste water samples for a 20 hour period. For purposes of
the comparative cytotoxicity studies, cellular ATP was used
as the principal cytotoxicity endpoint.

Toxicity of the effluent samples to CHO cells was
evaluated by examination of clonal growth after a six-day
incubation period. As illustrated in Figure 16, assay of
ATP in the RAM and WI-38 cultures indicated that these cells
were approximately equivalent in their response to the liquid
effluents. The sensitivity of the CHO clonal assay was equal
to or greater than that of the other systems. However, one
sample, Sample R, found to be relatively nontoxic in the

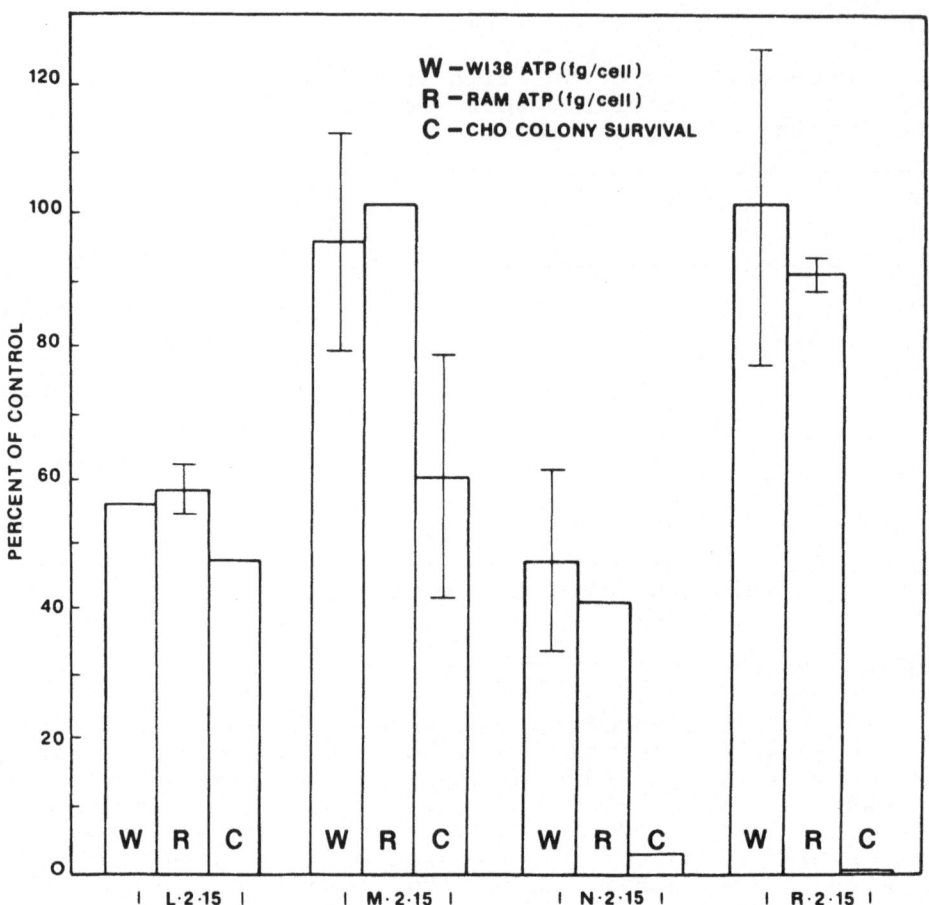

Figure 16. Comparative toxic effects of textile effluent
samples on WI-38 and RAM cell, ATP content, and on CHO cell
colony survival.

WI-38 and RAM, was highly toxic in the CHO system. The
result was reproducible and suggested that the clonal assay
may offer a different measure of relative toxicity than mass
cultures of dividing cells such as WI-38 or nondividing
cells such as RAM. Clonal growth should provide amplifica-
tion of the effect of toxicants which influence a specific
portion of the cell cycle. Such agents may go undetected

over an exposure interval that respresents only one or two cell cycles. Alternatively, other mechanisms may be involved; for example, surface active agents may dimish clonal growth. The agreement seen for a number of endpoints in the WI-38 and RAM systems suggests that similar mechanisms of cytotoxicity are operative in these two systems.

Since clonal growth is conventionally used to correct for cytotoxicity in mammalian cell mutagenicity and oncogenic transformation assays, the differing responses observed raise questions as to whether such a means of standardization relates directly to the acute cytotoxic response discussed in this paper. These issues may be resolved by additional comparative cytotoxicity studies using a variety of cell systems and toxicants.

CONCLUSION

In conclusion, in vitro cytotoxicity testing, while one of the oldest short-term test areas, is still in its infancy, especially as applied to complex environmental mixtures.

Several recommendations could be made for future work in the area.

- That increased effort go into the cultivation of cell types which are truly representative of the metabolically active target organs which receive direct and/or indirect exposure to environmental toxicants.

- That both directly and indirectly exposed target cell systems be fully characterized with respect to their ability to activate and detoxify a variety of classes of environmental agents.

- That increased effort be put forth to isolate and maintain in culture the mammalian and human cell epithelial types that are directly exposed as targets of environmental chemicals.

- That comparative in vitro and in vivo studies be undertaken with the same samples to provide direct information on the predictive capabilities of a variety of in vitro toxicity test systems.

- That the endpoints of somatic mutation and oncogenic transformation be pursued in epithelial cell
  systems, as well as in conventional cell systems,
  so that a better understanding of the relationship
  among the various manifestations of the cytotoxic
  response to environmental chemicals may be achieved.
  It is only through simultaneous measurement of the
  various biological activities of environmental
  chemicals that we will be able to fully characterize their toxicity and genotoxicity.

- That collaborative and comparative studies with
  existing cytotoxicity, mutagenicity, and oncogenic
  transformation bioassays be encouraged in order to
  define further their attributes and limitations in
  research on complex mixtures.

REFERENCES

1. Allison AC: Lysosomes and the responses of cells to
    toxic materials. Sci Basis Med Ann Rev 18-30, 1968

2. Allison AC: On the role of macrophages in some patho-
    logical processes. In: Mononuclear Phagocytes (Ralph
    Van Furth, ed.). Oxford, Blackwell Sci Pub, 1970, pp
    420-444

3. Aranyi C, Andres S, Ehrlich R, Fenters JD, Gardner DE,
    Waters MD: Cytotoxicity to alveolar macrophages of
    metal oxides absorbed on fly ash. In: Pulmonary
    Macrophage and Epithelial Cells, Conf-760972 (Sanders
    C, Schneider RP, Dagle GE, Ragan HA, eds.). ERDA Sym-
    posium Series 43, Technical Information Center, Energy
    Research and Development Administration, 1977, pp 58-65

4. Aranyi C, Miller FJ, Wiehle SA, Ehrlich R, Fenters J,
    Gardner DE, Waters MD: Cytotoxicity to alveolar macro-
    phages of trace metals absorbed on fly ash. In prepara-
    tion

5. Autian J, Guess W: A new field of plastics toxicology-
    methods and results. CRC Critical Review in Toxicology
    1:39, 1973

6. Berky JJ, Sherrod PC (eds.): Short term in vitro
    testing for carcinogenesis, mutagenesis and toxicity.
    Philadelphia, The Franklin Institute Press, November
    1977

7.   Bissell DM, Guzelian PS: Microsomal functions and
     phenotypic change in adult rat hepatocytes. In:
     Primary Monolayer Culture in Gene Expression and Car-
     cinogenesis in Cultured Liver (Gerschenson LE, Thompson
     EB, eds.). Academic Press, NY, 1975, pp 119-136

8.   Bonney RJ: Adult liver parenchymal cells in primary
     culture: characteristics and cell recognition stan-
     dards. In Vitro 10:130-142, 1974

9.   Bonney RJ, Maley F: Some characteristics and functions
     of adult rat liver parenchymal cells in primary culture.
     In: Gene Expression and Carcinogenesis in Cultured
     Liver (Gerschenson LE, Thompson EB, eds.). Academic
     Press, NY, 1975, pp 24-45

10. Christian RT, Cody TE, Clark CS, Lingg R, Cleary EJ:
     Development of a biological chemical test for the
     potability of water. AIChE 70:15-21, 1973

11. Christian RT, Cooke J, Elia VJ, Cody TE: The effects
     of aqueous and simple extracts of coal on cultured
     mammalian cells. Proceedings of the Second National
     Conference on Complete Water Reuse. Water's Interface
     with Energy, Air and Solids, sponsored by AIChE and
     EPA Technology Transfer, Chicago, Illinois, May 4-8,
     1975

12. Christian RT, Nelson J: Coal: responses of cultured
     mammalian cells corresponds to the incidence of coal
     workers pneumoconiosis. Environ Res, in press

13. Cody TE, Christian RT, Elia VJ, Clark CS: Cell culture
     as a toxicity bioassay for potable reuse water. Pro-
     ceedings of the Second National Conference on Complete
     Water Reuse. Water's Interface with Energy, Air and
     Solids, sponsored by AIChE and EPA Technology Transfer,
     Chicago, Illinois, May 4-8, 1975

14. Coffin DL, Gardner DE, Holzman RS, Wolock FJ: Influence
     of ozone on pulmonary cells. Arch Environ Health 16:633-
     636, 1968

15. Cross CE, Ibrahim AB, Ahmed M, Mustafa MG: Effect of
     cadmium ion on respiration and ATPase activity of the
     pulmonary alveolar macrophage: a model for the study
     of environmental interference with pulmonary cell func-
     tion. Environ Res 3:512-520, 1970

16. Dawson M: Cellular pharmacology, the effects of drugs
     on living vertebrate cells in vitro. Charles C.
     Thomas, Publisher, Springfield, 1972

17. Gardner DE, Holzman RS, Coffin DL: Effects of nitrogen
     dioxide on pulmonary cell population. J Bacteriol 98:
     1041-1043, 1969

18. Garrett NE, Campbell JA, Huisingh JL, Waters MD: The
     use of short-term bioassay systems in the evaluation of
     environmental particulates. Proceedings of the "Sym-
     posium on the Transfer and Utilization of Particulate
     Control Technology," Denver, Colorado, July 24, 1978,
     in press

19. Guess WL, Rosenbluth SA, Schmidt B, Autian J; Agar
     diffusion method for toxicity screening of plastics on
     cultured cell monolayers. J Pharm Sci 54:1545, 1965

20. Huisingh JL, Campbell JA, Waters MD: Evaluation of
     trace element interactions using cultured alveolar
     macrophages. In: Pulmonary Macrophage and Epithelial
     Cells Conf-760972, (Sanders CL, Schneider RP, Dagle GE,
     Ragan HA, eds.). ERDA Symposium Series 43, Technical
     Information Center, Energy Research and Development
     Administration, 1977, pp 346-357

21. Huisingh JL, Inmon JP, King LC, Williams K, Waters MD:
     The use of primary rat liver parenchymal cells in eval-
     uating cellular responses to toxic metals and carcino-
     genic polycyclic aromatic hydrocarbons. In Vitro, 1977

22. Huisingh JL, Nesnow S, Selkirk JK, Inman JP, Bergman H,
     Russell B, Waters MD: Comparison of benzo[a]pyrene
     metabolite formation in liver microsomes and cultured
     primary rat liver parenchymal cells. In Vitro, 1978

23. Hurst DJ, Gardner DE, Coffin DL: Effect of ozone on
     acid hydrolases of the pulmonary alveolar macrophage.
     J Reticuloendothelial Soc 8:288-300, 1970

24. LaBelle CW, Breiger H: The fate of inhaled particles
     in the early post exposure period. II. The role of
     pulmonary phagocytosis. Arch Environ Health 1:423-427,
     1960

25. Lainhart WS: Roentgenographic evidence of coal workers' pneumoconiosis in three geographic areas in the United States. J Occup Med 11:399-408, 1969

26. Lee RE, Von Lehmden DJ: Trace metal pollution in the environment. APCA J 23:853-857

27. Leuchtenberger E, Leuchtenberger R: Cytologic and cyto-chemical effects on primary mouse kidney tissue and lung organ cultures after exposure to whole, fresh smoke and its gas phase from unfiltered charcoal-filtered, and cigar tobacco cigarettes. Can Res 29:862-872, 1969

28. Litterst CL, Lichtenstein EP, Kajiwara K: Effects of insecticides on growth of HeLa cells. J Agr Food Chem 17:1199-1203, 1969

29. Mahar H: Evaluation of selected methods for chemical and biological testing of industrial particulate emis-sions. Environmental Protection Technology Series EPA-600/2-76-137, May 1976

30. Marks J, Mason MA, Nagleschmidt G: A study of dust toxicity using a quantitative tissue culture technique. Brit J Industr Med 13:187-191, 1956

31. Marks J, Mason MA: A quantitative technique for study-ing the effect of dust on phagocytic cells in vitro. Brit J Industr Med 13:192-195, 1956

32. Melamid MR, Kamentsky LA, Boyse EA: Cytotoxic test automation: a live-dead cell differential counter. Science 163:285-286, 1969

33. Myrvik QN, Leake ES, Farris B: Studies on pulmonary alveolar macrophages from the normal rabbit: a tech-nique to procure them in a high state of purity. J Immunol 86:128-132, 1961

34. Nardone RM: Toxicity testing in vitro. In: Growth, Nutrition, and Metabolism of Cells in Culture, vol. 3. (Rothblat G and Cristosalo V, eds.). New York, Academic Press, 1978, pp 471-495

35. Natusch DFS, Wallace JR, Evans CN Jr: Toxic trace ele-ments: preferential concentration in respirable parti-cles. Science 183:202-204, 1973

36. Ormsbee RA, Cornmen I: The place of tissue culture in
     a cancer chemotherapy screening program. Can Res 8:
     384-385, 1948

37. Pearsall NN, Weiser RS: The macrophage. Philadelphia,
     Lea & Febiger, 1970

38. Rolfe PC: Tissue culture and toxicology. Fd Cosmet
     Toxicol 9:683-696, 1971

39. Rosener VM, Jacobson W: Tissue culture in pharmacology.
     In: Cells and Tissues in Culture, Vol. 3 (Willmer EN,
     ed.). London, Academic Press, 1966, pp 351-396

40. Rounds DE, Awa A, Pomerat CM: Effect of automobile
     exhaust on cell growth in vitro. Arch Environ Health
     5:319-324, 1962

41. Rounds DE: Environmental influences on living cells.
     Arch Environ Health 12:78-84, 1966

42. Schepartz S, MacDonald M, Leiter J: The use of cell
     culture as a presumptive screen for antitumor agents.
     Proc Amer Assoc Cancer Res 3:265, 1961

43. Schindler R: Use of cell culture in pharmacology.
     Ann Rev Pharmacol 9:393-406, 1969

44. Schmidt JL, McIntire FC, Martin DL, Hawthorne MA,
     Richards RK: The relationships among different in vivo
     properties of local anesthetics and toxicity to cell
     cultures in vitro toxicol. Appl Pharmacol 1:454-461,
     1959

45. Shelburne JD, Ingram P, Hawkins HK, Waters MD: X-ray
     mapping of lysosomal inclusions. Scanning Electron
     Microscopy/1976 (Part III) Proceedings of the Workshop
     on Techniques for Particulate Matter Studies in SEM,
     IIT Research Institute, Chicago, Illinois, April, 1976

46. Trump BF, Valigorsky SM, Dees JH, Mergner WJ, Kim KM,
     Jones RT, Pendergrass RF, Garbus J, Cowley RA: Cellular
     change in human disease. A new method of pathological
     analysis. Human Pathol 4:89, 1973

47. Walton JR, Buckley IK: Cell models in the study of mech-
     anisms of toxicity. Agents and Actions 5:69-88, 1975

48. Waters MD, Abernethy DR, Garland HR, Coffin DL: Toxic effects of selected metallic salts on strain WI-38 human lung fibroblasts. In Vitro 10:342, 1974

49. Waters MD, Garnder DE, Coffin DL: Cytotoxic effects of vanadium on rabbit alveolar macrophages in vitro. Toxicol Appl Pharmacol 28:253-263, 1974

50. Waters MD, Vaughan TO, Campbell JA, Miller FJ, Coffin DL: Screening studies on metallic salts using the rabbit alveolar macrophage. In Vitro 10:342-343, 1974

51. Waters MD, Vaughan TO, Campbell JA, Stead AG, Coffin DL: Adenosine triphosphate concentration and phagocytic activity in rabbit alveolar macrophages exposed to divalent cations. J Reticuloendothelial Soc 18:296, 1975

52. Waters MD, Gardner DE, Aranyi C, Coffin DL: Metal toxicity for rabbit alveolar macrophages in vitro. Environ Res 9:32-47, 1975

53. Waters MD, Vaughan TO, Abernethy DJ, Garland HR, Cox CC, Coffin DL: Toxicity of platinum (IV) salts for cells of pulmonary origin. Environ Health Perspectives 12:45-56, 1975

54. Weissbecker L, Carpenter RD, Luchsinger PC, Osdene TS: In vitro alveolar macrophage viability-effect of gases. Arch Environ Health 18:756-759, 1969

55. Yamate G, Ashley H: Preparation and characterization of finely divided particulate environmental contaminants for biological experiments. IITRI Report No. C6321-5, IIT Research Institute, Chicago, Illinois, September, 1975

SECTION 2

COLLECTION AND
CHEMICAL ANALYSIS OF
ENVIRONMENTAL SAMPLES

ATMOSPHERIC GENOTOXICANTS— WHAT NUMBERS DO WE COLLECT?

Eugene Sawicki
Environmental Sciences Research Laboratory
U.S. Environmental Protection Agency
Research Triangle Park, North Carolina

To advance the estimation of human environmental risk, especially cancer prevention, from the body-count phase to realistic extrapolation, we need to carry out more sophisticated carcinogenic studies with animals and submammalian species mimicking the human condition. We have seemingly unsurmountable difficulties with our perspective in the study of human carcinogenesis because the paucity of our environmental data and the simplicity of our carcinogen models of purebred animals and "pure" chemicals so misleads us as to obscure reality. Before we can begin to understand human chemical carcinogenesis, we need to know the genetic background of the individual, the key genotoxicant(s) to which the individual is heavily exposed, and the families of genotoxicants in the individual's environment. We are exposed to genotoxicants because of our particular life style (e.g., cigarette smoking, drugs, medicines, and cosmetics), and because the chemicals are present in the life-supporting environment (e.g., food, water, and air) we share with other people.

Since nearly all of our polluted environment has never been satisfactorily investigated, much data need to be collected. But first, necessary analytical instrumentation and methodology have to be developed and/or perfected. Research is needed now on:

> The expansion of the methodology to many other types of pollutants.

- The unequivocal identification of atmospheric chemicals.

- The quantitation of the genotoxicants.

- The optimization of the accuracy and precision.

- The recognition of the limitations of the methods.

If this research is not done, then the premature and indiscriminate introduction into routine practice of the new poly-pollutant monitoring methods for atmospheric chemicals not only will result in some damage to present industrial activity, but also will compromise a completely researched application of these methods to the prevention of cancer and other genotoxic problems.

The main barriers to estimating human environmental risk stem from a lack of knowledge of the chemical composition of our environment, the failure to use the information we have, and our indecision as to what to measure. An example is the current inadequacy in measuring the key genotoxicants in carcinogenic coke oven effluents. These problems probably stem from the fact that only an inadequate fraction of the massive resources committed to cancer programs has been allocated to identifying carcinogens and their many cofactors, determining their environmental concentrations, and estimating exposures of high risk groups.

Let's just look at what needs to be done to determine the chemical composition of the polluted atmosphere, an area we know so little about. The different materials that need to be investigated are summarized in Table 1. The available sampling methods for the gaseous constituents of the air are inadequate. We could use separate specific cartridges for the collection of gases (b.p. <40°), polar vapors, non-polar vapors (b.p. 40°–250°), and high boiling vapors (b.p. 250°–400°). The extraction procedure for airborne particles has never been developed and routinely used for total extraction of the organic material followed by subfractionation into aliphatic, aromatic, neutral oxygenated, weak acid, strong acid, basic, and water-soluble fractions. In addition, an extraction-analytical procedure should be developed for the water soluble inorganic and organic cations and anions. These extraction methods would involve Polytron or ultrasonic extraction methods at lowered temperatures. The organic material would be analyzed by GC-MS-COMP with help from HPLC-MS-COMP and FTIC. The anions and cations would be analyzed by ion chromatography.

Table 1

Polypollutant Methods of Assay for Air Pollutants

Material	Sampling	Analysis[1]

Particulate organics
  Aliphatic hydrocarbons
  PAH[2]
  Acids
  Bases
  Phenolic compounds
  Neutral oxygenates
  Water solubles
	Hi-Vol glass fiber filter	LC GC MS COMP

Particulate PAH		LC HPLC SP, SPF
Organic gases	Cartridge	GC MS COMP
Organic vapors, nonpolar[3]	Tenax GC cartridge	GC MS COMP
Organic vapors, polar[3]	Cartridge	GC MS COMP
Particulate metals	Teflon filter	XRF, AA, etc.
Particulate anions and cations	Hi-Vol quartz filter, Lo-Vol teflon filter	IC

[1] AA = atomic absorption, COMP = computer, GC = gas chromatography, HPLC – high pressure liquid chromatography, IC = ion chromatography, LC = liquid chromatography, MS = mass spectrometry, SP = UV absorption spectrophotometry, SPF = spectrophotofluorimetry, XRF = X-ray fluorescence

[2] PAH = polynuclear aromatic hydrocarbons

[3] And also cartridges for the high boiling vapors of this type and for using GC–MS–COMP

Since there are such a tremendous number of organic
compounds in our environment, a much better separation of
these compounds is desirable before their analysis. One
method is by increasing the reliability and resolution of
capillary gas chromatography. The possibilities as reported
by G. Grob and K. Grob are shown in Table 2.

Table 2

Resolution of GC Peaks

(Water extract - OV-1 Columns - Grob and Grob)

I.D. (m x mm)	Column	No. Peaks
3 x 2	Packed	118
60 x 0.6	Glass capillary	320
35 x 0.28	Glass capillary	490

A reliable routine automated system of qualitative and
quantitative analysis of the highly complicated mixtures in
our polluted environment could be developed from a promising
method, termed HISLIB, that compares combined gas chromato-
graphic/mass spectrometric profiles of new environmental mixtures with historical libraries of GC/MS data on related mixtures (34). The presence of several components is established
by matching retention indexes and mass spectra after removal
of column bleed, contamination, and other types of background
and after resolution of overlapping GC components. The system
is quantified with the help of internal standards by comparing
relative concentrations of components.

The simplification and standardization of routine screening methods for the key genotoxicants are desirable. These
key chemicals are in high production, are present in the
atmosphere in high concentration, or are carcinogenic to humans.
Examples of some of the genotoxicants carcinogenic to humans
are given in Table 3, and some of the carcinogens found in
the polluted atmosphere are listed in Table 4.

Although a fairly large amount of information is available
on the carcinogenic and mutagenic activities, the reaction with
DNA, and the metabolic properties of the atmospheric carcinogens, little work has been done with the various families of

Table 3

Genotoxicants Carcinogenic to Humans

Genotoxicant	Target	Pathway
4-Aminobiphenyl	Bladder	Inhalation, oral
Arsenic	Lung	Inhalation
Asbestos	Lung, pleural cavity	Inhalation
Auramine	Bladder	Inhalation, oral, skin
Benzene	Bone marrow	Inhalation, skin
Benzidine	Bladder	Inhalation, oral, skin
Bis-chloromethyl ether	Lung	Inhalation
Cadmium	Prostate	Inhalation, oral
Chimney soot	Scrotum	Skin
Chloroprene	Lung, skin	Inhalation, skin
Chromate	Lung	Inhalation
Coal hydrogenation vapors	Skin	Skin
Coal tar pitch	Lung, skin	Inhalation, skin
Coke oven effluents	Bladder, lung	Inhalation
Creosote oils	Lung, skin	Inhalation, skin
Cutting oils	Lung, scrotum	Inhalation, skin
Hematite	Lung	Inhalation
Isopropyl oil	Nasal cavity, larynx	Inhalation, skin
Mineral oil	Scrotum	Skin
Mustard gas	Lung, larynx	Inhalation, skin
β-Naphthylamine	Bladder	Inhalation, oral, skin
Nickel	Nasal cavity, lung	Inhalation
Petroleum wax	Scrotum	Skin
Radium	Lung	Inhalation
Radon and radon daughters	Lung	Inhalation
Rubber plant effluents	Brain	Inhalation
Shale oil	Lung	Inhalation
Soots, tars and oils	Lung, scrotum	Inhalation, skin
Vinyl chloride	Brain, liver, lung	Inhalation, skin
Wood dust	Nasal cavities	Inhalation

Table 4

Carcinogenic Air Pollutants

Compound[1]	Remarks[2]
Acrylonitrile	C,h(inh),r(oral)
Aldrin	C,m,r(oral)
Anthanthrene	C,m(skin)
Arsenic (III)	C,h(oral,skin)
Asbestos	C,h,r(inh)
BaA	C,m(oral,skin,sc)
BaCAR	C,m(sc,skin),r(skin)
BaP	C,9 species(it,oral,skin)
BbFT	C,m(sc,skin)
BcACR	C,m(skin),r(bi)
BcCAR	C
Benzene	C,h(inh),m(inh),r(oral,inh)
Benzyl chloride	C,r(sc)
Be (II)	C,mk(inh),r(inh),rb(iv)
BeP	C,m(skin)
BHC	C,m(oral)
Bis-chloromethyl ether	C,h(inh),m(inh,sc,skin)
BjFT	C,m(skin)
C	C,m(sc,skin)
Carbon tetrachloride	C,ha,m,r(inh,oral)
Cd (II)	C,h(inh),r(im,sc)
Chloroform	C,m(oral),r(oral,sc)
Chloromethyl methyl ether	C,h(inh),m(skin),r(inh,skin)
Chloroprene	C,h(inh,skin)
Chromium (VI)	C,h(inh),r(im,ipl)

[1]A = anthracene, ACR = acridine, B = benzo, BHC = α-benzene hexachloride, C = chrysene, CAR = carbazole, DB = dibenzo, DDD = 1,1-dichloro-2,2-bis(p-chlorophenyl) ethane, DDE = 1,1-dichloro-2,2-bis(p-chlorophenyl) ethylene, DDT = 1,1,1-trichloro-2,2-bis(p-chlorophenyl) ethane, FT = fluoranthene, IND = indeno, P = pyrene, and PEP = pentaphene. Thus, BaP = benzo(a)pyrene while IND 1,2,3-cdP = indeno(1,2,3-cd) pyrene.

[2]bi = bladder implantation, C = carcinogenic, d = dog, gp = guinea pig, h = human, ha = hamster, im = intramuscular, inh = inhalation, ip = intraperitoneal, ipl = intrapleural, it = intratracheal, iv = intravenous, m = mice, pn = prenatal exposure following iv injection in pregnant female, r = rat, and sc = subcutaneous.

Table 4 (continued)

Dimethylnitrosamine	C, 16 species
Dimethyl sulfate	C,r(inh,pn,sc)
p-Dioxane	C,gp(oral),r(oral)
Ethylene dibromide	C,m(oral),r(oral)
Hematite	C,m(inh)
Heptachlor	C,m(oral)
IND 1,2,3,-cdP	C,m(skin)
Kepone	C,m,r(oral)
Lead (II)	C,m,r(oral)
Lindane	C,m(oral)
Methyl iodide	C,m(ip),r(sc)
Mirex	C,m(oral)
Nickel (III)	C,h(inh),m,r(im)
Perchloroethylene	C,m(oral)
Propylene oxide	C,r(sc)
Quinoline	C,r(oral)
Styrene oxide	C,m(skin)
o-Toluidine	C,m,r
p-Toluidine	C,m
Trichloroethylene	C,m(oral)
Vinyl chloride	C,h(inh)
Vinylidene chloride	C,m(inh)
DBaeP	C,m(sc,skin)
DBahA	C,6 species(oral,it,sc,skin)
DBahACR	C,m(sc,skin)
DBahP	C,m(im,skin),r(im,skin)
DBaiP	C,ha(sc,skin),m(sc,skin)
DBajACR	C,m(sc,skin)
DBalP	C,m(sc)
DBcgCAR	C,d(bi),ha(inh,it),m(bi,im, ip,iv,oral,sc,skin),r(im,sc)
DBh,rstPEP	C,m(sc)
DDD	C,m
DDE	C,m
DDT	C,m(oral)
Dieldrin	C,m(oral)
Diethylnitrosamine	C, 16 species
1,1-Dimethylhydrazine	C,m(oral)

genotoxicants that usually are associated in the polluted
atmosphere (Table 5). Analytical methodology for these
families as a family of one (e.g., total aliphatic hydrocarbons) or as a family of individuals, needs to be developed
further, perfected, and used in cancer prevention studies.
In the same way, the carcinogenic and mutagenic activities,
reaction with DNA, and the metabolic properties of these
families need to be determined, especially in ways that mimic
the human situation.

Table 5

Families of Genotoxicants in Air

Aldehydes
Aliphatic amines
Aromatic amines and precursors
Asbestos
Azaarenes (mono and dicyclic)
Azaarenes (polycyclic)
Benzene derivatives
Epoxides
Halogenated alkanes
Halogenated alkenes
Halogenated ring compounds
Long chain aliphatic acids
Long chain aliphatic alcohols
Long chain aliphatic hydrocarbons
Long chain aliphatic esters
Metals and their compounds
Nitrosamines
NO_x
Olefins
Oxidants (O_3, NO_2, PAN)
PAH (Di- and tricyclic)
PAH (tetra-, penta- and hexacyclic)
SO_x

The most extensive and important interfacing that man
has with his environment is through his respiratory membrane.
Each day this membrane, which has a surface area as large as
a tennis court, is exposed to a volume of contaminated air that
would fill a 15 m swimming pool (18). From the genotoxic
viewpoint, it would be of value to know the relative amounts
of atmospheric organic gases, vapors, and particles in contact

with this epithelial tissue.   (The reason for dividing air
pollutants into these three states is primarily because of
sampling protocol.)

     Normally, gases such as SO_2 are almost completely
assimilated by the nose.   When these gases are absorbed on
respirable particles, they can penetrate the lower respiratory tract more easily.   In addition, any circumstance dis-
posing to breathing through the mouth is likely to increase
exposure of the lung to pollutants.   Respirable particles
constitute that portion of the inhaled particles which penetrate to the non-ciliated portions of the lung.   The vapors,
being liquids or solids in the neat state, are more readily
retained in the respiratory tract than are the gases, other
conditions being equal.   In dogs, the respiratory retention
of inhaled benzene and toluene in the upper and lower respiratory tracts was very high (15).

     Regarding atmospheric vapors, we certainly need a
reordering of priorities in our research on environmental
genotoxicity.   Organic vapors can be present in many industrial atmospheres in four orders of magnitude larger than
the total airborne particulates (33).   Some of the carcinogens in the vapor state can be present in five to six orders
of magnitude larger than the carcinogens in the solid state.
The high amount of genotoxic vapors produced in the United
States is shown in Table 6 (3).   This means that genotoxic
vapors such as toluene, trichloroethane, trichloroethylene,
vinyl chloride, and benzene are emitted in much larger
amounts compared to benzo(a)pyrene, and BaP is emitted into
the air in greater amounts as compared to DDT or the PCB.

     With an increase in coal consumption to 665 million
tons in 1976 and a projected increase to 1.27 billion tons
in 1985 (6), the pollutants in airborne particulates contaminated with coal combustion products may become of greater
importance.   The same tenuosity and delicacy that qualify
the air-blood barrier in our lungs for the rapid exchange
of oxygen and carbon dioxide reduce its effectiveness as a
barrier to inhaled genotoxic gases, vapors, and particles.
Particles greater than 10µ are taken out by the filter
system of the nose.   Particles of 2-10µ settle on the walls
of the trachea, the bronchi, and the bronchioles.   Particles
in the range of 0.3-2µ reach the alveolar ducts and alveoli
while those less than 0.3µ, if not taken up by the blood,
are cleansed from the lungs with air.   The deposition can be
as low as 10-15% for particles in the size range 0.5-1.0µ
diameter.   Above 1µ deposition is stated to increase quite
rapidly to become effectively quantitative at 5 of 6µ (16).

Table 6

USA Production of Some Genotoxicants in 1976 (Anderson)

Ranking	Chemical	10^9 lbs	Production Increase %, 1966-1976
5	Ethylene	22	7.0
13	Benzene	10.6	4.3
14	Propylene	9.8	7.6
15	Toluene	8.2	6.9
16	Ethylene dichloride	7.9	8.2
18	Xylene	7.3	11.9
19	Styrene	6.3	7.1
21	Ethylbenzene	6.1	6.6
22	Vinyl chloride	5.7	8.7
23	Formaldehyde	5.6	4.2
26	Ethylene oxide	4.2	6.1
31	p-Xylene	3.2	19.9
33	Cumene	2.7	11.6
37	Phenol	2.2	5.4
41	Propylene oxide	1.80	9.7
45	Acrylonitrile	1.52	7.8

Large dust particles containing genotoxicants are usually filtered out by the nose. An example of such a material is wood dust. Those who work closely with wood dust have a higher than normal incidence of cancer of the nose and sinuses (1). The somewhat smaller particles deposit themselves in the bronchi, then ride the ciliary escalator to exit from the lungs within hours of deposition. In most cases, these particles are then swallowed and tend to end up in the stomach. This is essentially what has been postulated to explain a high incidence of gastric cancer in Carbon and Emory Counties, Utah (24). The coal miners who are affected breathe in large particles of coal dust in the mines and coal soot at home where coal is used as a fuel. Among other characteristics, the coal soot has a high PAH content. With respirable particles, their alveolar clearance is a much slower process consisting of an initial phase lasting twenty-four hours involving phagocytosis, an intermediate phase of continued transport for three to ten days, and a prolonged phase lasting one hundred days or longer (17). Because of this longer period of contact, the situation can be conducive to carcinogenesis.

This is probably the reason why lung cancer in uranium miners originates from deposition of finer dust in the conducting airways or in the acini, the respiratory units that together constitute the pulmonary compartment of the lung.

On the basis of a significant amount of circumstantial evidence, it would appear that 80-90% of human cancers are derived from contacts with environmental factors. One school of thought believes that most human cancers are associated with personal pollution. The two main etiological factors are believed to be cigarette smoking and diet, which are believed to account for most of the cancers in the digestive tract, the respiratory tract, and the endocrine-sensitive and reproductive organs.

There are several difficulties with this type of belief. The cancer patients who smoked cigarettes and/or ate "badly" did not otherwise live in a chemical vacuum. If we hurdle this body-count type of reasoning, we can face the fact that most cigarette smokers do not get lung cancer. This is probably because other factors (genetic and environmental) are involved in this carcinogenesis as shown by the tip of the "cocarcinogenic" iceberg--the effects of asbestos and radiation on the cancer rate of cigarette smokers. Evidence has been presented which indicates that lung cancer in cigarette smokers is derived from the families of carcinogens and cocarcinogens in cigarette smoke, a genetic factor(s), and an urban factor(s) (37). The author suggests a strong synergistic interaction between cigarette smoking and the constitutional host susceptibility to lung cancer. In addition, it is premature to argue that the carcinogenicity of the polluted environs (air, water, industrial, and food pollution) compared to the personal type of pollution as denoted by cigarette smoke is relatively negligible since the production of chemicals is continually increasing at a rapid rate as are our exposures to these chemicals. Because of a latency period of about twenty to sixty years, the results of this type of pollution have not, as yet, hit us with full force. This situation will be aggravated further with increasing industrialization in other countries and with increasing world population (Table 7). This is shown by the data in Table 6 and the report that the production of organic chemicals in the non-communist world increased from seven million tons in 1960 to sixty-three million tons in 1970 and is predicted to increase to 250 million tons in 1985 (30).

The areas that have had the highest priority in our research studies on polluted atmospheres have usually had

Table 7

Doubling of World Population

Year, AD	No. Years	Billions of People
1		0.25
1600	1600	0.5
1850	250	1.0
1930	80	2.0
1975	45	4.0
2010	35	8.0

large numbers of chemical and petroleum refining industries concentrated near a large body of water. Examples of such areas are the Kanawha River in the Kanawha Valley of West Virginia; the Arthur Kill in the Rahway, Newark, Jersey City area; the Delaware River in the Philadelphia-Camden area; the Mississippi River in the New Orleans-Baton Rouge area; the Gulf Coast in the Texas, Louisiana, Mississippi State areas; the Niagara Falls area; and San Francisco Bay area. The areas of prime interest usually have a high order of pollution, a high production of chemicals and derived products, a high cancer rate, or an emergency hazardous chemical(s) situation.

One of the families of compounds found in such areas is the group of benzene derivatives. The activity of some of them is shown in Table 8. However, the data on the genotoxic activity of this family is very sparse. Thus far, there is no reliable carcinogenicity bioassay system for these compounds. This is particularly disturbing because of the large variety of benzene derivatives which are found in the polluted atmosphere (Table 9).

A combination of genetic factors and chronic exposure to benzene are vital factors in the etiology of leukemia (2). However, there are probably other key chemical factors involved in this problem. This is because people exposed to benzene are also exposed to other chemicals in their environment, and the benzene which they are in contact with is usually impure or may even be impure toluene or xylene.

Table 8

Genotoxic Benzene Derivatives

Compound	Species Exposure	Genotoxic Effect	References
Benzene	Rat Inhalation	Clastogen[1]	10
	Rabbit Inhalation	Clastogen[1]	21
	Human Inhalation	Clastogen	39
	Rat, Mouse Inhalation	Leukemogen[2]	4
	Human Inhalation	Leukemogen	39
	Rat Ingestion	Carcinogen[3]	23
Toluene	Rat Inhalation	Clastogen[1]	10
Styrene	Yeast Host-mediated	Mutagen	22
Hexachlorobenzene	Rat Intraperitoneal[4]	Comutagen	12
	Hamster Ingestion[5]	Carcinogen	9

[1]Chromosome lesions in bone marrow cells.

[2]Tentative link of high doses of benzene to leukemia in rats and mice.

[3]Male and female rats fed benzene down to 50 mg/kg of body weight resulted in some zymbal gland and dermal tumors. Preliminary data which need confirmation.

[4]Induction of 2,4-diaminoanisole mutagenicity in vitro.

[5]Resulting in hepatomas, haemangioendotheliomas, thyroid adenomas, and a shortened lifespan.

Table 9

Some Atmospheric Benzene Derivatives

Benzene	Dichlorobenzenes
Acetophenone	Diethylbenzenes
Anisole	Diethyl Phthalate
Benzaldehyde	Dimethyl Phthalate
Benzonitrile	Dipropyl Phthalate
Benzophenone	Ethylbenzene
Benzyl Bromide	Ethyltoluenes
Benzyl Chloride	Fluorobenzene
Benzyl methyl ether	Hexachlorobenzene
Biphenyl	Hexafluorobenzene
Biphenylene	Hexylbenzenes
Biphenyl ether	Methylstyrenes
Bromobenzene	Pentylbenzenes
Bromotoluenes	Perfluorotoluene
Bromoxylenes	Phenylethanol
Butylbenzenes	Propylbenzenes
Chlorobenzene	Styrene
Chlorotoluenes	Toluene
Chloroxylenes	Trichlorobenzenes
Cumene	Trimethylbenzenes
	Xylenes

Large aliphatic hydrocarbons are another family of compounds classified as cocarcinogens present in the environment
in fairly high concentrations. Methods are available for
their analysis as a family or as individuals. They and the
polynuclear aromatic hydrocarbons have been discussed in previous papers (31,32). The importance of cocarcinogenicity is
demonstrated in those reports that large aliphatic hydrocarbons
can increase the carcinogenicity of some PAH a thousand-fold;
they can cause lung tumors when painted on mice whose pregnant
parent had been previously injected with BaP, and they can
cause some noncarcinogenic PAH to become carcinogenic (30).

The large number of halogenated aliphatic (about 100
found so far) and ring (about 50 found so far) compounds in
the polluted atmosphere means that we have to redirect some
of our bioassay studies into investigations of integrated
genotoxic effects of both the individual members of a carcinogen family and the various families. We must look for additive, multiplicative, and initiation-promotion types of

insults. To complicate further the situation, some members
of the family can have more than one cancer pathway or different organotropic effects.

Another family of genotoxic compounds found in the polluted atmosphere is the aromatic amines. At least three
possible cancer pathways can be deduced for these compounds,
through ring epoxidation, N-hydroxylation, and nitrosation
with NO_x to form the diazonium salt. The genotoxic activities
of some monocyclic aromatic amines are given in Table 10. The
high bladder cancer rate in some American counties may be due
to the presence of aromatic amines and their azo dye precursors
in the polluted environment. In preliminary work we have found
some of these amines in the polluted atmospheres of some counties that have high bladder cancer rates. The analysis, atmospheric concentrations, and genotoxic properties of many of
these pollutants have been considered (30-33).

Doll (11) has discussed various industrial genotoxicants,
many of which have been found in the polluted atmosphere in
large numbers and sometimes in fairly high concentrations. The
genotoxicants are important for a number of reasons. Their
hazard to humans is derived from their effect on sentinel
individuals (workers who are a high risk group because of this
contact and a chemical and genetic background which is conducive
to carcinogenesis). They could cause potential problems to
the workers concerned. Many of these agents find their
way into the atmosphere through leakage, accident, dumping,
or use, so that large numbers of people are exposed to
them.

It is commonly thought that a carcinogen either causes
cancer (100% effect) or it doesn't (0% effect). If it takes
several exposures to cause cancer, what effect would too few
exposures have on the body? There could be other mutagenic
effects (besides the carcinogenic one) leading to an accumulation of metabolic errors, thus causing a decrease in the
quality of life and/or a life-shortening effect. One effect
of contacts with an industrial carcinogen could result in a
germinal mutagenic effect that would be passed on to future
generations.

It is highly unusual for one chemical or mixture to cause
cancer in humans, unlike carcinogenesis in animals. Man must
come in contact with huge amounts of a carcinogen for the risk
of cancer to be 100%. A 100% risk was reported for lung cancer
in miners of radioactive ore and for bladder cancer among some
aromatic amine producers (19). For example, in one small group
of 19 men employed in distilling 2-naphthylamine, the risk proved

Table 10

Genotoxic Properties of Anilines

Amine	Species	Genotoxic Effect[1]	References
Aniline	Mouse, rat, man	NC[2]	
	S. typhimurium	M[3]	25
o-Aminoacetophenone	Mouse	C	41
2,4-Diaminoanisole	S. typhimurium[4]	M	13
	Rodents	C?	5
4,4'-Diaminodiphenyl ether	Mouse, rat	C	14
4,4'-Diaminodiphenyl-methane	Rat	C	35
2,4-Diaminotoluene	Rat	C	20
	Drosophila melanogaster	M	7
8-Methoxylkynurenic acid	Mouse	C	8
4,4'-Methylene-bis-(2-chloroaniline)	Mouse, rat	C	27,36
4,4'-Methylene-bis-(2-methylaniline)	Rat	C	36
2-Nitro-p-phenylene-diamine	Mouse	C	38
	S. typhimurium	M	38
4-Nitro-o-phenylene-diamine	Mouse	C	38
	S. typhimurium	M	38
Phenacetin	Human	C	26
o-Toluidine	Rat	C	28
	S. typhimurium	M[3]	25
p-Toluidine	Rat	C	28

[1]C = carcinogenic, M = mutagenic, NC = noncarcinogenic
[2]See discussion of negative results in body of paper
[3]In presence of norharman and S-9 mixture
[4]Using liver microsomal fraction from rats pretreated with hexachlorobenzene

to be 100% (11). In the majority of cases, other factors are necessary; this applies to vinyl chloride, asbestos, coke oven effluents, benzidine, and even cigarette smoke. Thus, the epidemiological study of industrial carcinogens has shown the importance of cocarcinogenic factors. Because of our reliance on state-of-the-art animal models and our ignorance of the human chemical environment, epidemiological investigations are controversial and are sometimes doomed to failure. An example of such a controversial investigation involves the relationship between industrial vinyl chloride and cancer (40). Even with the most "thoroughly" investigated human carcinogens, asbestos fibers, and cigarette smoke, the epidemiological data for their effect on humans is said to be inadequate for choosing between the additive, multiplicative, or cocarcinogenic asbestos models to explain the results (29).

Because of the vital importance of industrial chemicals to our civilization, the boundaries of carcinogenicity and noncarcinogenicity of our atmospheric and other environmental pollutants should be determined. This means that we need much more knowledge on the carcinogenicity, cocarcinogenicity, and anticarcinogenicity of the numerous chemicals, families of chemicals, and mixtures in our environments. The first step, once we know what chemicals are in our environments, would be to decrease exposure to the potent human carcinogens and to those genotoxic chemicals or families present in highest concentrations in our environment. The carcinogenicity of any chemical, whether it is classified as a human or animal carcinogen, is actually potential carcinogenicity for which boundaries have not yet been determined. Although negative results always have a question of uncertainty about them, so do positive results. Newer carcinogenic results supersede all previous negative results, but negative data should not be discarded. It is still meaningful since it tells us something about the boundaries of the carcinogenicity and noncarcinogenicity of that particular chemical. Finally, it is significant to remember that while industrial chemicals can be a curse, they are humanity's best hope for increasing the meaning and the quality of life.

The background against which we study mutagenic problems is particularly complicated, annoying, and frustrating because of our continually changing chemical environment and the consequent alterations in the relative amounts and relative importances of the various types of cancers and other genotoxic manifestations. We need to know much more about the genotoxic properties of pure chemicals, key environmental mixtures, key families of chemicals, and non-ionizing and ionizing radiations.

To accumulate this knowledge on these genotoxicants, further improvements need to be made and data accumulated on rodent bioassays for carcinogenicity and short-term bioassays for carcinogenicity, germinal mutation, teratogenicity, athero-sclerosis, and aging. To show some of the possibilities in the investigations of genotoxic materials, the short-term bioassays for carcinogenicity should be considered. They can be used in the following:

- Predicting the carcinogenicity of pollutants of unknown activity that are highly toxic and/or are present in high concentrations in the environment.

- Setting priorities for chemicals to be tested in mammals.

- Identifying active fractions and chemicals in environmental mixtures.

- Identifying mutagenic metabolites in human body fluids.

- Determining the possible carcinogenicity of atmo-spheric, aqueous, or other mixtures with which humans are in contact.

- Identifying mutagenic metabolites in plants.

- Determining the cocarcinogenicity (or comutagenicity) of environmental and natural chemicals.

- Determining the anticarcinogenicity (or antimuta-genicity) of environmental and natural chemicals.

- Investigating the mechanism of carcinogenesis.

- Determining the effect of pollution control activities.

- Selecting relatively safe chemicals to replace the hazardous ones that are currently of great importance to our modern industrial society.

ACKNOWLEDGMENT

Much of the analytical data on the organic vapors was obtained mainly from the studies of a research group led by Dr. Edo Pellizzari at the Research Triangle Institute, Research Triangle Park, NC.

REFERENCES

1. Acheson ED, Hadfield EH, Macbeth RG: Carcinoma of the nasal cavity and accessory sinuses in woodworkers. Lancet 1:311-312, 1967

2. Aksoy M, Erdem S, Erdogan G, Dincol G: Combination of genetic factors and chronic exposure to benzene in the etiology of leukemia. Human Hered 26:149-153, 1976

3. Anderson EV: Top 50 chemicals regain output lost in 1975. Chem Eng News 55:37, 1977

4. Anonymous: More benzene data. Chem Week 17, August 17, 1977

5. Anonymous: Hair dyes a hazard? Chem Week 17, Dec 21, 1977

6. Anonymous: Increased use of coal deemed safe through 1985. Chem Eng News 56:22, 1978

7. Blijleven WGH: Mutagenicity of four hair dyes in Drosophila melanogaster. Mutat Res 48:181-186, 1977

8. Bryan GT: Neoplastic response of various tissues to the systemic administration of the 8-methyl ether of xanthurenic acid. Cancer Res 28:183-185, 1968

9. Cabral JRP, Shubik P, Mollner T, Raitano F: Carcinogenic activity of hexachlorobenzene in hamsters. Nature 269: 510-511, 1977

10. Dobrokhotov VB: The mutagenic action of benzene, toluene and a mixture of these hydrocarbons in a chronic test. Gig i San No. 1:32-34, 1977

11. Doll R: Strategy for detection of cancer hazards to man. Nature 265:589-596, 1977

12. Dybing E, Aune T: Hexachlorobenzene induction of 2,4-
     diaminoanisole mutagenicity in vitro. Acta Pharmacol
     Toxicol 40:575-583, 1977

13. Dybing E, Thorgeirsson SS: Metabolic activation of 2,4-
     diaminoanisole, a hair dye component: I. Role of cyto-
     chrome P-450 metabolism in mutagenicity in vitro. Bio-
     chem Pharmacol 26:729-734, 1977

14. Dzhioev FK: On carcinogenic activity of 4,4'-diaminodi-
     phenyl ether. Vopr Onkol 21:69-73, 1975

15. Egle Jr JL, Gochberg BJ: Respiratory retention of inhaled
     toluene and benzene in the dog. J Toxicol Environ Health
     1:531-538, 1976

16. Giacomelli-Maltoni G, Melandri C, Prodi V, Tarroni G:
     Deposition efficiency of monodisperse particles in human
     respiratory tract. Am Ind Hyg Assoc J 33:603-610, 1972

17. Gibb FR, Morrow PE: Alveolar clearance in dogs after
     inhalation of an iron-59 oxide aerosol. J Appl Physiol
     17:429, 1962

18. Green GM, Jakab GJ, Low RB, Davis GS: Defense mechanisms
     of the respiratory membrane. Am Rev Resp Disease 115:
     479-514, 1977

19. Hueper WC, Conway MD: Chemical carcinogenesis and cancers.
     Springfield, Charles C. Thomas, 1964, p 74

20. Ito J, Hiasa Y, Yoniski Y, Marugami M: The development
     of carcinoma in livers of rats treated with m-toluylene-
     diamine and synergistic and antagonistic effects with
     other chemicals. Cancer Res 29:1137-1145, 1969

21. Kissling M, Speck B: Further studies on experimental
     benzene induced aplastic anemia. Blut 25:97, 1972

22. Loprieno NA, et al.:   Mutagenicity of industrial com-
     pounds: styrene and its possible metabolite styrene
     oxide. Mutat Res 40:317-324, 1976

23. Maltoni C in Anonymous: Research "bombshell" hits ben-
     zene. Chem Week 33, November 2, 1977

24. Matolo NM, Klauber MR, Gorishek WM, Dixon JA: High inci-
     dence of gastric carcinoma in a coal mining region.
     Cancer 29:733-737, 1972

25. Nagao M, Yahagi T, Honda M, Seino Y, Matsushima T, Sugimura T: Comutagenic action of norharman and harman. Proc Japan Acad 53B:95-98, 1977

26. Rathert P, Melchior H, Lutzeyer W: Phenacetin: a carcinogen for the urinary tract? J Urol 113:653-657, 1975

27. Russfield AB, Homburger F, Boger E, Van Dongen CG, Weisburger EK, Weisburger JH: The carcinogenic effect of 4,4'-methylene-bis-(2-chloroaniline) in mice and rats. Toxicol Appl Pharmacol 31:47-54, 1975

28. Russfield AB, Homburger F, Weisburger EK, Weisburger JH: Further studies on carcinogenicity of environmental chemicals including simple aromatic amines. Toxic Appl Pharm 25:446-447, 1973

29. Saracci E: Asbestos and lung cancer: An analysis of the epidemiological evidence on the asbestos-smoking interaction. Int J Cancer 20:323-331, 1977

30. Sawicki E: The genotoxic environmental pollutants. In: Proceedings of the First Symposium on Management of Residues from Synthetic Fuels Production (Schmidt-Collerus JJ, Bonomo FS, eds.). Denver, University of Denver, 1976, pp 122-165

31. Sawicki E: Analysis of atmospheric carcinogens and their cofactors. In: Environmental Pollution and Carcinogenic Risks (Rosenfeld C, Davis W, eds.). Lyon, IARC Scientific Publications No. 13, 1976, pp 297-354

32. Sawicki E: Chemical composition and potential genotoxic aspects of polluted atmospheres. In: Air Pollution and Cancer in Man (Mohr U, Schmahl D, Tomatis L, eds.). Lyon, IARC Scientific Publications No. 16, 1977, pp 127-157

33. Sawicki E: Analysis of atmospheric pollutants of possible importance in human carcinogenesis. Presented at the conference on "Modern Measurement of Environmental Pollutants" at the University of Rochester Medical School, May 23, 1977 in press, 1978

34. Smith DH, Achenbach M, Yeager WJ, Anderson PJ, Fitch WL, Rindfleisch TC: Quantitative comparison of combined gas chromatographic/mass spectrometric profiles of complex mixtures. Anal Chem 49:1623-1632, 1977

35. Steinhoff D, Grundmann E: Zur cancerogenen wiskung von
     4,4'-diaminodiphenylmethan und 2,4'-diaminodiphenylmethan.
     Naturwissenschaften 5:247-248, 1970

36. Stula EF, Sherman H, Zapp Jr JA, Clayton Jr JW: Experi-
     mental neoplasia in rats from oral administration of
     3,3'-dichlorobenzidine, 4,4'-methylene-bis(2-chloro-
     aniline), and 4,4'-methylene-bis(2-methylaniline).
     Toxicol Appl Pharmacol 31:159-176, 1975

37. Tokuhata GK: Cancer of the lung: Host and environmental
     interaction. In: Cancer Genetics (Lynch HT, ed.).
     Springfield, Charles C. Thomas, 1976, 213-232

38. Venitt S, Searle CE: Mutagenicity and possible carcino-
     genicity of hair colourants and constituents. IARC
     Scientific Publications No. 13, INSERM 52:263-272, 1976

39. Vigliani EC, Fornia A: Benzene and leukemia. Env Res
     11:122-127, 1976

40. Wagoner JK, Infante PF, Saracci R, Duck BW, Carter JT:
     Vinyl chloride and mortality? Lancet ii:194-195, 1976

41. Zharova EI: Characteristics of blastomogenesis induced
     with tryptophan metabolites. Patol Fiziol Ekspter 17:
     54-58, 1973

STATE-OF-THE-ART ANALYTICAL TECHNIQUES FOR AMBIENT VAPOR PHASE ORGANICS AND VOLATILE ORGANICS IN AQUEOUS SAMPLES FROM ENERGY-RELATED ACTIVITIES

Edo D. Pellizzari
Chemistry and Life Sciences Group
Research Triangle Institute
Research Triangle Park, North Carolina

The presence of organic components in the ambient air is a fact of life in a modern society, since volatile organic compounds are ubiquitous. Automobile exhaust, fossil fuel burning, and the chemical industry contribute many organic compounds to the air. It is not unreasonable to expect that products from reactions of these chemicals with NO_2 and SO_2, by photochemical (1-5) or other processes, will be also observed in the atmosphere (6). However, many organic constituents are suspected to enter the environment directly by industrial pollution (7). Carcinogenic and mutagenic compounds find frequent use as intermediates in organic synthesis, e.g., in the preparation and use of plastics, fabrics, dyes, resins, cosmetics, pharmaceuticals, etc. Organic·solvents, heavily used in industry, are also sources of high levels of organic vapors.

Comprehensive studies on levels of carcinogenic agents in air and correlation of this information with health effects in humans are mandatory if we are to understand better the current genetic diseases, as well as problems in carcinogenesis and mutagenesis. While immediate and life-threatening effects of some of these compounds are obvious, the consequences of chronic low levels of exposure are often not known for many years. Qualitative/quantitative analysis of the atmosphere is vital to establish the etiology of cancers and other diseases. It is essential to understand the organic composition of the atmosphere because of the existence of anti- and cocarcinogenic factors. Statistical studies demonstrate that the incidence of cancer aassociated with the respiratory system is elevated where high air pollution occurs. Thus,

an analytical technique that provides information on the
identity and quantity of organic constituents of ambient
air is highly desirable.

An assortment of methods are described in the literature
for the collection and analysis of volatile organics from the
atmosphere. In fact, the variety of methods is a problem
because most techniques are too restrictive, i.e., they focus
only on one or a "few" substances at any given time; only a
"narrow window" is examined. More recently we have developed
and perfected techniques that provide for a polypollutant
approach and yield a more representative and quick chemical
analysis of the surrounding atmosphere (8-18). The polypollutant method is based upon the use of a solid sorbent followed by capillary gas chromatography/mass spectrometry/
computer analysis for qualitative and quantitative determinations.

COLLECTION AND ANALYSIS

Because organic constituents of the air are present
usually at ppt to ppm levels in a vast amount of a diluting
medium (air and water vapor), it is generally not practical
to perform in situ analyses of all these organic compounds.
There is no widely applicable method of detection that could
distinguish each compound from all others at such low concentrations. Therefore, to register enough sensitivity, vapors
must be concentrated from a large volume of air.

There are four basic steps necessary to successfully
analyze organic vapors in air. They are:

- The collection/concentration of vapors.

- Their transfer to an analytical system.

- Their separation and identification.

- The ability to measure the quantities of each of
  the components of interest.

Concentration Techniques

With regard to collection/concentration, there are several
possible techniques (Table 1). Cryogenic sampling is excellent
for extremely volatile compounds such as acetylene, NO_x, SO_x,

Table 1

Methods of Concentrating Organic Vapors from Air

Method	Mechanism	Selectivity Based On:
Cryogenic Traps	Low Temperature Condensation	Temperature
Solvent Impingers	Dissolution in Liquid	Solvent Polarity, Temperature
Sorbent Cartridges	Adsorption on Solid Surface	Structure of Sorbent, Temperature

or freons, using liquid nitrogen, oxygen, or solid CO_2/acetone
as the cooling medium.   However, there are several drawbacks.
This method condenses considerable quantities of water vapor.
The reactive gases NO_x, SO_x, and ozone can cause artifacts by
reacting with amines, oxygenates, olefins, etc.   Also, in
inaccessible field locations, setting up and maintaining
cryogenic traps can be difficult, and cryogenic samples are
difficult to store or ship.   Furthermore, the use of oxygen
(or nitrogen which condenses oxygen) is rather dangerous.
Solvent impingers are also used, but handling and shipping
volumes of solvent also presents problems.   Artifacts during
collection are also prevalent.   The use of a solid sorbent,
such as Tenax GC, can achieve reasonably effective concentrations of organic components.   Sorbent cartridges by contrast, can be made clean, lightweight, and compact.   All the
vapors collected from a large volume of air can be delivered
to an analytical system by thermal desorption of the trapped
vapors.   During collection of vapors on a sorbent, each
adsorbed compound is in equilibrium with its vapor in the
air stream, so that it moves slowly through the sorbent bed.
The partition ratio between the air stream and the sorbent
surface for each compound is unique, depending on the temperature and the structure of the sorbent.   The ratio determines selectivity for the compound because it controls the
rate at which the compound moves through the bed.   Compounds
are quantitatively collected until sufficient air has passed
through the bed to elute them.   This elution volume, or breakthrough volume, must be known as a function of temperature
for any compound which is to be collected for quantitative
analysis by this technique.

Recovery Methods

Once collected, the concentrated vapors must be recovered from the collection system and delivered to an analytical instrument. One method employs cryogenic traps and solvent impingers. If cryogenic traps are used, a substantial quantity of water is collected. Solvent impingers use liquid as the collection medium. In either case, the collected compounds must be separated from a substantial volume of liquid. This could be done by inert gas purge, by solvent extraction and concentration, or, for highly volatile compounds, by low temperature vacuum distillation. These procedures are tedious, sample throughput is low, and substantial losses of the collected compounds may occur. Generally, only a small aliquot of the sample can be analyzed, and thus the sensitivity is poor.

The other method employs sorbent cartridges. The vapors can be recovered from a sorbent bed by solvent extraction or by thermal desorption. The disadvantages of solvent extraction have already been mentioned. Thermal desorption is done by simply heating the sorbent in an inert gas stream.

Sampling System

After considering all of the above factors, a "polypollutant" technique was developed using the sorbent cartridge and pumping apparatus shown in Figure 1. In sampling, air is drawn first through a glass fiber filter to remove particulates, and then through the cartridge. A manifold can be used to collect replicate cartridges. A 12-volt DC pump is used which is powered by an automobile battery, or, if a 110 volt AC is available, by a built-in battery charger. Thus, the system is portable, weighing about 20 pounds. The air stream exhaust passes through a rotameter and a gas meter. Typical sampling rates vary between 1 and 10 liters per minute, although the volume sampled generally is between 20 and 200 liters.

Choosing a Sorbent. There are several criteria which a sorbent must fulfill if it is to be acceptable for ambient air pollution studies. These criteria are:

> The sorbent must withstand repeated use without deterioration and must not contaminate the sample.

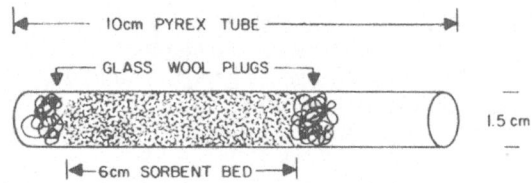

2.2g 35/60 MESH TENAX GC CARTRIDGE

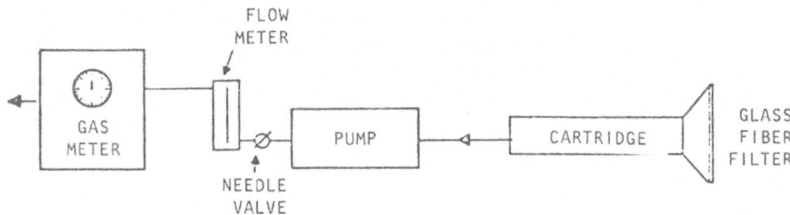

Figure 1. Organic vapor collection system.

- In order to collect a compound, it must adsorb substantially all the vapor passing into it.

- The adsorbed vapors must be completely released upon thermal desorption.

- The sorbent should possess a sufficient breakthrough volume or retention volume if the method is to be used for quantitative analysis.

- The sorbent must not catalyze in situ reactions on its surface during and after vapors have been adsorbed on its surface. In other words, it should not be involved in hydrolysis, rearrangement, synthesis, or decomposition of compounds.

Before Tenax was selected, a number of sorbents--carbon, porapaks, chromosorbs, etc.--were evaluated according to the first four criteria. Tenax has been found to be, at least for the time being, the best compromise. Thermal and storage

stability of Tenax were observed by repeatedly desorbing it
at 270°C after varying storage intervals. The vapors thus
produced were analysed chromatographically. Ethylene oxide
and styrene were often observed, but the amounts were too
small to interfere with use of Tenax for sampling organic
vapors from air.

Collection efficiences for sorbents for representative
vapors have been evaluated by purging a small quantity of
vapor from a 2 liter bulb as shown in Figure 2. An air
stream carries the vapors from the bulb to the flame ionization detector. A cartridge is interposed between the bulb
and the detector.

The detector response depicts an exponential decrease
in the amount of vapor coming from the bulb when no cartridge
is in place. This is depicted in Figure 3. Complete collection has been demonstrated for a variety of representative
chemical classes.

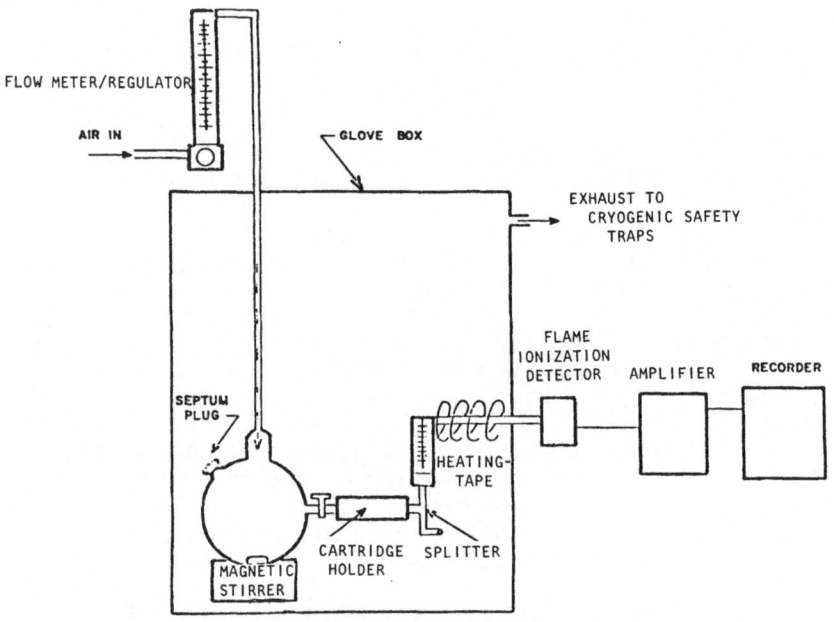

Figure 2. Monitoring system for vapors in cartridge
effluents.

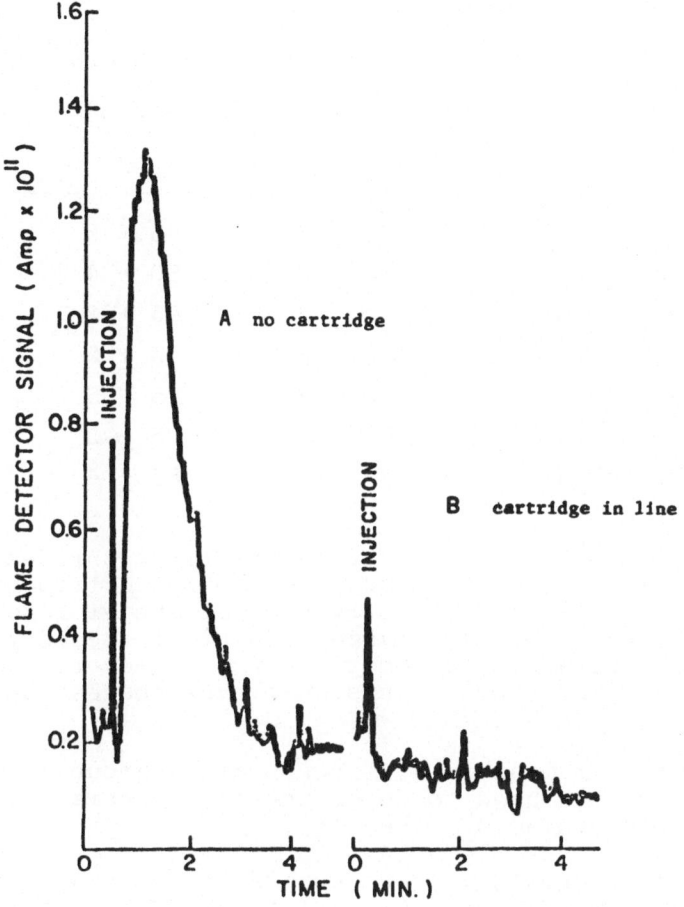

Figure 3.   Elution profile of cartridge effluent.

        Percent recovery of collected vapors has been determined
by comparing gas chromatographic responses when cartridges
loaded with a known amount of vapor were desorbed with the
responses obtained by direct injection into the chromatograph
of the same amount of vapor.   Examples of some results are
given in Table 2.   Other data have also been published for a
variety of representative compounds.   In each case the results
of immediate analysis were compared with the results obtained

Table 2

Percent Recovery of Vapors After Storage

Storage Period (Weeks)

	0	3	5
1-Nitropropane	95 ± 2	93 ± 3	50 ± 9
Chlorobenzene	95 ± 2	80 ± 4	50 ± 8
Phenyl methyl ether	95 ± 2	95 ± 2	70 ± 8
N-Ethylaniline	95 ± 2	95 ± 2	70 ± 6
Nitrobenzene	95 ± 2	95 ± 3	50 ± 9
Aniline	95 ± 2	95 ± 2	80 ± 5
4'-Fluoroacetophenone	95 ± 2	80 ± 4	90 ± 4

after a period of storage. During the first three-week storage period the cartridges were subjected to round trip shipment by air freight to test the effectiveness of the storage containers at high altitude. Little change in percent recovery has been seen after three weeks for most vapors. However, two weeks of further storage results in significant losses for some compounds.

In a similar manner, a representative group of model compounds has been used to determine four general methods of breakthrough volumes. They are:

- Cartridges have been purged into a monitoring system as discussed earlier, and the response of the flame ionization detector was observed.

- Disappearance of vapor from cartridges during purging under laboratory and field conditions has been determined.

- Appearance of vapor in backup cartridges during purging also has been determined. One loads cartridges with a compound and then samples air that does not contain that compound. Backup cartridges are changed periodically during sampling. The volume of air required for half of the vapor in the backup cartridges to appear is then determined.

This approach has been used to test the possibility of premature breakthrough or displacement chromatography when sampling in the presence of high ambient levels of hydrocarbons, such as occurs with auto exhaust.

● Elution volumes may be determined on a gas chromatographic column packed with a known quantity of sorbent. The breakthrough volumes determined by all of these methods generally are in substantial agreement.

Table 3 lists the breakthrough volumes for a few organic compounds. We have determined this characteristic for approximately 125 compounds and extrapolated the results to more than 400.

Table 3

Breakthrough Volumes at 26.7°C for Tenax GC 35/60M

	Per gram of Tenax	Per 6 cm Cartridge
Benzene	17 ℓ	37.4 ℓ
Carbon Tetrachloride	7 ℓ	15.4 ℓ
Dimethyl Amine	16 ℓ	35 ℓ
Dimethyl Nitrosamine	74 ℓ	163 ℓ
1,2-Dichloroethane	10 ℓ	22 ℓ

Finally, the sorbent must not catalyze _in situ_ reactions. One such possiblity of forming dimethylnitrosoamine (DMN) on the surface of Tenax shall be discussed.

In order to examine this reaction, a gas flow system as shown in Figure 4 has been used. Clean air passes through an oxide of nitrogen/ozone generator where it mixes with selected concentrations of nitric oxide, nitrogen dioxide and ozone, then is humidified. The air then passes through a glass reaction tube. At the upstream end of this tube dimethylamine (DMA) can be introduced from a thermostatted chamber containing a permeation tube. At the downstream end, the air is analyzed for oxides of nitrogen and ozone and sampled for organic vapors.

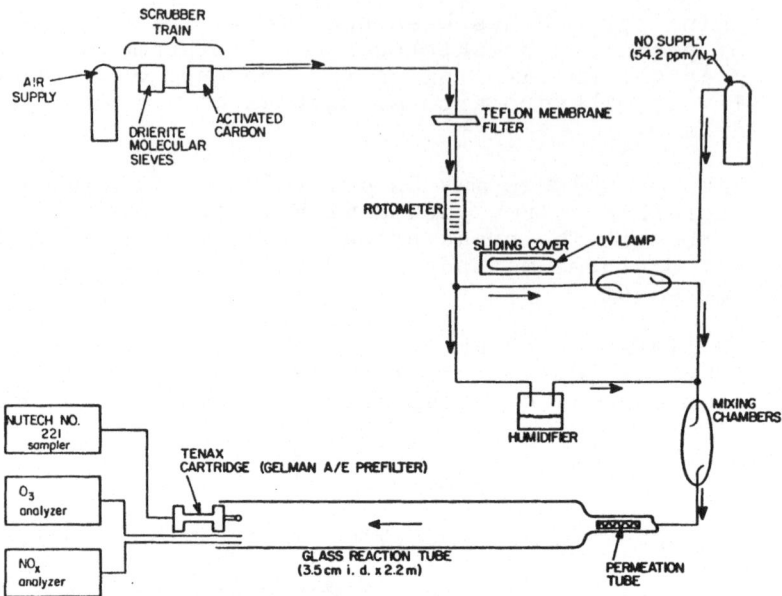

Figure 4.   Schematic of apparatus for studying artifacts on
solid sorbents.

       Table 4 presents experiments that have been conducted.
The sampling cartridges were preloaded with DMA and used to
sample air containing oxides of nitrogen, ozone, and 151 ppb
of deuterated DMA.   DMN could be formed only on the cartridge
in this case, while deuterated DMN could form in the flow tube
and on the cartridge.   Each experiment was repeated in reverse
fashion, with the cartridges being preloaded with deuterated
DMA and used to sample 131 ppb of DMA in the tube.   The identity of the amine which was preloaded on the cartridge is
indicated in Table 4 by the letter d and h by the amount of
DMN formed only on the cartridge.

       The ratio of DMN formed from the amine which passed down
the tube to the DMN formed from the amine preloaded on the
cartridge is consistently larger than one.   This means that
some of the DMA must have been converted before reaching the
sorbent bed.   The length of the flow tube was changed from
15.8 cm to 130 cm without producing any significant change
in the amount of amine attributable to formation in the tube.
This would suggest that the reaction occurred on a surface in

Table 4

Relative Amount of DMN Formation
Flow Tube vs. Sorbent Bed

Flow Tube Length x Diameter (cm)	Concentration (ppb)			Total DMA (nM)	DMN (nM)		
					Formed only on Cartridge	Formed in Flow Tube and on Cartridge	Ratio $\frac{C + T}{C}$
	O_3	NO	NO_2				
Flow Tube 25°C							
15.8 x 3.5	95	0	530	404	0.212d	0.310	1.46
	25	10	545	404	0.379d	0.459	1.21
	105	0	515	404	0.189h	0.437	2.31
	2	60	530	404	0.081h	0.412	5.09
130 x 3.5	90.	0	540	404	0.200d	0.256	1.28
	7	30	505	404	0 d	0	–
	90	0	545	404	0.159h	0.287	1.52
	7	30	480	404	0.148h	0.312	2.11
Flow Tube 70°C							
	4	35	485	404	0.243h	0.625	2.57
Flow Tube 25°C							
15.8 x 7.5	3	260	225	404	0.137d	0.391	2.85
130 x 7.5	3	270	230	404	0.100d	0.256	2.56
70 x 7.5	2	275	240	404	0.125d	0.391	3.13

the inlet to the cartridge such as on the glass fiber filter
which was used. The flow tube was replaced with one having a
larger diameter. The flow rate of the air was increased to
maintain the linear velocity at 520 cm/sec, and the upstream
DMA permeation tube was replaced with a faster one in an
attempt to maintain the same DMA concentration in the tube.
The new permeation tube had a higher rate than expected, and
the resulting DMA concentration was actually 466 ppb or 3.5
times that in the smaller tube. This would be expected to
lead to a three-fold increase in the production of DMN in the
tube if it were formed in a homogeneous, gas phase reaction.
Then the ratio of DMN formed in the tube and cartridge to the
DMN formed only on the cartridge would become substantially
larger than three since the ratio was already larger than one
before the change was made. If the reaction were heterogeneous
then less DMN would be produced in the tube, since the surface/
volume ratio was decreased from 1.14 to 0.53 in going from the
smaller to the larger tube. This would cause the ratio to
decrease, since the amount of amine from the tube drawn into
the cartridge was kept the same.

The results actually obtained indicated a less than three-
fold increase in the ratio. Furthermore, the ratio was substantially independent of the length of the tube. This means that
while some conversion of amine may have occurred in the flow
tube, most of the DMN formation that happened in front of the
cartridge must have taken place on the glass fiber filter and
the glass fiber plug which was used to anchor the Tenax in
the cartridge. Fortunately, the percent conversion of DMA to
DMN was very small.

Similar experiments have been conducted with molecular
chlorine and olefins, specifically ethylene. No chlorinated
products have yet been detected. Many other types of artifact
reactions still need to be examined to precisely define the
limitations of Tenax or any other sorbent as a collection
medium for ambient air pollutants. In fact, these experimental concepts which have been outlined must be applied to
any collection technique before it is used in this capacity.

Once a collection device has been thoroughly tested,
the next step is to interface it to an analytical system.

Instrumental Analysis

An inlet-manifold has been used to thermally desorb the
vapors at 270°C from cartridges in a helium stream from which

vapors are trapped at -196°C (Figure 5). Then the trap is switched into the carrier gas stream of a capillary gas chromatograph and rapidly heated to 250°C. Figure 6 depicts a schematic of the gas chromatography/mass spectrometer/ computer system that was used for this analysis. The mixture is resolved by using glass capillary columns. The detector, a mass spectrometer, is coupled to a computer which stores full scan data on disc or magnetic tapes. After data acquisition is complete, the system plots normalized mass spectra indexed to a total ion current chromatogram. Mass fragmentograms may also be derived from the acquired mass spectra.

The advantages of this collection and analysis system are:

● It is easy to transport and operate the samplers in the field, even under adverse weather conditions.

● Little water is collected.

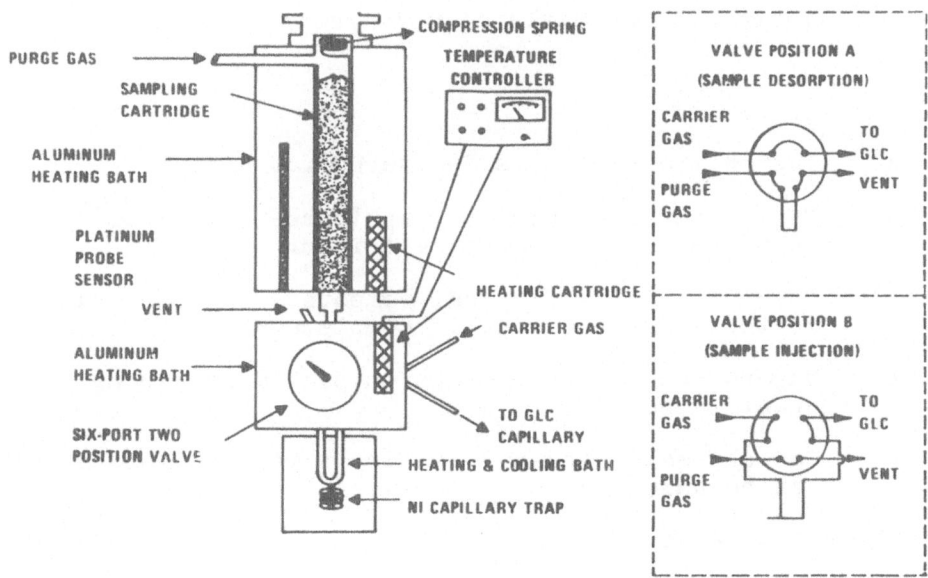

Figure 5. Inlet-manifold for recoverying vapors from sampling cartridges.

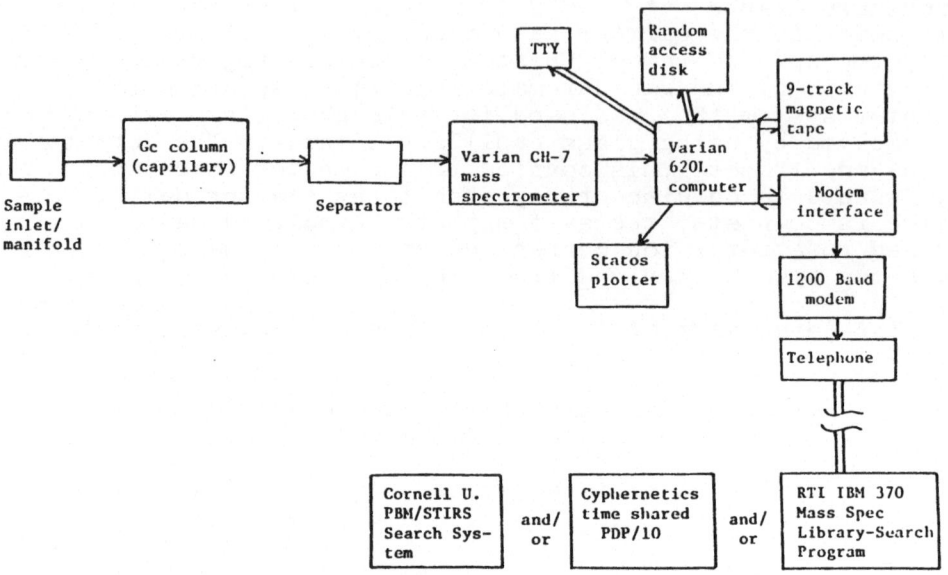

Figure 6.   Schematic of gc/ms/comp system.

- The entire sample is delivered for analysis.

- Early complete chromatographic resolution of individual compounds is attained.

- Resolution of individual compounds by mass fragmentography is virtually complete.

- Vapors present in the ppb to ppt range can be quantitated.

- Gas chromatographic retention times and mass spectra can be used for positive identification.

The disadvantages are:

- The volume of data produced is very large - 1000 mass spectra per sample, or 8,000 per day, or 40,000 mass spectra per week.. Fast data processing systems are needed to assist in the identification of the components represented by their mass spectra. Current computer mass spectral search systems are costly and inaccurate. However, efforts to improve these problems are in progress in several laboratories.

- Some compounds are not seen because they are too volatile to be collected effectively on Tenax GC. Other sorbents such as XAD, Porapaks, Chromosorbs, etc., while they have higher adsorption affinities, do not meet the criteria outlined earlier, i.e., they collect too much water, exhibit artifact reactions, have poor thermal stability, or may yield poor recoveries.

APPLICATION OF POLYPOLLUTANT METHOD

Qualitative Analysis

Figure 7 depicts an example of a volatile organic vapor profile of ambient air taken in an industrial area in Deer Park, Texas. Even though glass capillary columns were used for effecting separation of components, complete resolution was not achieved. The arrows in this figure indicate components in the sample that have been identified as halogenated hydrocarbons. The composition of this profile is given in Table 5. The list of components are in their order of elution from the gas chromatographic column (from left to right in Figure 7). A large assortment of chemical classes are represented: oxygenated and halogenated hydrocarbons, alkanes, alkenes, and alkyl aromatics. The highly volatile organics such as ethylene, acetylene, ethane, and pentane are not efficiently collected by the Tenax cartridge sampler, nor are compounds such as methyl chloride, methyl bromide, vinyl chloride, formaldehyde, or acetaldehyde. Thus, the highly volatile end of this profile is skewed. Other techniques of collection would have to be employed to trap these substances. Altogether, 150 compounds were identified. This particular sample was taken during the summer when the ambient air temperature was about 98°F. This condition favors the vaporization of relatively high molecular weight compounds. In this case a seventeen-carbon hydrocarbon was detected. Presumably other materials present in the air were associated with the particulate fraction which was filtered out by the glass fiber filter in front of the Tenax cartridge. Therefore, some skewing of the upper end of the window is observed which is influenced by the ambient air temperature. The

Table 5

Volatile Organic Vapors in Ambient Air From Deer Park, TX

Chromatographic Peak No.	Elution Temp. (°C)	Compound	Chromatographic Peak No.	Elution Temp. (°C)	Compound
1	40	CO_2	17A	74	n-pentanal
2	42	dichlorodifluormethane	17B	75	trichloroethylene + C_7H_{14} isomer
3A	45	1-butene			
4	46	chloroethane + acetaldehyde	18	76	n-heptane
4A	47	isopentane	19	78	C_8H_{18} isomer
5	48	trichlorofluormethane	19A	79	C_8H_{16} isomer
6	50	acetone	20	80	methylcyclohexane
7	52	isopropanol + dichloromethane	21	82	4-methyl-2-pentanone
8	53	freon 113 (BKG) + chloropropene isomer	22	83	C_8H_{18} isomer
			22A	84	C_8H_{16} isomer
8A	55	C_4H_8O isomer	22B	85	C_8H_{16} isomer
8B	56	dichloroethylene	23	85	1,1,2-trichloroethane
8C	56	isobutanal	24	87	toluene
9	57	1,1-dichloroethane	24A	88	C_8H_{16} isomer
9A	57	2-methylpentane	25	89	C_8H_{18} isomer
9B	58	dichloropropene (tent.) isomer	25A	89	methylethylcyclopentane isomer
9C	58	3-methylpentane	26	90	methylethylcyclopentane isomer
9D	59	n-butanal	26A	92	C_8H_{16} isomer
10	60	hexafluorobenzene (eℱ)	27	93	n-hexanal + C_8H_{16} isomer
10A	60	methyl ethyl ketone	28	94	C_8H_{18} isomer
11	61	n-hexane	28A	95	C_8H_{16} isomer
12	61	chloroform	29	96	n-octane
12A	62	2-butanol	30	97	n-butyl acetate (tent.)
12B	63	C_4H_6O isomer (tent.)	30A	98	C_9H_{18} isomer
12C	64	perfluorotoluene (eℱ)	31	100	C_9H_{18} isomer
13	65	1,2-dichloroethane	31A	102	ethylcyclohexane
13A	67	methylcyclopentane	32	106	ethylbenzene
14	68	benzene	33	107	p-xylene
14A	69	carbon tetrachloride + C_7H_{16} isomer	34	108	C_9H_{20} isomer
			34A	108	C_9H_{18} isomer
14B	70	cyclohexane	35	109	C_9H_{20} isomer
15	71	2-methylhexane	35A	110	$C_7H_{14}O$ isomer
15A	71	2,3-dimethylpentane	36	110	styrene
16	72	3-methylhexane	36A	111	$C_{10}H_{22}$ isomer
16A	73	dichloropropane + C_7H_{14} isomers	37	111	o-xylene
			37A	112	n-heptanal
17	74	dichloropropene isomer	38	112	C_9H_{18} isomer

Chromatographic Peak No.	Elution Temp. (°C)	Compound	Chromatographic Peak No.	Elution Temp. (°C)	Compound
39	113	dichloropropene	59	148	C_4-alkyl benzene isomer
40	114	n-nonane	60	150	n-nonanal
40A	116	C_9H_{18} isomer	60A	151	$C_{11}H_{12}$ isomer
40B	116	$C_{10}H_{20}$ isomer	60B	152	C_4-alkyl benzene isomer
40C	117	isopropylbenzene + $C_{10}H_{22}$ isomer	61	153	n-undecane
41	118	$C_{10}H_{22}$ isomer	61A	154	C_5-alkyl benzene isomer
41A	119	$C_8H_{10}O$ isomer (tent.)	61B	155	n-pentylbenzene
41B	120	$C_{10}H_{20}$ isomer	62	155	tetramethylbenzene isomer
41C	121	C_9H_{18} isomer	62A	156	$C_{12}H_{24}$ isomer
42	122	$C_{10}H_{22}$ isomer	62B	156	$C_{11}H_{20}$ isomer
43	123	benzaldehyde	63	157	$C_{12}H_{25}$ + C_5-alkyl benzene isomers
43A	123	n-propylbenzene	64	157	$C_{12}H_{26}$ isomer
43B	124	$C_{10}H_{22}$ isomer	64A	158	$C_{11}H_{22}$ isomer
44	125	p-ethyltoluene	65	159	methylindan isomer
44A	126	$C_{10}H_{22}$ isomer	65A	159	C_4-alkyl benzene isomer
45	127	1,3,5-trimethylbenzene	65B	160	C_5-alkyl benzene isomer
46	128	$C_{11}H_{24}$ isomer	66	161	$C_{12}H_{26}$ isomer
47	129	$C_{10}H_{22}$ isomer	66A	163	C_5-alkyl benzene isomer
47A	129	$C_{10}H_{20}$ isomer	67	164	2-decanone + naphthalene
48	130	$C_{10}H_{22}$ isomer	67A	165	$C_{12}H_{24}$ isomer
49	131	o-ethyltoluene + n-octanal	68	166	n-decanal
49A	132	$C_{10}H_{20}$ isomer	68A	167	$C_{12}H_{24}$ isomer
50	134	n-decane + dichlorobenzene isomer (tent.)	69	168	n-dodecane
50A	135	C_4-alkyl benzene isomer	69A	170	$C_{12}H_{24}$ isomer
51	135	$C_{10}H_{20}$ isomer	70	171	$C_{13}H_{28}$ isomer
52	137	1,2,3-trimethylbenzene	72	181	$C_{13}H_{28}$ isomer
52A	137	C_4-alkyl benzene isomer	72A	182	$C_{11}H_{22}O$ isomer
52B	138	$C_{11}H_{24}$ isomer	73	185	$C_{13}H_{26}$ isomer
53	139	$C_{11}H_{24}$ isomer	74	187	n-tridecane
54	141	$C_{11}H_{22}$ isomer	76	191	$C_{12}H_{24}$ isomer (tent.)
55	142	C_4-alkyl benzene + $C_{11}H_{24}$ isomers	77	199	$C_{14}H_{28}$ isomer
56	143	acetophenone	78	200	m-tetradecane
56A	144	C_4-alkyl benzene isomer	78A	212	$C_{15}H_{32}$ isomer
56B	144	$C_{11}H_{22}$ isomer	79	222	diethyl phthalate
57	145	$C_9H_{18}O$ isomer (tent.)	81	226	$C_{16}H_{34}$ isomer
57A	146	$C_{10}H_{18}$ isomer	82	227	n-hexadecane
58	147	C_4-alkyl benzene isomer	82A	228	$C_{15}H_{32}$ isomer
			82B	238	$C_{15}H_{30}O$ isomer (tent.)
			83	240	n-hexadecane
			84	240	$C_{18}H_{38}$ isomer

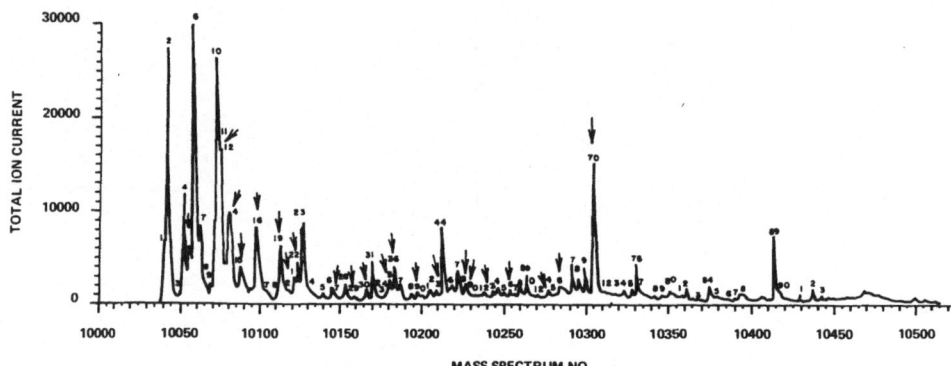

Figure 7.   Profile of volatile organics in ambient air from Deer Park, TX.

ambient air temperature, of course, would determine those compounds which would partition between the vapor and aerosol state. A different procedure is needed for the collection of organics associated with particulates.

This phenomenon notwithstanding, the collection and analysis procedure presents a wider window unlike the mono-pollutant methods. The procedure is extremely versatile and applicable to many chemical classes with the exception of the peroxides and the hydroperoxides.

Quantitative Analysis

Once the compounds have been identified, their quantities need to be determined. There are basically two approaches to the extraction of quantitative information. One is to prepare standard curves relating instrument response vs concentration for each compound of interest. The primary deterrent to extensive use of this method is that a calibration curve for each of the compounds listed for this sample would be prohibitive because of the amount of instrument/operator time required. The sample throughput would be extremely low. Thus, from several points of view, it would be most desirable to extract quantitative information from full-scan data, i.e., mass fragmentography exchanging a level of accuracy for breadth of information. Before analysis, approximately 200 ng of two standards, perfluorobenzene and perfluorotoluene, are loaded on all the cartridges.

In the second approach, it is necessary to determine Relative Molar Response (RMR) factors. Successful use of the RMR method requires a knowledge of the exact amount of reference standard added and the exact amount of compound added.

$$RMR_{unknown/standard} = \frac{R_{unknown}/Moles_{unknown}}{R_{standard}/Moles_{standard}}$$

R is a system response; it may be a peak area (a total ion current peak with a value determined by either integration or triangulation), a peak height, or the area of the peak produced by a particular ion. The ionic peak areas are especially useful in those situations where even 100 m of SCOT column cannot resolve chromatographic peaks. The value of the RMR is determined from at least three independent analyses.

Since

$$Moles_{compound} = g_{compound}/GHW_{compound}$$

where GMW = gram molecular weight, the number of grams of unknowns can be calculated from the RMR factors and values observed in the sample analysis in a straightforward manner:

$$g_{unknown} = \frac{R_{unknown} \cdot GMW_{unknown} \cdot g_{standard}}{R_{standard} \cdot GMW_{standard} \cdot RMR_{unknown/standard}}$$

Usually, two or three characteristic ions are selected for a given compound to avoid overlap with the ions of other compounds, since the ratio of one ion to another is known from the mass spectrum (either from a compendium or determined in the laboratory), RMR factors can be calculated quite readily in those cases where the most intense ion of the spectrum is saturated. Quantitative data for all the organic vapors in an air sample can be obtained in a single sample analysis once the response factors are known.

Table 6

Ambient Air Levels of Halogenated and Other Organics in Houston, Pasadena, Deer Park, Freeport, and La Porte, TX Areas

Chemical Class	HL1	HL2	HL3	PL1	PL2	DSL1	DSL2	DDL1	DTL1	DTL2	DTL3	DTL4	FL2	FL3	LL1	LL2	LL3
HALOGENATED HYDROCARBONS																	
2-chloro-1,3-butadiene (chloroprene) (tent.)	4,000[a]	266	-	-	-	-	-	-	-	-	-	-	-	-	-	-	-
dichloropropene isomer (tent.)	-	-	-	-	-	-	-	-	-	-	-	-	-	-	-	-	-
chloroform	11,539	-	11,538	T	T	53,846	6,420	7,692	1,923	8,846	T	15,384	280	T	8,461	8,850	7,692
vinylidene chloride	-	-	-	-	-	555	-	-	-	-	-	-	-	531	-	-	-
1,1-dichloroethane	T	-	-	158	-	-	-	-	-	-	-	-	-	-	-	-	-
1,2-dichloroethane	-	-	-	-	158	66,300	-	6,722	-	4,055	-	-	3,300	4,500	778	-	-
dibromoethane	-	-	-	-	-	-	-	-	-	-	-	-	-	T	-	-	-
1,1,1-trichloroethane	522	-	900	T	T	846	144	-	-	1,000	400	400	16,600	15,200	3,889	27,700	T
carbon tetrachloride	238	114	T	T	146	T	T	T	-	T	276	69	11,538	1,478	T	1,230	T
dichloropropane isomer	-	-	-	-	-	-	-	-	-	-	-	-	69	200	-	-	-
trichloroethylene	75	-	39	76	5,071	321	-	2,535	-	2,586	-	-	107	-	-	43	-
dichlorobutane isomer	-	-	-	-	-	-	-	-	-	-	-	-	-	-	-	-	-
dichlorobutene isomer	262	700	-	-	-	-	-	-	-	-	-	-	-	-	-	-	-
1,1,2-trichloroethane	-	-	-	-	-	-	-	-	-	T	-	-	-	3,821	-	-	-
tetrachloroethylene	29	T	21	20	18	6,700	T	2,019	75	3,334	-	72	94	1,585	17	83	T
dichlorobutane isomer	52	-	-	-	-	-	-	-	-	-	-	-	-	-	-	-	-
dichloropropene isomer(s)	-	-	-	-	-	180, 90, 90	-	-	T,T	241, 1,293, 72, 1,293, 345	-	T	-	-	-	-	-
1,1,2-tetrachloroethane	-	-	-	-	-	-	-	-	-	-	-	-	-	21	-	-	-
bis-(2-chloroisopropyl)-ether	-	-	-	-	-	-	-	-	-	-	-	-	-	-	-	T	T
hexachloro-1,3-butadiene	-	-	-	-	-	-	-	334	-	2,066	-	25	27	333	-	T	T
1,2-dibromopropane	-	-	-	-	-	-	-	-	-	-	-	-	13.3	8.3	-	-	-
1,1,2,2-tetrachloroethane	-	-	-	-	-	-	-	-	-	19	-	-	-	33	-	-	-
1,2,3-trichloropropane	-	-	-	-	-	-	-	-	-	-	-	-	-	298	-	-	-
pentachloroethane	-	-	-	-	-	-	-	-	-	-	-	-	-	3,984	-	-	-
perchloroethane	-	-	-	-	-	-	-	-	-	-	-	-	-	2,903	-	-	-
tetrachlorobutadiene isomer	-	-	-	-	-	-	-	-	-	T	-	-	-	-	-	-	-
pentachlorobutadiene isomer(s)	-	-	-	-	-	-	-	67	-	T, 100	-	-	-	-	-	-	-
vinyl chloride	-	-	-	-	-	-	-	T	-	-	-	-	-	-	-	-	-
trichloropropene isomer(s)	-	-	-	-	-	-	-	-	-	T,T	-	-	-	-	-	-	-

[a] Values are in ng/m³.

Table 6 (continued)

Chemical Class	HL1	HL2	HL3	PL1	PL2	DSL1	DSL2	DDL1	DTL1	DTL2	DTL3	DTL4	FL2	FL3	LL1	LL2	LL3
OXYGENATED COMPOUNDS																	
isobutyl isobutyrate	T	1,233	-	-	?	-	T	-	-	33	T	-	-	-	-	-	T
2-butyl-n-butyrate	330	-	-	-	-	-	-	-	-	-	-	-	1,291	1,586	-	1,334	2,066
n-butyl-n-butyrate	T	600	-	-	-	-	T	230	-	-	T	670	1,435	1,010	3,334	7,300	4,167
dimethyl phthalate	-	-	-	-	-	-	-	-	-	-	-	-	-	-	100	-	-
diethyl phthalate	-	113	-	T	-	134	T	-	-	-	T	330	-	-	500	1,000	-
amyl benzoate	-	T	-	-	-	-	-	-	-	-	-	-	-	-	-	-	-
dibutyl phthalate	-	-	-	-	-	-	-	547	T	-	-	-	-	-	567	-	-
methyl methacrylate	-	-	-	-	-	-	-	333	1,334	-	-	-	-	-	-	-	-
isobutyl methacrylate	-	-	-	-	-	-	-	-	-	67	-	-	-	-	-	-	-
n-butyl methacrylate	-	-	-	-	-	-	-	-	2,380	-	-	-	-	-	-	-	-
sec-butyl acrylate	-	-	-	-	-	-	-	167	-	-	-	-	-	-	-	-	-
n-butyl acrylate	-	-	-	-	-	-	-	2,670	-	-	-	-	-	-	-	-	-
n-hexyl acrylate	-	-	-	-	-	-	-	3,000	-	-	-	-	-	-	-	-	-

Using the latter approach to quantitation, this method has been used to obtain quantitative data for several samples taken in the Houston, Pasadena, Deer Park, LaPorte, and Freeport, Texas areas. Examples of these data are shown in Table 6. The range of concentrations were from a few ng/m to several μg/m³. An inspection of the vertical columns of data quickly reveals which samples were downwind from industrial facilities. These data also can be categorized into two general groups of pollutants, those that are ubiquitous, occurring in upwind and downwind samples and those that are "site" specific.

Table 7 summarizes the total weight of halogenated and oxygenated compounds found in the Texas study.

ANALYSIS OF VOLATILE ORGANICS IN AQUEOUS SAMPLES

Many of the concepts and instrumental techniques used in ambient air analysis can also be used in the identification and quantitation of volatile organics (VOA) in aqueous effluent samples from energy-related activities. The VOA technique employs an inert gas, helium, which is bubbled through the sample to transfer the volatile compounds from the aqueous phase to the gaseous phase and then trapped on a Tenax cartridge (19,20). The sample is heated between 40° to 95°C during the purging. Figure 8 depicts one of the many configurations which has been used.

Several other configurations have been reported. Regardless, foaming of the sample remains a serious problem. A paucity of data has been published on the percent recovery of chemicals from aqueous samples using these devices. The recovery of carbon-14 labeled acetone, acetonitrile, benzene, and toluene has been examined for several aqueous samples from energy-related processes such as in situ coal gasification (19). For compounds that are highly soluble in water, e.g., acetonitrile, the recovery was very low, about 10%. On the other hand, the recovery of hydrocarbons, aromatics, and alkyl-aromatics was > 80%. In general, the purging of volatile organics from an aqueous medium utilizing an inert gas is quantitative for compounds with boiling points <210° and <2% solubility, and for compounds with boiling points of <150° with a solubility of <10% in water.

The VOA method has been applied to the analysis of aqueous samples from energy-related activities using the instrumental methods described earlier (Figure 6). Figure

Table 7

Estimated Minimum Total Ambient Air Levels
of Volatile Organic Chemical Classes

Chemical Class	Halogenated hydrocarbons	Oxygenated compounds
HL1	16,737[1]	370
HL2	1,100	1,966
HL3	12,518	–
PL1	294	20
PL2	5,433	–
DSL1	128,948	134
DSL2	6,604	60
DDL1	19,409	6,947
DTL1	2,078	3,734
DTL2	25,398	120
DTL3	716	60
DTL4	15,990	1,020
FL2	32,028	2,726
FL3	40,026	2,596
LL1	13,165	4,601
LL2	37,926	9,634
LL3	7,832	6,253

[1]Values are in ng/m^3.

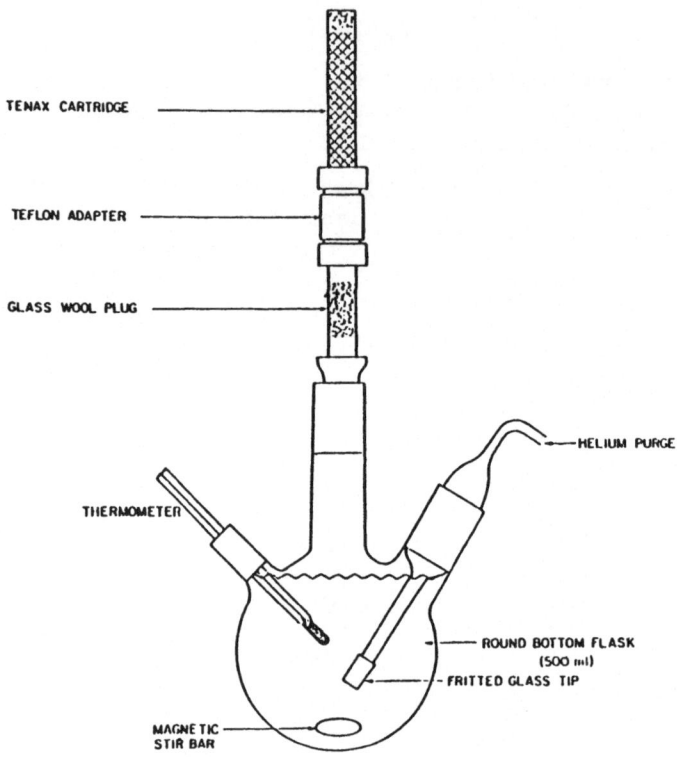

Figure 8. VOA apparatus.

9 depicts a profile of volatile organics from a low Btu
gasification process for coal. The window representing
only the volatile organics is quite complex. The gas chromatographic resolution is insufficient. However, when mass
fragmentography is employed, complete deconvolution is
achieved (Figure 10). As with ambient air analysis, the use
of ion chromatography is very important for quantification.

     The identity and concentrations of the components in
this sample are shown in Table 8. Many alkanes, alkyl-
aromatics, thiophenes, pyridines, indanes, indenes, furnas,
etc., were present. In this case approximately 134 chemicals were identified and quantified. The range of component
concentrations in this sample was from a few ppb to 817 ppb
for naphthalene.

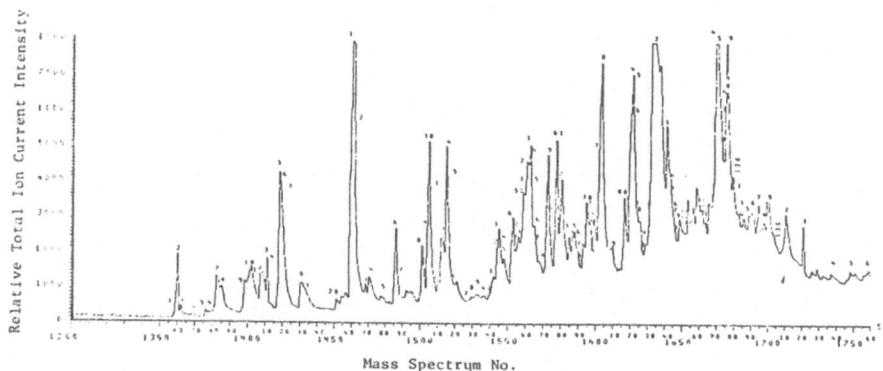

Figure 9. Profile of volatile organics in aqueous sample from low btu gasification of coal (MERC, ERDA).

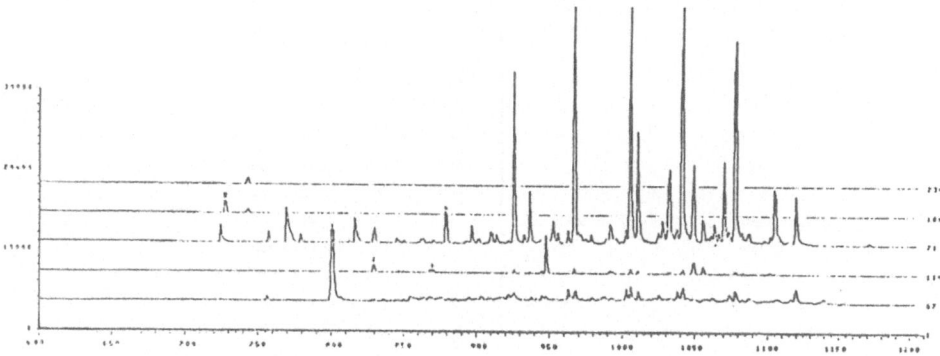

Figure 10. Ion chromatograms of sample in Figure 9.

CONCLUSIONS

Even though significant strides have been made in the development of techniques for volatile organic analysis of ambient air and aqueous samples; improvements are still warranted. A master scheme would be useful to serve as a guide to these analyses. Standardization is needed but it should be tempered with flexibility for modification of methods as improvements are made.

Table 8

Volatile Organics in Aqueous Condensate (-2L) from Low BTU Gasification of Rosebud Coal (MERC, ERDA)

Chromatographic Peak No.	Elution Temp. (°C)	Compound	ppb
1	49	$N_2 + O_2$	-
2	50	CO_2	-
4	54	C_4H_8 isomer	NQ
6	59	C_4H_{10} isomer	NQ
8	63	acetone	100±20
9	70	perfluorobenzene (e§)	
10	71	C_5H_{12} isomer	154±66
12	74	perfluorotoluene (e§)	
14	77	C_5H_{12} isomer	T
15	79	benzene	308±134
16	81	thiopene	4.5±1
17	81	C_6H_{14} isomer	57±10
18	85	C_6H_{14} isomer	90±17
19	87	C_6H_{14} isomer	84±33
20	96	C_7H_{16} isomer	T
21	100	toluene	381±83
22	101	methylthiophene isomer	T
23	104	C_7H_{16} isomer	T
24	106	C_7H_{16} isomer	33±17
25	110	C_7H_{14} isomer	T
27	116	C_8H_{18} isomer	20±13
28	121	ethylbenzene	38±7
29	123	C_8H_{16} isomer	40±20
30	124	m- and p-xylene	210±67
31	124	dimethylthiophene isomer	32±24
32	126	C_8H_{18} isomer	T
33	127	styrene and/or cyclooctatetraene	12±3
34	128	o-xylene	16±8
35	129	C_8H_{16} isomer	23±5
36	131	dimethylthiophene isomer	47±10
37	132	C_9H_{20} isomer	T
38	133	anisole	4.6±1.4
39	136	C_3-alkyl benzene isomer	T
40	137	C_9H_{20} isomer	T
41	141	C_3-alkyl benzene isomer	T
42	143	benzaldehyde	T
43	143	C_3-alkyl benzene isomer	107±10
44	144	C_9H_{18} isomer ⎱	67±33
45	144	C_9H_{20} isomer ⎰	

Chromatographic Peak No.	Elution Temp. (°C)	Compound	ppb
46	147	C_3-alkyl benzene isomer	T
47	147	methylpyridine isomer	47±10
48	148	C_9H_{18} isomer	33±20
49	148	C_9H_{20} isomer	T
50	149	benzofuran	59±40
51	149	methylpyridine isomer	38±11
52	150	C_3-alkyl benzene isomer	144±18
54	152	methylpyridine isomer	16±7
55	154	$C_{10}H_{22}$ isomer	T
56	154	C_9H_{18} isomer	T
57	155	unknown	T
58	155	methylanisole isomer	T
59	158	C_4-alkyl benzene isomer	301±104
60	158	C_4-alkyl benzene isomer	33±20
61	160	indan	112±53
62	161	indene	114±40
63	162	C_4-alkyl benzene isomer	T
64	164	C_4-alkyl benzene isomer	17±8
65	165	$C_{10}H_{20}$ isomer	10±2
66	165	$C_{10}H_{22}$ isomer	13±3
67	166	C_4-alkyl benzene isomer	T
68	167	$C_{10}H_{20}$ isomer	T
69	168	C_4-alkyl benzene isomer	T
70	168	$C_{10}H_{22}$ isomer ⎱	
71	168	C_4-alkyl benzene isomer ⎬	82±44
72	168	C_4H_7-benzene isomer ⎰	
73	170	C_4H_9-benzene isomer	T
74	170	C_4H_7-benzene isomer	T
75	171	$C_{11}H_{22}$ isomer	73±60
76	171	methylbenzofuran isomer	127±50
77	172	$C_{11}H_{24}$ isomer	177±37
78	173	C_5H_9-benzene isomer	367±74
79	176	C_4H-benzene isomer	T
80	179	methylindan isomer	19±7.4
81	180	C_5H_9-benzene isomer ⎱	37±6
82	180	$C_{11}H_{22}$ isomer ⎰	
83	180	methylindan isomer	39±19
84	181	methylindene isomer	57±37
85	182	methylindan isomer	24±20
86	183	C_5H_9-benzene isomer	80±37

Table 8 (continued)

Chromatographic Peak No.	Elution Temp. (°C)	Compound	ppb	Chromatographic Peak No.	Elution Temp. (°C)	Compound	ppb
87	184	methylindene isomer	55±37	111	203	C_6H_{13}-benzene isomer	T
88	184	ethylphenol isomer	NQ	112	204	$C_{11}H_{14}$ isomer	117.3±6.7
89	185	C_5-alkyl benzene isomer	T	113	205	C_6H_{13}-benzene isomer	T
90	186	C_5-alkyl benzene isomer	T	114	206	β-methylnaphthalene	350±83
91	187	$C_{11}H_{24}$ isomer	T	115	206	$C_{13}H_{28}$ isomer	38±5
92	188	naphthalene	817±445	116	207	C_6H_{13}-benzene isomer	T
93	189	dimethylindan isomer	47±24	117	208	C_6H_{13}-benzene isomer	T
94	189	C_5-alkyl benzene isomer	84±23	118	208	C_7H_{15}-benzene isomer	T
95	189	$C_{12}H_{24}$ isomer	96±35	119	210	α-methylnaphthalene	143±67
96	190	$C_{12}H_{26}$ isomer	74±34	120	210	C_6H_{11}-benzene isomer	T
97	190	dimethylindan isomer	82±44	121	211	C_6H_{11}-benzene isomer ⎫	
98	191	C_6H_{13}-benzene isomer	T	122	211	C_7H_{15}-benzene isomer ⎬	10±4.7
99	192	dimethylbenzimidazole (tent.)	35±11	123	211	C_7H_{13}-benzene isomer ⎭	
100	193	C_6H_7-benzene isomer ⎫		124	214	C_8H_{17}-benzene isomer	T
101	196	C_6H_{13}-benzene isomer ⎬	38±11	125	216	$C_{13}H_{18}$ isomer ⎫	
102	196	C_6H_{13}-benzene isomer ⎪		126	217	$C_{14}H_{26}$ isomer ⎬	39±11
103	196	dimethylindan isomer ⎭		127	219	$C_{13}H_{18}$ isomer ⎫	
104	196	C_6H_{13}-benzene isomer	T	128	220	$C_{14}H_{24}$ isomer ⎬	11±6
105	199	C_4H_7-benzene isomer	6.7±3.3	129	221	$C_{14}H_{30}$ isomer ⎭	
106	200	C_5H_9-benzene isomer	T	130	224	$C_{15}H_{24}$ isomer	T
107	200	C_6H_{13}-benzene isomer	3.3±2.0	131	225	C_8H_{17}-benzene isomer	T
108	201	dimethylindole (tent.)	32±31	134	240	dibenzofuran	T
109	201	$C_{11}H_{14}$ isomer	17±7				
110	202	C_7H_{15}-benzene isomer	3.3±2.0				

REFERENCES

1. Calvert JG, Pitts JN: Photochemistry. New York, John Wiley and Sons, Inc, 1966, pp 366-557

2. Leighton PA, Perkins WA: Air Pollution Foundation. Los Angeles, Rept 14, 1956

3. Leighton PA, Perkins WA: Air Pollution Foundation. Los Angeles, Rept 24, 1958

4. Leighton PA: Photochemistry of Air Pollution. New York, Academic Press, 1961, pp 1-200

5. Gould RF: Photochemical smog and ozone reactions. In: Advances in Chemistry Series 113, Washington, DC, Amer Chem Soc, 1976, p 285

6. Matz J: Z Ges Hyg Ihre Grenzebiete 18:903, 1972

7. Fishbein L: Chromatography of Environmental Hazards. New York, Elsevier Pub Co, 1972, p 499

8. Pellizzari ED: Development of method for carcinogenic vapor analysis in ambient atmospheres. Research Triangle Park, Environ Prot Agency, EPA-650/2-74-121, 1974, pp 148

9. Pellizzari ED: Development of analytical techniques for measuring ambient atmospheric carcinogenic vapors. Research Triangle Park, Environ Prot Agency, EPA-600/2-75-075, 1975, pp 187

10. Pellizzari ED: The measurement of carcinogenic vapors in ambient atmospheres. Research Triange Park, Environ Prot Agency, EPA-600/7-77-055, 1977, pp 288

11. Pellizzari ED: Analysis of organic air pollutants by gas chromatography and mass spectroscopy. Research Triangle Park, Environ Prot Agency, EPA-600/2-77-100, 1977, pp 104

12. Pellizzari ED: The measurement of carcinogenic vapors in ambient atmospheres. Research Triangle Park, Environ Prot Agency, Contract No 68-02-1228, in preparation

13. Pellizzari ED, Bunch JE, Berkley RE, Bursey JT: Identification of n-nitrosodimethylamine in ambient air by capillary gas-liquid chromatography-mass spectrometry-computer. Biomed Mass Spec 3:196-200, 1976

14. Pellizzari ED, Carpenter B, Bunch JE, Sawicki E: Collection and analysis of trace organic vapor pollutants in ambient atmospheres - a technique for evaluating the concentration of vapors by sorbent media. J Environ Sci Tech 9:552-555, 1975

15. Pellizzari ED, Bunch J, Carpenter B, Sawicki E: Collection and analysis of trace organic vapor pollutants in ambient atmospheres - studies on thermal desorption of organic vapor from sorbent media. J Environ Sci Tech 9:556-560, 1975

16. Pellizzari ED, Bunch JE, Berkley RE, McRae J: Collec-
     tion and analysis of trace organic vapor pollutants in
     ambient atmospheres - the performance of a Tenax GC
     cartridge sampler for hazardous vapors. Anal Letters
     9:45-63, 1976

17. Pellizzari ED, Bunch JE, Berkley RE, McRae J: Deter-
     mination of trace hazardous organic vapor pollutants in
     ambient atmospheres by gas chromatography/mass spectrom-
     etry/computer. Anal Chem 48:803-807, 1976

18. Pellizzari ED, Bunch JE, Bursey JT, Berkley RE, Sawicki
     E, Krost K: Estimation of n-nitrosodimethylamine levels
     in ambient air by capillary gas-liquid chromatography/
     mass spectrometry. Anal Letters 9:579-594, 1976

19. Pellizzari ED: Identification of components of energy-
     related wastes and effluents. Athens, Environ Prot
     Agency, Contract No 68-03-2368, in preparation

20. Pellizzari ED, Castillo NP, Willis S, Smith D, Bursey
     JT: Identification of organic constituents in aqueous
     effluents from energy-related processes. Fuel Chemistry
     23:144-155, 1978

STRATEGY FOR COLLECTION OF DRINKING WATER CONCENTRATES

Carl C. Smith
Department of Environmental Health
College of Medicine
University of Cincinnati
Cincinnati, Ohio

In the first chapter of a recent book on water analysis, Aaron Rosen reviewed the state of water analysis in 1950 and pointed out the need for a new and entirely different approach to sampling (1). Thus in 1950, Braus, Middleton, and Walton (2), using a column of activated carbon to filter 5000 gallons or more of water, were able to recover from 2 to 4 g of organic pollutants by extracting the carbon with ethyl ether and an additional 2 to 10 g by a second elution with ethanol. The process was subsequently scaled up eighty-fold (3); the ether was replaced with chloroform and water-free extracts of 150-1700 g were obtained. As recently as 1974, this same procedure was used to prepare the starting material for a detailed analysis of the organic pollutants in New Orleans drinking water (4).

Once large samples of organic pollutants became available, various specialized analytical procedures were applied. These included infrared spectroscopy and, in particular, various chromatographic procedures including column, thin-layer, and gas/liquid chromatography. The introduction first of gas chromatography followed by coupled gas chromatography-mass spectrometry led to the detection of an increasing array of compounds that have been identified in various drinking water samples. The list published by EPA in 1976 included 398 compounds and the current number exceeds 700 (5).

The problems encountered in using activated carbon are detailed in Rosen's review and include pore size, pore volume, surface groups, content of extractable organic matter, and desorption with different organic solvents. The bases for

these problems were studied at length and led to the final
evolution of a standard material and a standard elution
procedure using chloroform and ethanol (6). In spite of
some inherent difficulties with the procedure, the U.S.
Public Health Service issued Drinking Water Standards -
1962, which specified a limit of 200 µg/1 for the concentration of carbon chloroform extractable or CGE compounds
in order to "avert, if possible, the health hazard of
unidentified and unnumbered industrial organic pollutants"
(7).

At the same time other methods for concentrating and
fractionating various contaminants in water supplies were
being developed. One of the earliest approaches, and one
still employed with many variations, is the sparging or
purge and trap procedure of Bellar and Lichtenberg (8).
Other methods for isolating the purgeable components from
drinking water included those of Rook (9), who was the
first to demonstrate the origin of the trihalomethanes in
drinking water, Mieure et al. (10), who continously swept
the head space gas through a porous polymer trap, and the
method of Zlatkis (11), in which the volatile compounds
in the sample were thermally extracted into the head space.
The latter procedure was used by Dowty et al. (12,13) in
analyzing drinking water. Kopfler et al. (14) applied a
revision of the Bellar and Lichtenberg method to the determination of volatiles for the National Organics Reconnais-
sance Survey (NORS). The GC/MS techniques for identifying
and quantifying the volatile organics in this study are
described by Lingg et al. (15).

A liquid-liquid procedure for determining halomethanes
in drinking water was developed by Glaze and coworkers (16).
In this procedure, 120 ml samples were collected in serum
bottles in a manner which excluded any head space. When
the sample was returned to the laboratory, 3 to 5 ml of
pentane was added to the bottle using two syringes (one to
add pentane, the other to accept the displaced water) and
the bottle was shaken for 15 minutes at 500 rpm on a
gyratory platform shaker. Glaze et al. suggested that this
procedure offers several advantages over the Bellar and
Lichtenberg procedures: (1) multiple samples can be
processed; (2) the GC procedure is shorter because thermal
desorption is eliminated; and (3) the isothermal GC procedure cuts analysis time.

Another approach for detecting and/or quantifying volatile impurities in water samples is the closed loop stripping procedure developed by Grob and coworkers (17). Some difficulties were experienced in certain laboratories in applying this method until some additional information on the equipment and procedure was published in 1976. The first paper (18) stressed the need to carefully control certain parameters including extraction (water) temperature; filter characteristics and amount; stripping flow rate, duration and temperature; and, finally, desorption from the special filter. The second paper (19) discussed the application of narrow and wide bore capillary columns for the GC analyses and contrasted the typical separation (118 peaks) obtained with a 3 m/2 mm column packed with OV-1 on Gaschrom Q to the greatly increased number of peaks detected (490) when a 35 m/0.28 mm glass capillary column coated with an OV-1 film was used.

Investigators at the U.S. Environmental Protection Agency have prepared a slightly larger model of the Grob device (20). It will accept a 4 l water sample and should provide sufficient materials for analysis and limited biological testing.

Another approach for monitoring contaminants in water samples is that described by Junk et al. (21). The Ames, Iowa, group has primarily used the XAD-2 resin. This resin is considered to have low polarity, and consists of smooth white spheres, 0.25-0.5 mm in diameter with surface area of 300 m^2/g and average pore size of 90 Å. The general flow chart used is shown in Figure 1.

For grab samples the design shown in Figure 2 is used. Clear, decanted water samples are fed through the column at a rate of 25 to 50 ml/min. Sediment, if there is some, is transferred to the reservoir with several rinses of organic-free water. The column is eluted with ethyl ether, and 1 to 5 µl of the dried ether extract is injected into a GC. A typical run from the extract of the Ames, Iowa, municipal water supply is shown in Figure 3.

Junk et al. (21) also employ another type of resin column, shown in Figure 4, to process 55 gal samples from numerous cities throughout the country. This apparatus provides more material and, of course, has the great advantage of being rugged, relatively stable, and can be shipped for subsequent analysis.

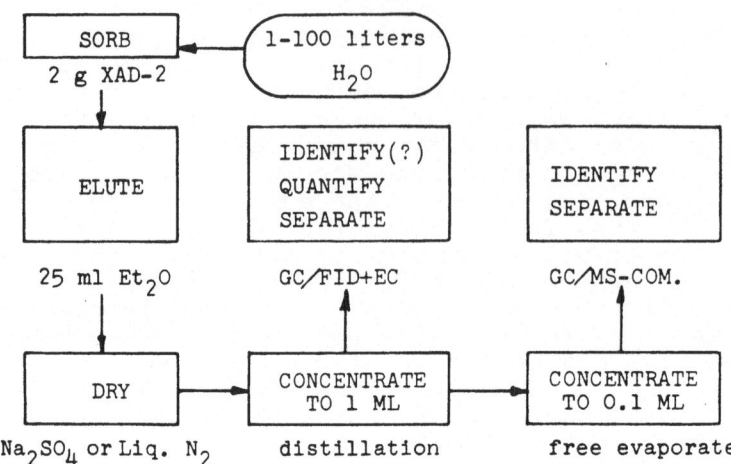

Figure 1. Flow chart of general resin sorption scheme of Junk et al. (21).

This group has also been testing raw and finished water samples from 14 major U.S. water utilities on a monthly basis for the presence of mutagenic materials, using the Ames spot test (22). The ether extracts from their XAD columns are concentrated to 1 ml, 0.1 ml of DMSO is added and the solution allowed to evaporate until the ether is gone. Dr. Bonita Glatz, a member of their group, has applied these DMSO solutions to discs which are applied to dishes containing Ames strains TA98, TA100, TA1535, TA-1537 and TA1538 with and without the addition of liver microsomes from Aroclor 1254-induced rates (S-9). The most sensitive of the Ames strains for detecting mutagens in water samples was TA100, a finding confirmed by Loper, Lang, and Smith (23). Also, they found almost no increase or decrease in activity in the presence of the rat liver S-9 fraction. This finding is also in general agreement with the results presented previously by Loper et al. (23).

The authors obtained a positive mutagenic response with extracts equivalent to 15 liters of water sample. This finding is somewhat surprising, since from our results one would expect a doubling of the mutagenic activity with either TA98 or TA100 when the extract of about 1.5 liters of water is applied to the plate. Some possible explanations for these apparent differences will be considered in the comments on the reverse osmosis procedure.

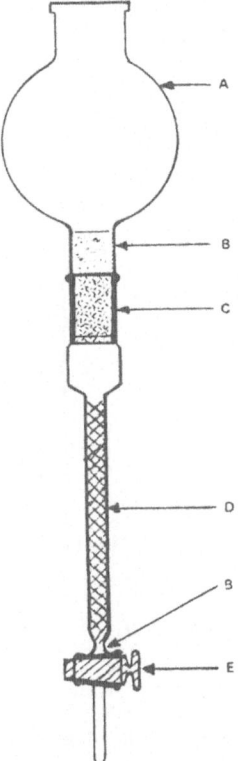

Figure 2.   Device for extracting organics from grab samples
(21).   A: 5-1 reservoir, scaled down; B: glass wool plugs;
C: 24/40 ground glass joint with PTFE sleeve; D: 8 x 140
mm glass tube packed with ~2 g, 40-60-mesh resin; E: PTFE
stopcock.

These methods developed by Junk, Fritz, Svec, and Cris-
well, including the application of the Ames Test, are of
great interest and are described in more detail in another
chapter in this symposium (22). With the acquisition of
larger (2000 1) samples, more quantitative Ames tests and
extensive fractionation procedures are possible.

The last procedures to be described for concentrating
and fractionating contaminants in potable water samples are
the reverse osmosis methods developed by Kopfler and coworkers
at the Cincinnati Laboratory of the USEPA (14).

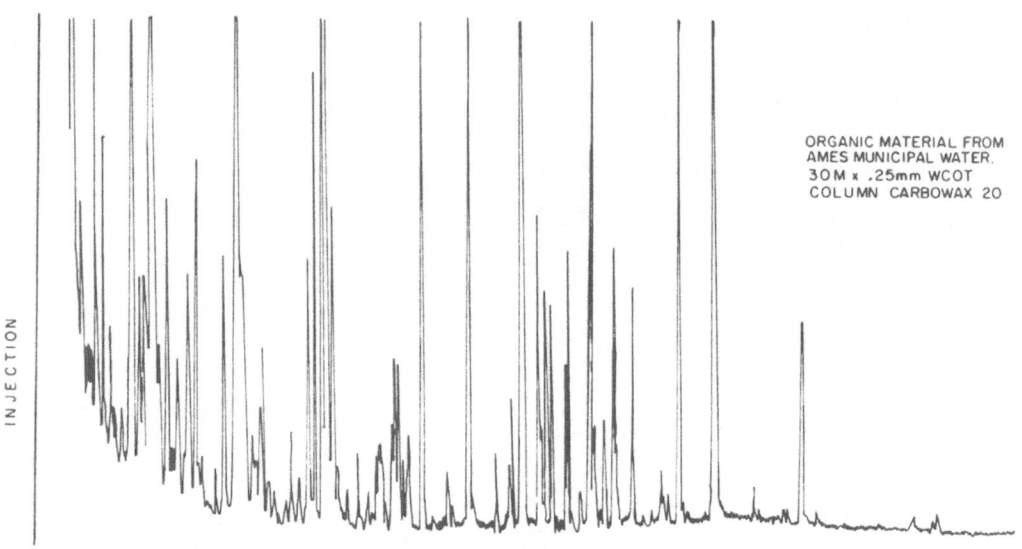

Figure 3.   Organic material from Ames municipal water (21).

The need for these procedures can be appreciated when one remembers that at present only about 10-15% of the organic material in drinking water has been identified. The volatiles such as chloroform and the other halomethanes, although easily detected and measured either by the Purge-and-Trap or the Liquid-Liquid extraction procedures, still represent less than 10% of the total organic compounds in water.   Because of the shortcomings of all the previous methods [see Kopfler et al. (14)], the reverse osmosis (RO) procedure was developed.   It appeared to be the only method that would provide sufficiently large samples of the whole array of non-volatile constituents to support the various biological and chemical screens (Figure 5) devised by Dr. Robert Tardiff and other members of the EPA staff (24). Some idea of the total amount of organic compound present

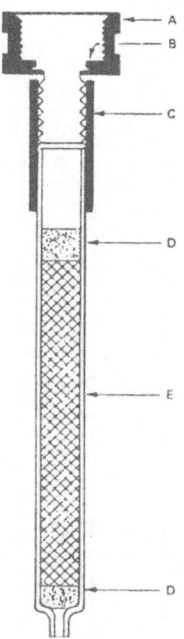

Figure 4. Device for extracting organics from composite
samples (21). A: standard garden hose coupling; B: PTFE
washer; C: 12.7 mm ID PTFE tubing; D: glass wool plugs;
E: 12.7 mm OD x 9 cm long glass tube packed with ~2 g,
40-60-mesh resin.

can be obtained by determining total organic carbon (TOC).
In Cincinnati, this varies typically from 1 to 2 mg/l
[Figure 6 (25)]. Considering carbon to be about 50% of
the weight of the organic contaminants present, 1500 liters
will contain 3 to 6 g of organic material. The scheme
shown in Figure 7 has been applied repetitively to water
samples from 5 cities; these were chosen from the group
studied by Keith et al. (26) and shown in Table 1.

    The salt brines obtained by lyophilization of the reject
from each membrane were extracted as shown in the diagram.
Originally, only the total ROC-OE and XAD-eluate were provided but with the demonstration of mutagenicity in these
extracts as described by Loper et al. (23), the separate
fractions were provided and these have been examined for

PROTOCOL FOR BIO-SCREEN OF ORGANIC CONCENTRATES FROM TAP WATER

Assay	Sample/City at 2 month intervals					
	1	2	3	4	5	6
RANGE-FINDING (LD$_{50}$ MOUSE)	X					
MUTAGENESIS (SALMONELLA)		X-f			X-f	
MAMMALIAN CELL TRANSFORMATION		X-f			X-f	
IN VIVO CARCINOGEN BIO-ASSAY (NEONATE)			X			?
TERATOGEN ASSAY (RAT)				X		?
CHEMICAL CHARACTERIZATION (GC/MS)	?	?	?	?	?	X-?

Figure 5.   Bioscreen of Tardiff et al. (24).

mutagenicity, cell death (toxicity) and in a few instances
for capability of transforming clones of the BALB/3T3 cell
line (27).

    From the data in Table 2 one can see that the concentra-
tions of ROC-OE materials as well as XAD adsorbable components
vary from city to city. Sometimes they are unexpectedly high
as in the case of Miami. In all cases recovery appeared to
account for 35%-40% of the total organic carbon (TOC). Although this represents the best recovery reported so far
using large initial water samples, there are some drawbacks
in terms of an ideal methodology:

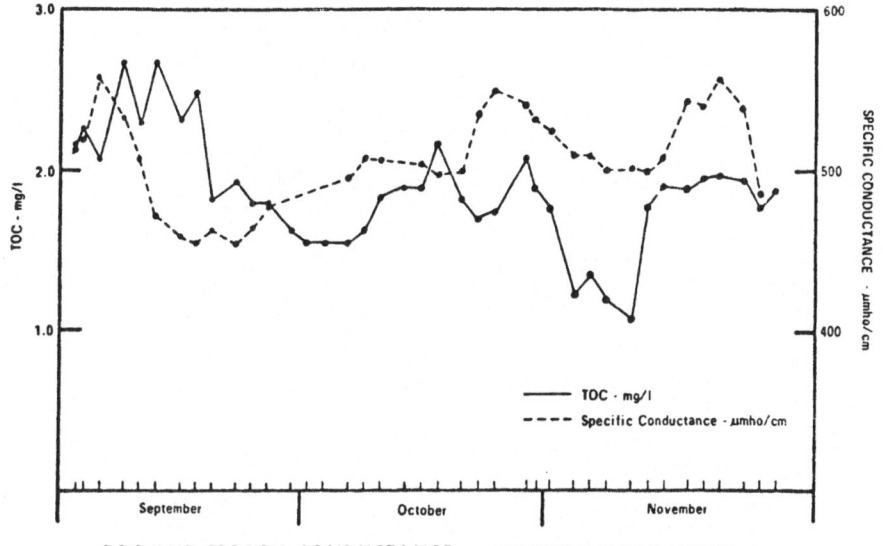

TOC AND SPECIFIC CONDUCTANCE - CINCINNATI TAP WATER

Figure 6. Variation over a 3 month period of the TOC and specific conductance of Cincinnati tap water (25).

        1. Most, if not all, the volatile organics are removed by the procedure.

        2. At present there is no way of ascertaining how well the various compounds in the original water sample are represented in the final concentrates.

        3. It has been shown that some of the components of the final concentrates are derived from the metal and plastic which comprise the plumbing, membranes, etc., required by the process.

        4. Stability of the extracts is not thought a problem. Previous studies within EPA and elsewhere have shown that Cl_2 must be removed to stop further reaction to synthesize trihalomethanes.

        5. Although analyses on successive samples appear to agree, too little information has been accumulated so far to determine this beyond doubt.

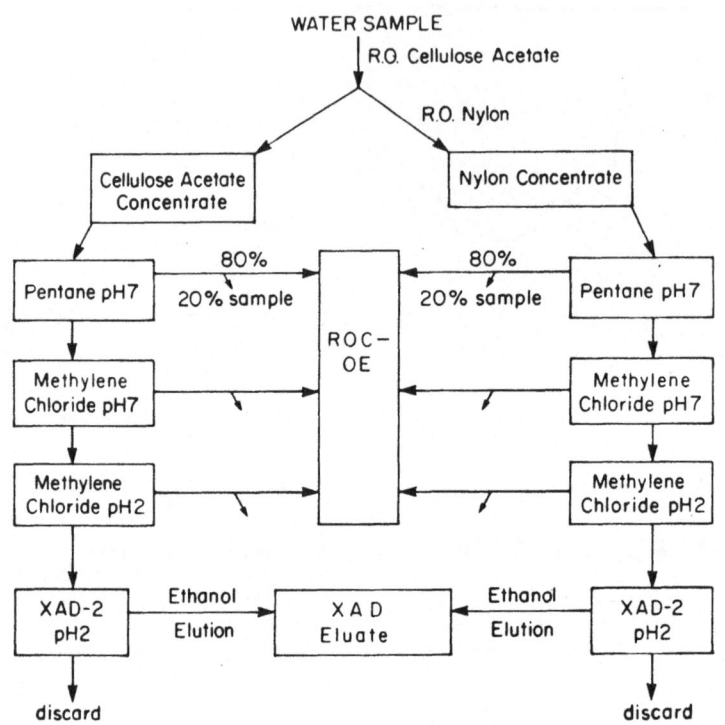

Figure 7.   Diagram of the reverse osmosis (ROC–OE) and XAD
eluate devised by Kopfler et al. (14).

    6.   In spite of the large samples provided by the pro-
cedure, the projected studies in which we hope to fractionate
in greater detail the ROC–OE and XAD eluate using the Ames
Test as a detector are still limited by sample availability.

    In view of our findings (27) that all of the city water
supplies examined to date appear to contain significant
amounts of mutagenic material, several areas need further
study:

    1.   Methods need to be developed which will permit one
to determine the relative mutagenic risk to the human population consuming the water of the (1) volatile and (2) non-
volatile fractions of the organic contaminants.   At present
EPA, understandably, has focused its initial attention on

Table 1

Type and Location of Various Types of Water Supplies (26)

Source Character	Water Plants
Mississippi River at New Orleans	1. Carrollton (City of New Orleans) 2. Jefferson Parish #1 (east bank) 3. Jefferson Parish #2 (west bank)
Uncontaminated upland water	1. Seattle, WA 2. New York, NY
Ground water	1. Miami, FL 2. Tucson, AZ
Contaminated by agricultural runoff	1. Ottumwa, IA 2. Grand Forks, ND
Contaminated by industrial waste	1. Cincinnati, OH 2. Lawrence, MA
Contaminated by municipal waste	1. Philadelphia, PA 2. Terrebonne Parish, LA

(NORES brackets the five non-Mississippi categories)

Table 2

Composition of Organic Concentrates from Five Cities

City & Sample	Tap Water Processed, Liters	ROC-OE, G*	XAD Eluate, G
New Orleans 1B	7,800	1.0	5.6
Miami 2	2,270	1.0	8.8
Philadelphia 1	5,800	1.8	3.9
Ottumwa 1	6,000	0.9	6.4
Seattle 1	13,000	0.4	3.0

*Reverse osmosis concentrate, organic extract

the halomethanes and has issued proposed controls on these
suspected carcinogens. The proposed standard for total chloroform and related trihalomethanes ($CHCl_2Br$, $CHClBr_2$, and $CHBr_3$)
is 0.1 mg/liter or 100 parts per billion. The Environmental
News for January 25, 1978 contained the preliminary data
which later appeared in the Federal Register (28). On the
basis of limited but repeated studies (National Organic
Monitoring Survey) some 36 cities were named as either having
a possible problem with trihalomethanes (THM) or synthetic
organic compounds (SOC) in their drinking waters. Also listed
are some 40 cities already employing granular activated carbon
(GAC) in their water treatment plants.

It is obvious that there is need to develop reproducible
methodology for collecting and measuring various contaminants
in drinking water. Although the strategy for collecting
samples, the best designs for trapping and eluting organics,
and the most efficient and economical procedures for analyzing these samples are yet to be determined, it should be
pointed out that the EPA is working diligently on these
problems and has engaged the help of groups with expertise
to (1) examine many water sources for a significant number
of organics; (2) to collect in a computerized file GC-MS
data on all known organic water contaminants; and (3) to
develop methodology to determine the chemical identity of
as many of the presently unidentified components as possible.
Studies on mutagenicity of individual halo-organics and other
synthetic organic compounds detected in drinking water are
being carried out as well as the attempts to identify active
components of the non-volatile and presently unknown com-
poments isolated by the ROC-XAD procedures described above.

New and improved procedures for isolating the THM and
quantifying them usually by improved GC procedures are
being described in scientific journals and at various
meetings such as the 1978 Pittsburgh Conference on Analytical
Chemistry and Applied Spectroscopy in which no less than ten
papers in the general area are being presented.

Finally, Leland J. McCabe recently presented (29) a
very thought provoking and penetrating analysis of the
overall drinking water problem in relation to the current
information on trihalomethanes and mutagenicity of drinking water concentrates. His comments, compilations and
suggestions deserve serious consideration by those concerned
with the impact of the Safe Drinking Water Act of 1974.

ACKNOWLEDGMENTS

Special thanks are due to Dr. Robert Tardiff, Dr. Fred Kopfler, and Ms. Geraldine Wolfe for their assistance with the manuscript and to Dr. Colin Chriswell for permission to use certain figures.

REFERENCES

1. Rosen AA: The foundations of organic pollutant analysis. In: Identification and Analysis of Organic Pollutants in Water (Keith LH, ed.). Ann Arbor, MI, Ann Arbor Science Publishers Inc., 1976, pp 3-14

2. Braus H, Middleton FM, Walton G: Organic chemical compounds in raw and filtered surface waters. Anal Chem 23:1160-64, 1951

3. Middleton FM, Pettit HH, Rosen AA: The megasampler for extensive investigation of organic pollutants in water. Proceedings 17th Industrial Waste Conference, Engineering Ext Ser 112, 454-460, Purdue University, Lafayette, IN May 1-3, 1962

4. US Environmental Protection Agency: New Orleans area water supply study. Draft analytical report, Lower Mississippi River Facility, Slidell, LA, Nov, 1974

5. Kopfler FC, Coleman WE: Personal communication

6. Middleton FM, Rosen AA, Burttschell RH: Manual for recovery and identification of organic chemicals in water. Part II. US Public Health Service, Robert A. Taft Sanitary Engineering Center, Cincinnati, OH, May, 1957

7. US Public Health Service: Drinking Water Standards - 1962. Washington, DC

8. Bellar TA, Lichtenburg, JJ: Determining volatile organics at microgram-per-liter levels by gas chromatography. J Am Water Works Assoc 66:739-744, 1974

9. Rook JJ: Formation of haloforms during chlorination of natural waters. Water Treatment Exam 23:234-243, 1974

10.   Mieure JP, Mappes GW, Tucker ES, Dietrich MW: Separa-
       tion of trace organic compounds from water. In: Identi-
       fication and Analysis of Organic Pollutants in Water
       (Keith LH, ed.). Ann Arbor, MI, Ann Arbor Science
       Publishers, Inc., 1976, pp 113-133

11.   Zlatkis A, Lichtenstein HA, Tishbee A: Concentration
       and analysis of trace volatile organics in gases and
       biological fluids with a new solid adsorbent. Chromato-
       graphia 6:67-70, 1973

12.   Dowty B, Carlisle D, Laseter J: Halogenated hydrocarbons
       in New Orleans drinking water and blood plasma. Science
       187:75-77, 1975

13.   Dowty B, Carlisle D, Laseter J: New Orleans drinking
       water sources tested by gas chromatography - mass
       spectrometry (Occurrence and origin of aromatics and
       haologenated aliphatic hydrocarbons). Environ Sci
       Technol 9:762-765, 1975

14.   Kopfler FC, Coleman WE, Melton RG, Tardiff RG, Lynch SC,
       Smith JK: Extraction and identification of organic
       micropollutants: reverse osmosis method. Ann N Y Acad
       Sci 298:29-30, 1977

15.   Lingg RD, Melton RG, Kopfler FC, Coleman WE, Mitchell
       DE: Quantitative analysis of volatile organic com-
       pounds by GC-MS. J Am Water Works Assoc 69:605-612,
       1977

16.   Henderson JE, Peyton GR, Glaze WH: A convenient
       liquid-liquid extraction method for the determination
       of halomethanes in water at the parts-per-billon level.
       In: Identification and Analysis of Organic Pollutants
       in Water (Keith LH, ed.). Ann Arbor, MI, Ann Arbor
       Science Publishers Inc., 1976, pp 105-111

17.   Grob K, Grob G: Organic substances in potable water
       and in its precursor. II. Applications in the area of
       Zurich. J Chromatogr 90:303-313, 1974

18.   Grob K, Grob G:' Glass capillary gas chromatography in
       water analysis: how to initiate use of the method.
       In: Identification and Analysis of Organic Pollutants
       in Water (Keith LH, ed.). Ann Arbor, MI, Ann Arbor
       Science Publishers, Inc., 1976, pp 75-85

19. Grob K, Zurcher F: Stripping of trace organic sub-
     stances from water equipment and procedure. J Chroma-
     togr 117:285-294, 1976

20. Kopfler FC: Personal communication

21. Junk GA, Richard JJ, Fritz JS, Svec HJ: Resin sorption
     methods for monitoring selected contaminants in water.
     In: Identification and Analysis of Organic Pollutants
     in Water (Keith LH, ed.). Ann Arbor, MI, Ann Arbor
     Science Publishers, Inc., 1976, pp 135-153

22. Chriswell CD: Chemical and mutagenic analysis of water
     samples. In: Application of Short-Term Bioassays in
     the Fractionation and Analysis of Complexes in Mixtures
     (Waters M, ed.). Williamsburg, Virginia, Feb. 21-22,
     1978

23. Loper JC, Lang DR, Smith CC: Mutagenicity of complex
     mixtures from drinking water. In: Water Chlorination:
     Environmental Impact and Health Effects, Vol. 2 (Jolley
     RL, Gorchev H, Hamilton DH, Jr., eds.). Ann Arbor,
     MI, Ann Arbor Science Publishers, Inc., 1978, pp 433-
     450

24. Tardiff RG, Carlson GP, Simmon V: Halogenated
     organics in tap water: a toxicological evaluation.
     Proceedings of the Conference on the Environmental
     Impact of Water Chlorination (Jolley RL, ed.). Oak
     Ridge National Laboratory, Oak Ridge, TN, Oct. 22-24,
     1975, pp 213-227

25. Kopfler FC, Melton RG, Mullaney JL, Tardiff RG: Human
     exposure to water pollutants. In: Fate of Pollutants
     in the Air and Water Environments. Part 2 (Suffet IH,
     ed.). New York, John Wiley & Sons, 1977, pp 419-433

26. Keith LH, Garrison AW, Allen FR, Carter MH, Floyd TL,
     Pope JD, Thruston AD, Jr.: Identification of organic
     compounds in drinking water from thirteen US cities.
     In: Identification and Analysis of Organic Pollutants
     in Water (Keith LH, ed.). Ann Arbor, MI, Ann Arbor
     Science Publishers, Inc., 1976, Ch 22, pp 329-373

27. Loper JC, Lang DR, Schoeny RS, Richmond BB, Gallagher
     PM, Smith CC: Residue organic mixtures from drinking
     water show in vitro mutagenic and transforming activity,
     in press

28.   USEPA: Interim primary drinking water regulations:
      Control of organic chemical contaminants in drinking
      water.   Federal Register Vol. 43 (No. 28):   pp 5756-
      5780, Feb. 9, 1978

29.   McCabe LJ: Health effects of organics in drinking
      water.   Water Quality Technology Conference.   Kansas
      City, MO, Am Water Works Assoc, pp 1-11, Dec. 4-6, 1977

SECTION 3

CURRENT RESEARCH

SHORT-TERM BIOASSAY OF COMPLEX ORGANIC MIXTURES: PART I, CHEMISTRY

M.R. Guerin, B.R. Clark, C.-h. Ho,
J.L. Epler, and T.K. Rao
Analytical Chemistry and Biology Divisions
Oak Ridge National Laboratory
Oak Ridge, Tennessee

INTRODUCTION

A multidisciplinary program to elucidate the health
and ecological effects of advanced fossil fuels use is
currently underway at the Oak Ridge National Laboratory.
The development of short-term bioassays applicable to the
characterization of complex mixtures is an important aspect
of the program. Studies reported here have been designed to
identify constituents responsible for the mutagenicity of
energy related materials. Observations made in the course
of these studies may prove useful for designing biotesting
methods suitable for the quantitative routine application to
complex mixtures.

PREPARATION OF COMPLEX MIXTURES FOR SHORT-TERM BIOTESTING

The preparation of complex mixtures for _in vitro_
bioassay is complicated by at least two concerns: (a) the
relevance of the material applied to the test system and,
(b) the compatibility of the material with the test system.
Chemical "relevance" is achieved when the test system is
dosed with a material whose chemical composition mimics
that which reaches the natural (man, plant, animal, soil,
water, etc.) point of impact. Difficulties with "compatibility" are encountered when the material being bioassayed
contains constituents which interfere with the test
organisms' ability to respond to the effect of interest.
Physical properties of the test material may inhibit the
release, for example, of its mutagenic constituents to the

test bacteria. High concentrations of mildly toxic constituents or small quantities of highly toxic constituents
can mask the more subtle effect of mutagenic constituents.

Complex mixtures are generally entities of poorly defined and continuously changing chemical composition. If,
by virtue of sample history (generation, storage, handling),
the material is altered in its content of bioactive constituents or undergoes other changes which affect the biological test system, the results of the bioassay may be
invalidated. Whole animal inhalation bioassays of cigarette
smoke, for example, are designed with considerable attention
(11,12,15,18) to the compatibility of the "smoke" being bioassayed to that which is freshly generated by the cigarette
under conditions comparable to those representative of the
human situation. Pellet implantation, while highly successful for the application of pure compounds to respiratory
tract epithelium, is questionably applicable to the study of
complex mixtures because (19) the constituents may be released
to the test organism at differing rates depending on their
chemical properties.

Methods for the preparation of complex materials for
bioassay must meet different requirements from those designed for chemical characterization. Methods for quantitative chemical analyses seek to recover 100% of the individual
constituent being determined without regard to the remaining
constituents. Differing degrees of recovery of several
constituents being quantitatively determined in a single
material are acceptable for the purpose as long as the
degrees of recovery are known. For bioassay purposes,
however, quantitative recovery is important only because
it provides the maximum amount of material for bioassay
from the starting material. The critical objective is
to produce a material whose constituents have been recovered
equally. Recovering the constituents to an equal degree
ensures that the composition of the material bioassayed is
the same as that which was sampled.

"Compatibility" with the test system becomes of greater
concern when the researcher moves from whole animal models
to in vitro bioassay systems. Mechanisms of selective adsorption, metabolism, and of detoxification, for example,
embodied in whole animal models may be absent in in vitro
test systems. Observation of a highly toxic reaction to a
test material need not imply the absence of constituents
capable of producing the more subtle biological effects.

The requirements of relevance and compatibility are themselves incompatible. Any steps taken to remove toxic constituents or to otherwise make the material compatible with the test system necessarily involves a change in physical or chemical nature of the test material. The objective of methods development here has been to prepare materials in a form suitable for biotesting but with a minimal or at least interpretable impact on the relevance of the test material.

The approach used most successfully to date is to separate the constituents of the mixture into a manageable number of chemically distinct fractions. Each fraction is subsequently bioassayed to determine whether mutagenic constituents of any type are present independent of synergism, toxic interferences, or other complicating factors due to the interaction of constituents of widely differing properties. If it is assumed (or experimentally demonstrated) that the mutagenicities of the fractions are additive, the activities of the fractions may be summed to estimate the activity of the starting material. The primary chemical requirement is that all of the constituents present in the starting material are recovered in the fractions to be biotested.

Acid-base extractive fractionation and gel chromotagraphic fractionation have been used in an attempt to identify constituents responsible for the mutagenicity of synthetic crude oils and to assess their utility for preparing the oils for bacterial mutagenesis testing.

CHEMICAL CLASS FRACTIONATION

Procedures

Acid-base fractionation involves the liquid-liquid partitioning of the sample between an immiscible organic solvent and an alkaline or acidic aqueous phase. The procedure (6,20) used here first subjects the sample to partitioning between ether and 1N NaOH. Acidic constituents of the sample concentrate in the aqueous phase while alkaline and neutral constituents concentrate in the organic phase. The organic phase is subsequently contacted with 1N HCl which preferentially extracts the alkaline constituents. The constituents of the aqueous phase can be backextracted into ether following pH adjustment. Four fractions (acids, bases, neutrals, and any insoluble residue which may be formed) thus result for biological testing.

Additional discrimination is possible by subjecting the primary fractions to further separation. The procedure (Figure 1) used here was chosen because of the extensive literature (1,2,3,4,14,24,25,26) available dealing with its application to chemical and biological characterizations of condensed tobacco smokes. Basic constituents are further separated into those which are water soluble at pH 11, those which are ether soluble, and those which are insoluble under these conditions of separation. Acidic constituents are divided into weakly acid (presumably phenolic), strongly acidic/water soluble, strongly acidic/ether soluble, and two insoluble residues. Neutral constituents are divided into any number of subfractions by Florisil column chromatography or other chromatographic methods.

The extractive fractionation procedure is advantageous in that most fractions contain chemically similar constituents of predictable types, the procedure is applicable to large sample sizes (25,26), and that it has been demonstrated (1,2,14) to be effective for elucidating the biological properties of a complex mixture. An additional advantage is that it can be applied to both hydrophilic (e.g., an aqueous effluent) and lipophilic (e.g., a crude oil) materials. A practical disadvantage is that the procedure is highly labor-intensive and therefore both costly and time-consuming when a high discrimination is required. A potentially more serious concern is that contact with highly acidic and highly alkaline environments can lead to chemical reactions which may alter the nature and quantities of bioactive constituents present, thus invalidating the bioassay.

A new procedure (Figure 2), optimized to fractionate materials which are predominantly lipophilic, has been developed (13) for comparison with the extractive method. The sample is dissolved in hexane and added to a column of Sephadex LH-20 gel previously equilibrated with methanol/water (85/15, vol/vol). Lipophilic constituents are eluted from the column using hexane. Hydrophilic constituents are subsequently eluted from the column using methanol and/or acetone. In the second step of the procedure, the lipophilic constituents are separated into "polymeric," "sieved," and hydrogen bonding fractions by eluting the lipophilic fraction from a column of Sephadex LH-20 using tetrahydrofuran. The "sieved" material is subsequently separated according to aromaticity by a column of Sephadex LH-20 eluted with isopropanol. Because the pore size of

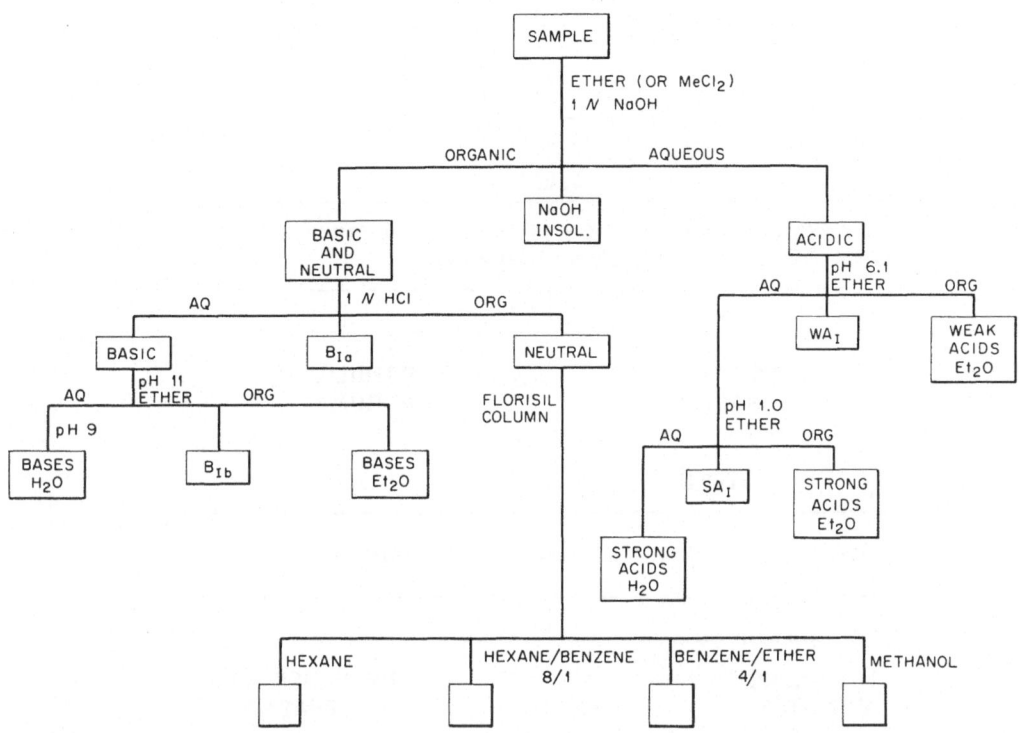

Figure 1. Acid-base extractive separation of complex mixtures (21).

the gel is somewhat smaller when equilibrated with isopropanol than tetrahydrofuran, additional "polymeric" material elutes prior to the aliphatic constituents in this third step of the procedure. The gel chromatographic procedure thus produces the following size fractions: hydrophilic, polymeric, hydrogen bonding, aliphatic, simple aromatic, and polyaromatic.

The primary advantage of the gel chromatographic procedure is that separation is affected by gentle mechanisms. The gel acts as an inert support for the methanol/water phase in the first step of the procedure yielding separation of the lipophilic constituents from the hydrophilic constituents by an essentially continuous liquid-liquid partition mechanism. Molecular sieving and hydrogen bonding are the primary mechanisms associated with the second step. Final separation of the "sieved" subfraction obtained in the second

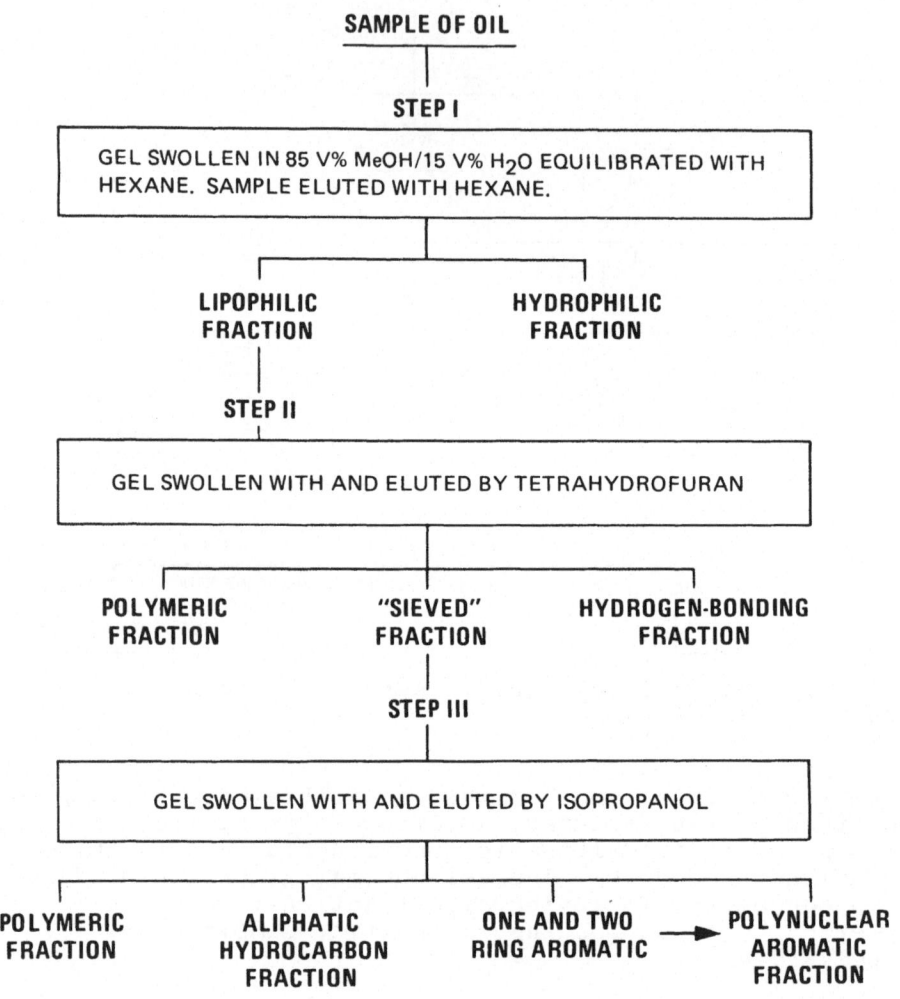

Figure 2.   Gel chromatographic separation of complex
mixtures (14).

step is effected by a combination of size exclusion, π-bonding to the gel matrix, and hydrogen bonding. Since this is a chromatographic procedure, additional precautions such as excluding light, deoxygenating solvents, and operating under inert gas blankets could be taken to minimize the likelihood of chemical changes occurring.

Discrimination and Reproducibility

Class fractionation procedures are optimized to recover all of the starting material rather than to isolate quantitatively one class of constituents. This factor, combined with the extreme chemical complexity of most natural or anthropogenic mixtures, generally results in fractions which themselves contain a variety of chemical types. As an example, the "aliphatic" fraction of Shale Oil B (10) obtained using the gel chromatographic procedure was further separated (8) into acids, bases, and neutrals by acid-base partition. The neutral constituents were then chromatographed on alumina (neutral, activity I) using stepwise elution with hexane, benzene, methylene chloride, and methanol. Essentially all (~98% by weight) of the "aliphatic" fraction was recovered in the "neutral" subfraction as would be expected. Only ~50% by weight of the "neutrals" were found in the hexane eluate from alumina with an additional ~10% by weight being found in the benzene eluate. Up to 40% by weight of the mass of the aliphatic fraction would, therefore, be expected to contain aromatic or polar moieties or to consist of very high molecular weight species.

Gas chromatographic retention, mass spectra, and reference to a detailed study (7) of shale oil composition suggest (8) the presence of a complex variety (Table 1) of hydrocarbons in the aliphatic fraction derived from Shale Oil B. Aromaticity is noted in the benzene elutables and might be expected to increase in proportion to aliphatic character in fractions eluted by methylene chloride and methanol. The aliphatic fraction would thus be considered primarily "aliphatic" in character but also contains constituents with aromatic moieties. Fractions are not distinctly different in composition but rather are "enriched" in a given class of constituents relative to the other fractions.

Preparative scale class fractionations have generally been carried out (20,25) in a semiquantitative manner. As is illustrated in Table 2 for studies carried out here,

Table 1

Some Constituents of the "Aliphatic" LH-20
Fraction of a Shale-Derived Oil (8)

Alumina/cyclohexane elutables of neutral subfraction

n C_{10} – n C_{30} (C_{18} – C_{23} predominating)

Phytane, Pristane

Miscellaneous 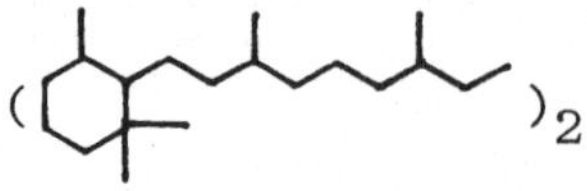   mw = 558

Terpanes

   tricyclics C_nH_{2n+1} n= 2-8, mw = 276-360

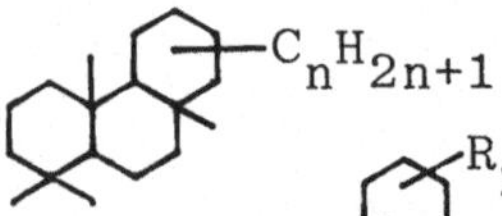

      R_2 mw = 398+

  pentacyclics R_1

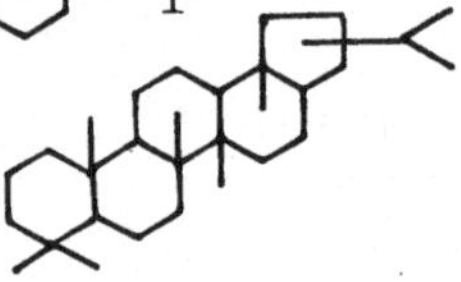

         mw = 412

Steranes   R

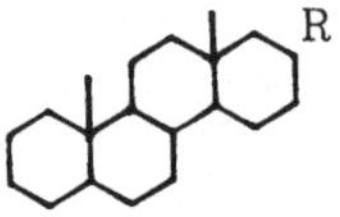

       mw = 288-400

Alumina/benzene elutables of neutral subfraction

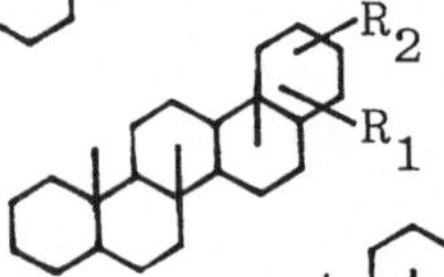

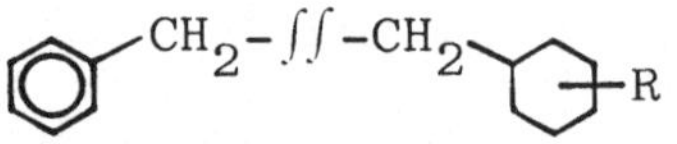

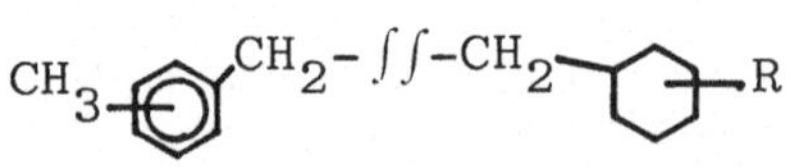

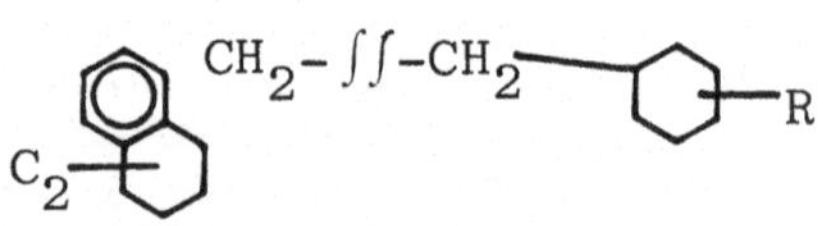

CH_2–∫∫–CH_2–R CH_3– CH_2–∫∫–CH_2–R

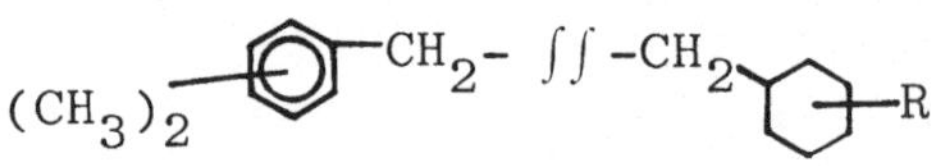

$(CH_3)_2$– CH_2– ∫∫ –CH_2–R CH_2–∫∫–CH_2–R

C_2–

R = C_1 – C_{10}

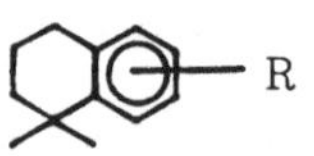

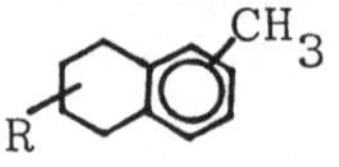

R_2

R CH_3 R_2

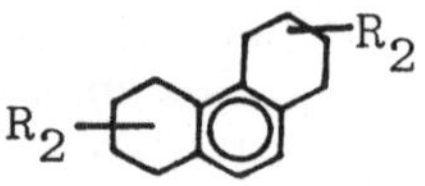

R = C_1, C_2

NMR Average $(CH_3)_2$– –CH_2–CH_2– –$C_{10}H_{21}$

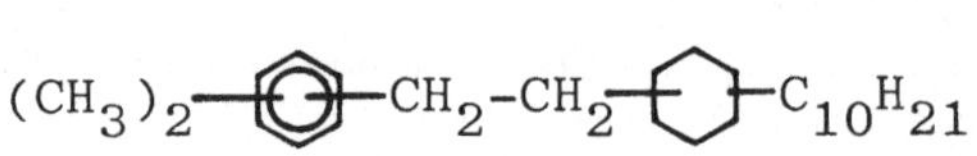

Table 2

Typical Reproducibility and Recoveries of Preparative
Scale Class Fractionation

Extractive Fractionation[1]			Sephadex Gel Chromatographic Fractionation		
Designation	Fraction % By Weight	Rel. Std. Deviation (%)	Designation	Fraction % By Weight	Rel. Std. Deviation (%)
NaOH Insol.	0.9	34	Lipophilic[2]	92.4	1.4
WA$_I$	0.2	86	Hydrophilic[3]	6.5	12
WA$_E$	1.9	18	Total	100.2	1.5
SA$_I$	0.2	41			
SA$_E$	1.1	50			
B$_{Ia}$	1.6	82			
B$_{Ib}$	0.2	29	Polymeric[4]	1.2	34
B$_E$	2.2	38	Sieved[5]	92.5	17
B$_W$	7.3	16	H-Bonding[5]	7.7	23
N/Hexane	74.2	2	Total[5]	101	19
N/Hex-Bz	4.9	47			
N/Bz-Ether	4.7	17			
N/MeOH	2.4	17			
Total	102	9			

[1]Coal-derived crude oil "D" (10), four repetitions, sample sizes of 4.4-11.9 grams (20).

[2]Shale-derived oil "B" (10), five repetitions, sample sizes of 17.5-302.4 grams (13).

[3]Shale-derived oil "B", three repetitions, sample sizes of 17.5-302.4 grams (13).

[4]From lipophilic fraction of coal-derived oil "D" (10), five repetitions, 4 gram samples (13).

[5]From lipophilic fraction of coal-derived oil "D" (10), eight repetitions, 4 gram samples (13).

quantitative reproducibility is particularly poor for
fractions constituting less than 5% by weight of the starting
material. It must be recalled, however, that the purpose
of these (1,2,6) studies has been to identify constituents
contributing to the biological activity of the starting
material rather than to quantify that activity. Little
attention has yet been given to improving the quantitative reproducibility of fractionation procedures. Reduced
sample sizes, simplified fractionation procedures, and a
reduction in the number of manual operations may contribute
to the development of a procedure sufficiently reproducible
for quantitative biotesting.

Mutagenicity

Methods used and significant observations concerning
the mutagenicity of fractionated samples are summarized
elsewhere (6) in these proceedings. Table 3 summarizes
the results of mutagenicity testing of fractions from the
extractive and chromatographic separations of Shale Oil B
for an evaluation of the efficacy of the fractionation procedures. The mutagenicities of the individual fractions
and their calculated contributions to the mutagenicity of
the starting material are tabulated. The basic fractions
are seen (Table 3) to exhibit the highest mutagenic activities for Shale Oil B separated by the extractive fractionation method. The neutral fraction is of substantially
lower specific activity but makes the largest contribution
to the calculated activity of the starting material because
it constitutes the largest part (87% by weight) of the
material. The gel chromatographic procedure produces three
fractions of high specific activity. These fractions are
those which would be expected to concentrate constituents
suggested as mutagenic by the extractive fractionation
procedure. Basic constituents would be expected to concentrate in the hydrophilic and hydrogen-bonding fractions
while polycyclic aromatic hydrocarbons and azaarenes would
be expected to accumulate in the polyaromatic fraction. The
aliphatic fraction, a somewhat refined version of the neutral
fraction provided by the extractive procedure, is seen to be
the major contributor to the calculated activity of the
starting material by virtue of its quantitative contribution
(60% by weight) to the mass of the material. The gel chromatographic data thus tends to confirm the utility of the
acid-base extraction fractionation procedure.

Table 3

Distribution of Mutagenic Activity by Fractionation Procedures

Extractive Fractionation			Sephadex Gel Chromatographic Fractionation		
Fraction[1]	Activity[2] rev/mg	Weighted Activity[3] rev/mg	Fraction[1]	Activity[2] rev/mg	Weighted Activity[3] rev/mg
NaOH Insol.	256	3	Hydrophilic	1300	78
Weak Acid 1 (WA_I)	185	<1	Polymeric 1	54	3
Weak Acid 2 (WA_E)	52	<1	H-Bonding	1040	52
Strong Acid 1 (SA_I)	0	0	Polymeric 2	0	0
Strong Acid 2 (SA_E)	159	<1	Aliphatic	180	108
Strong Acid 3 (SA_W)	160	<1	Mono/Diaromatic	24	3
Basic 1 (B_{Ia})	1377	3	Tri/Tetra-aromatic	132	7
Basic 2 (B_{Ib})	800	2	>Tetra-aromatic	1220	49
Basic 3 (B_E)	952	68			
Basic 4 (B_W)	223	1			
Neutral	112	97			
TOTAL		178			300
Unfractionated		-		2204,268,196	

[1] Shale-derived crude oil "B", separate aliquots 6 months apart.

[2] Revertants per milligram of fraction using TA98, S-9, and Aroclor 1254 (6).

[3] Revertants per milligram of starting material computed by multiplying activity of fraction by weight percentage of total recovered weight constituted by the fraction (6).

The studies yielding the results given in Table 3 were carried out for qualitative rather than quantitative purposes. The weighted activities are tabulated and summed, however, to illustrate the degree of additivity and quantitative reproducibility accompanying these procedures. Summation of weighted activities from the extractive method indicates that Shale Oil B exhibits a mutagenicity of 178 rev/mg. The gel chromatographic method yields a calculated result of 300 rev/mg as compared to an average of 223 rev/mg from three tests of the unfractionated oil carried out at the same time. Considering that the oil sample used for the gel chromatographic study was first subjected (13) to azeotropic distillation and other substantial differences in procedures, agreement between the methods is good.

The efficacy of summing mutagenicities of fractions to determine the mutagenicity of the starting material is yet to be systematically studied. Additivity observed (Table 4) to date has, however, been generally good.

Table 4

Comparison of Mutagenicity Calculated by Summing
Fractions With That Experimentally Determined

| Separation | Revertants per milligram | | | | | |
	A[1]	B[2]	C[3]	D[4]	E[5]	F[6]
Calculated	300	260	138	167	252	112
Determined	223	223	196	100	350	109

[1]Shale oil separated by gel chromatographic procedure (13).

[2]Shale oil separated into lipophilic and hydrophilic fractions.

[3]Lipophilic fraction separated into polymeric, sieved, hydrogen bonding fractions.

[4]Sieved fraction separated into polymeric, aliphatic, aromatic, polyaromatic fractions.

[5]Shale oil separated into isopropanol and acetone elutables from Sephadex LH-20.

[6]Neutral fraction separated into hexane, hexane/benzene, benzene/ether, and methanol elutables from Florisil.

IDENTIFICATION OF MUTAGENIC CONSTITUENTS

Acid-base fractionation has consistently shown (6) basic constituents of synthetic crude oils to exhibit a high mutagenicity relative to those from petroleum crude oils. The basic fractions of condensed cigarette smoke have also been reported (14) to exhibit high mutagenicities. Recent reports which demonstrate a high correlation of factors such as tobacco type (17), degree of fertilization (18), and stalk position (16) with the mutagenicity of the resulting cigarette smoke condensate further suggest (9) that nitrogenous constituents are important contributors to mutagenicity. Candidate nitrogenous constituents are known to be present in synthetic oils (21,22), cigarette smoke condensate (23), and airborne particulate matter (5).

The general approach of fractionation and mutagenicity testing has been used (6) to define better the nature of the constituents responsible for the observed mutagenic activity. A procedure (Figure 3) optimized to isolate mutagenic constituents of the ether soluble basic (B_E or ESB) fraction has been developed. Approximately one gram of the EAB from Shale Oil B and coal synthoil C ["synfuel A" (10)] were eluted from a basic alumina column using benzene followed by ethanol. For both oils, the benzene fraction contained (Table 5) from 75-80% by weight of the mass but no more than 2% of the mutagenic activity. The ethanol fraction, containing essentially all of the mutagenic activity, was then separated into isopropanol and acetone subfractions using Sephadex LH-20 gel. For both oils, the acetone subfraction was found to contain approximately 90% of the mutagenic activity of the ESB in approximately 10% by weight of its mass.

The acetone subfraction constitutes approximately 0.5% by weight of the original oils. The subfraction is extremely complex in spite of the 20-fold enrichment. The gas chromatogram (Figure 4) resulting from the analysis of acetone subfraction from coal synthoil C reveals the presence of at least one hundred individual constituents.

The compositions of the benzene, isopropanol, and acetone subfractions from both oils are being examined. Table 6 summarizes observations to date based on gas chromatographic and mass spectral analysis. All of the fractions are seen to contain nitrogen heterocyclics. Pyridines and quinolines predominate in the inactive benzene fraction while constituents of greater structural

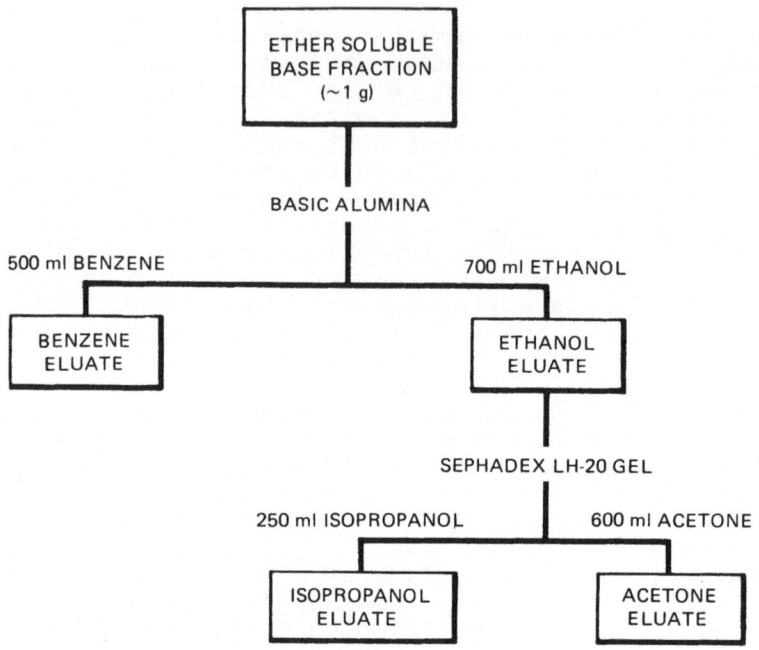

Figure 3.   Isolation of mutagenic constituents of ether soluble bases.

complexity are enriched in the isopropanol and acetone fractions. A general feature of these materials is that they contain a large variety of alkyl-substituted and partially hydrogenated derivatives of the parent heterocyclics.

The primary difference between the acetone, isopropanol, and benzene fractions is that the acetone (highly mutagenic) fraction contains higher molecular weight azaaarenes. Constituents such as benzacridines and azabenzpyrenes are found in the acetone fraction. Perhaps of importance is that the exceptionally mutagenic acetone fraction from the coal-derived oil contains a larger variety of these constituents than does the shale-derived acetone fraction. Mass spectral evidence suggests the presence of azacoronene in coal derived fraction.

Table 5

Distribution of Mutagenic Activity in ESB Subfractions

	Shale-Derived Oil B		Coal-Derived Oil C	
Subfraction	Average Weighted Activity[1] (%)	Average Relative Weight[2] (%)	Average Weighted Activity[1] (%)	Average Relative Weight[2] (%)
Benzene	0	78	2	76
Isopropanol	1	13	0	12
Acetone	92	9	88	12
TOTAL	93	100	90	100

[1]Percentage of mutagenic activity [TA98, S-9, Aroclor 1254 (6)] of the ether soluble bases (ESB) fraction accounted for in the subfraction.

[2]Percentage by weight of the ESB.

SUMMARY

Petroleum substitutes produced from coal and shale are among the materials requiring biological evaluation to assess environmental and health impacts of new energy technologies. Intractability and the presence of highly toxic constituents are among the physical and chemical properties of products and process streams which complicate short-term biotesting for subtle health effects. An effective approach is to separate the starting material into chemically well defined fractions. Bioassay results obtained for the separated fractions may be summed to estimate the biological activity of the starting material. Biological activities of individual fractions provide evidence as to the types of constituents responsible for the biological activity of the starting material.

Liquid-liquid partition from strongly acidic and alkaline solutions has proven viable for the testing of coal- and shale-derived oils. A theoretically more gentle separation procedure based on Sephadex LH-20 gel chromatography has been found viable for lipophilic materials.

Table 6

Constituents of the Ether Soluble Base Fraction

Shale Oil Subfractions			Synthoil Subfraction
Benzene	Isopropanol	Acetone	Acetone
C_1-C_{13} Pyridines C_2 Tetrahydroquinoline C_1-C_4 Quinolines C_2 Phenylpyridine C_2 Azafluorene C_1-C_3 Acridines	Azaindanes C_2-C_5 Dihydropyridines C_4-C_{11} Pyridines C_1 Diazaindane C_1-C_2 Azaindane C_1-C_2 Tetrahydroquinoline Tetrahydroquinoline C_2-C_4 Quinolines C_2 Phenylpyridine C_3-C_6 Dihydrodiazanaphthalene Azabenzo(def)fluorene C_5 Diazanaphthalene C_1 Acridines Diazabenzo(ghi)fluorene C_4 Tetrahydrodiazanaphthalenes Diazapyrenes C_1 Diazapyrenes	Azaindanes C_3-C_5 Dihydropyridine C_1-C_5 Pyridines C_1 Diazaindane C_1-C_2 Azaindane C_1 Tetrahydroquinoline Tetrahydroquinoline C_1-C_2 Quinolines Quinoline C_3-C_4 Dihydropyridine Vinylquinoline Phenylpyridine C_1-C_6 Phenylpyridines C_3 Dihydrodiazanaphthelene Dihydroazafluorene C_1-C_2 Azafluorene C_1-C_2 Acridines Acridine Tetrahydroacridine C_1-C_5 Diazafluorenes Azapyrenes C_1-C_3 Azapyrenes C_2-C_4 Diazapyrenes Benzacridine C_1-C_3 Benzacridines Azabenzopyrene C_1 Azabenzopyrene	C_1-C_4 Diazines C_2-C_6 Pyridines C_1-C_4 Tetrahydroquinoline Tetrahydroquinoline C_1-C_6 Quinolines C_2-C_4 Phenylpyridines Azadihydrofluorene C_1 Azahihydrofluorene C_1-C_2 Azafluorene C_1-C_2 Acridines Acridine C_1 Tetrahydroacridine C_1-C_3 Azabenzo(def)fluorene C_3 Diazobenzo(def)fluorene Azapyrenes C_1-C_3 Azapyrenes Benzacridine C_1 Benzacridines C_1-Indenoquinoline Azabenzopyrene Azadibenzopyrene Azabenzoperylene Azaindenopyrene Azacoronene

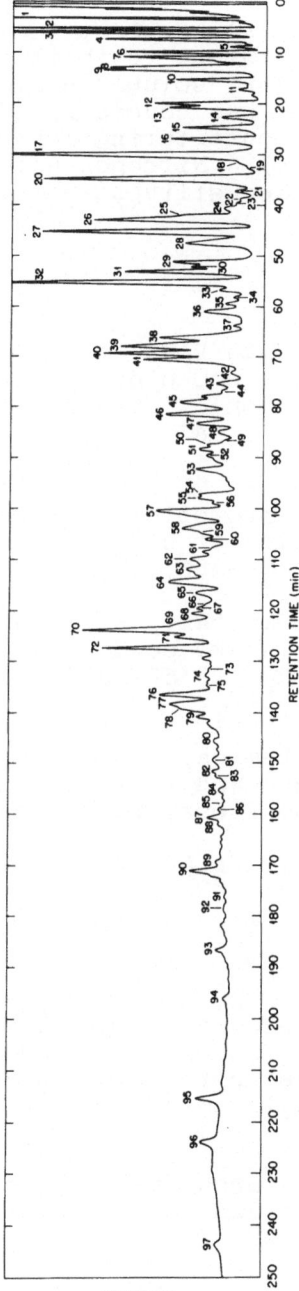

Figure 4. Gas chromatogram of a mutagenically enriched acetone subfraction. Coal synthoil C, 20' x 1/8" glass column of 3% Dexsil 400 on 100/120 mesh Chromosorb 750, programmed from 100°C to 320°C at 1°C/min following an eight minute isothermal hold at 100°C hydrogen flame ionization detection.

Studies suggest that alkaline constituents of petroleum substitutes are major contributors to their Ames Test activity. Subfractionation of ether soluble bases from a shale-derived and coal-derived oil has concentrated the bioactive constituents in a fraction constituting approximately 0.5% by weight of the starting oil. Nitrogen heterocyclics are found to be the predominant constituents of this active subfraction.

REFERENCES

1.   Bock FG, Swain AP, Stedman RL: Carcinogenesis assay of subfraction of cigarette smoke condensate prepared by solvent-solvent separation of the neutral fraction. J Natl Cancer Inst 49:477-483, 1972

2.   Bock FG, Swain AP, Stedman RL: Composition studies on tobacco. XLIV. Tumor-promoting activity of subfractions of the weak acid fraction of cigarette smoke condensate. J Natl Cancer Inst 47:429-436, 1971

3.   Bock FG, Swain AP, Stedman RL: Bioassay of major fractions of cigarette smoke condensate by an accelerated technic. Cancer Res 29:584-587, 1969

4.   Chamberlain WJ, Stedman RL: Fractionation of tabacco smoke condensate for chemical composition studies. In: The Chemistry of Tobacco and Tobacco Smoke (Schmeltz I, ed.). New York, Plenum Press, 1972, pp 99-105

5.   Dong MW, Locke DC, Hoffmann D: Characterization of aza-arenes in basic organic portion of suspended particulate matter. Environ Sci and Tech 11(6):612-618, 1977

6.   Epler JL, Clark BR, Ho C-h, Guerin MR, Rao TK: Short-term bioassay of complex organic mixtures: Part II, mutagenicity testing. Proc of Symposium on Application of Short-Term Bioassays in the Fractionation and Analysis of Complex Environmental Mixtures (these proceedings)

7.   Gallegos EJ: Terpane-sterane release from kerogen by pyrolysis gas chromatography-mass spectrometry. Anal Chem 47(9):1524-1528, 1975

8. Goeckner NA: Western Illinois University, Macombe, Illinois, Oak Ridge National Laboratory Faculty Participant, 1977-1978, private communication

9. Griest WH, Guerin MR: Influence of tobacco type on smoke composition. In: Recent Advances in Tobacco Science, Vol 3. Tobacco Smoke: Its Formation and Composition. Proc 31st Tobacco Chemists' Research Conference, 1977, pp 121-144

10. Guerin MR, Epler JL, Griest WH, Clark BR, Rao TK: Polycyclic aromatic hydrocarbons from fossil fuel conversion processes. Proc Second International Symposium on Polynuclear Aromatic Hydrocarbons (in press)

11. Guerin MR, Maddox WL, Stockely JR: Tobacco smoke inhalation exposure: concepts and devices. Proc Tobacco Smoke Inhalation Workshop on Experimental Methods in Smoking and Health Research. DHEW Pub No (NIH) 75-906:31-44, 1976

12. Guerin MR, Stokely JR: Proband-machine animal interactions in inhalation bioassays. Proc Tobacco Smoke Inhalation Workshop on Bioassay Models and Inhalation Toxicology. DHEW Pub (in press)

13. Jones AR, Guerin MR, Clark BR: Preparative-scale liquid chromatographic fractionation of crude oils derived from coal and shale. Anal Chem 49:1766-1771, 1977

14. Kier LD, Yamasaki E, Ames BN: Detection of mutagenic activity in cigarette smoke condensates. Proc Nat Acad Sci USA, Vol 71(10):4159-4163, 1974

15. McGill HC: The human model. Proc Tobacco Smoke Inhalation Workshop on Bioassay Models and Inhalation Toxicology, DHEW Pub (in press)

16. Mizusaki S, Okamoto H, Akiyama A, Fukuhara Y: Relation between chemical constituents of tobacco and mutagenic activity of cigarette smoke condensate. Mutat Res 48(3/4):319-325, 1977

17. Mizusaki S, Takashima T, Tomaru K: Factors affecting mutagenic activity of cigarette smoke condensate in Salmonella typhimurim TA1538. Mutat Res 48(1):29-36, 1977

18. Reist PC: Particle size and its role in physical and chemical interactions of tobacco smoke aerosol. Proc Tobacco Smoke Inhalation Workshop on Bioassay Models and Inhalation Toxicology. DHEW Pub (in press)

19. Rubin IB, Guerin MR: Chemical evaluation of the beeswax pellet implantation bioassay model for studies of environmental carcinogens. J Natl Cancer Inst 58(3): 641-644, 1976

20. Rubin IB, Guerin MR, Hardigree AA, Epler JL: Fractionation of synthetic crude oils from coal for biological testing. Envir Res 12:358-365, 1976

21. Schiller JE: Nitrogen compounds in coal derived liquids. Anal Chem 49(1):2292-2294, 1977

22. Shultz JL, White CM, Schweighardt FK, Sharkey AG: Characterization of the heterocyclic compounds in coal liquefaction products. Part I: nitrogen compounds. Pittsburgh Energy Research Center Report PERC/RI-77/7, 1977

23. Snook ME, Arrendale RF, Higman HC, Chortyk OT: Isolation of indoles and carbazoles from cigarette smoke condensate. Anal Chem 50(1):88-90, 1978

24. Swain AP, Bock FG, Cooper JE, Chamberlain WJ, Strange ED, Lakritz L, Stedman RL, Schmeltz I, Russell RB: Further fractionations of cigarette smoke condensate for bioassays. Weak acid and neutral subfractions and combinations of active fractions. Beitr Tabakforsch 7:1-7, 1973

25. Swain AP, Cooper JE, Stedman RL: Large scale fractionation of cigarette smoke condensate for chemical and biologic investigations. Cancer Res 29:579-583, 1969

26. Swain AP, Cooper JE, Stedman RL, Bock FG: Composition studies on tobacco XL. Large scale fractionation of the neutrals of cigarette smoke condensate using adsorption chromatography and solvent partitioning. Beitr Tabakforsch 5:109-114, 1969

SHORT-TERM BIOASSAY OF COMPLEX ORGANIC MIXTURES: PART II, MUTAGENICITY TESTING

J.L. Epler, B.R. Clark, C.-h. Ho,
M.R. Guerin, and T.K. Rao
Biology and Analytical Chemistry Divisions
Oak Ridge National Laboratory
Oak Ridge, Tennessee

INTRODUCTION

The feasibility of using short-term mutagenicity assays
to predict the potential biohazard of various crude and complex test materials has been examined in a coupled chemical
and biological approach. The principal focus of the research
has involved the preliminary chemical characterization and
preparation for bioassay, followed by testing in the Salmonella histidine reversion assay described by Ames (1). The
mutagenicity tests are intended to (a) act as predictors of
profound long-range health effects such as mutagenesis and/
or carcinogenesis, (b) act as a mechanism to rapidly isolate
and identify a hazardous biological agent in a complex mixture, and (c) function as a measure of biological activity
correlating baseline data with changes in process conditions.
Since complex mixtures can be fractionated and approached in
these short-term assays, information reflecting on the actual
compounds responsible for the biological effect may be accumulated. Thus, mutagenicity tests will (d) aid in identifying the specific hazardous compounds involved and in establishing priorities for further valid testing, testing in
whole animals, and more definitive chemical analysis and
monitoring.

Our work has emphasized test materials available from
the developing synthetic fuel technologies (2). However,
the procedures are applicable to a wide variety of industrial
and natural products, environmental effluents, and body
fluids. The general applicability of microbial test systems
has already been demonstrated with, for example, the use of
the assay as a prescreen for potential genetic hazards of

complex environmental effluents or products, e.g., tobacco
smoke condensates (3), natural products (4,5), hair dyes
(6), soot from city air (7), fly ash (8), and, in our work
with synthetic fuel technologies, oils and aqueous wastes
(9,10).

BIOASSAY METHOD

    For the study of application of mutagenicity testing
to environmental effluents and crude products from the synthetic fuels technology, we performed preliminary screening
with the highly sensitive Ames histidine reversion strains
known to repond to a wide variety of known mutagens/carcino-
gens. The working hypothesis was that sensitive detection
of potential mutagens in fractionated complex mixtures could
be used to isolate and identify the biohazard. In addition,
the information could be helpful in establishing priorities
for further testing, either with other genetic assays or
carcinogenic assays.

    The Salmonella strains used in the various assays are
listed below. All strains were obtained through the courtesy of Dr. Bruce Ames, Berkeley, California.

Salmonella typhimurium Strains

        TA1535 hisG46, uvrB, rfa (missense)
        TA100 hisG46, uvrB, rfa (missense plus R factor)
        TA1537 hisC3076, uvrB, rfa (frameshift)
        TA1538 hisD3052, uvrB, rfa (frameshift)
        TA98 hisD3052, uvrB, rfa (frameshift plus R factor)

    In the screening of fractionated materials, the two
strains TA98 and TA100 were generally employed. Standard
experimental procedures have been given by Ames et al. (7).
Briefly, the strain to be treated with the potential muta-
gen(s) is added to soft agar containing a low level of histidine and biotin along with varying amounts of the test substance. The suspension containing approximately 2×10^8
bacteria is overlaid on minimal agar plates. The bacteria
undergo several divisions with the reduced level of histidine,
thus forming a light film of background growth on the plate
and allowing the mutagen to act. Revertants to the wild-
type state appear as obvious large colonies on the plate.
The assay can be quantitated with respect to dose (added
amount) of mutagen and modified to include "on-the-plate"
treatment with the liver homogenate required to activate
metabolically many compounds.

Fractions and/or control compounds to be tested were
suspended in dimethyl sulfoxide (supplied sterile, spectro-
photometric grade from Schwarz/Mann) to concentrations in
the range of 10-20 mg/ml solids. The potential mutagen was
in some cases assayed for general toxicity (bacterial survival) with strain TA1537. Normally, the fraction was
tested with the plate assay over at least a thousand-fold
concentration range with the two tester strains TA98 and
TA100. Revertant colonies were counted after 48 h incubation. Data were recorded and plotted versus added concentration, and the slope of the induction curve was determined.
It is assumed that the slope of the linear dose-response
range reflects the mutagenic activity. Positive or questionable results were retested with a narrower range of
concentration. All studies were carried out with a parallel
series of plates plus and minus the rat liver enzyme preparation (7) for metabolic activation. Routine controls demonstrating the sterility of samples, enzyme or rat liver S-9
preparations, and reagents were included. Positive controls
with known mutagens were also included in order to recheck
strain response and enzyme preparations. All solvents used
were nonmutagenic in the bacterial test system.

SAMPLES

Samples tested and their sources are listed below:

- Coal-liquefaction product from a process under
  development, courtesy of the Pittsburgh Energy
  Research Center (Synfuel A), or Coal A from ORNL
  repository.

- Coal-liquefaction product from the COED Pyrolysis
  Process, courtesy of FMC (Synfuel B), or Coal B
  from ORNL repository.

- Louisiana-Mississippi sweet crude oil, courtesy
  of Dr. J.A. Carter of the Analytical Chemistry
  Division, Oak Ridge National Laboratory.

- Composite crude oil sample from materials obtained through the courtesy of Dr. Dee Latham of
  the Laramie Energy Research Center.

- A crude shale-oil sample (B) from the above-
  ground simulated in situ oil shale retorting
  process.

- The aqueous product water consisting of the centrifuged water of combustion from the same process
  (both samples 5 and 6 courtesy of the Laramie
  Energy Research Center).

- A coal-gasification aqueous condensate from a
  process under development, courtesy of Pittsburgh
  Energy Research Center.

- A separator liquor from a coal-liquefaction process, courtesy of FMC.

The authors recognize the possibility that these samples
may bear no relationship to the process as it may exist in
the future, nor should it be construed that these materials
are representative of all natural crudes, synthetic, or shale
oil processes. They are used here simply as appropriate and
available materials for the research.

BIOASSAY RESULTS

Class Fractionation

Oil Samples. The bulk of the samples listed above were
subjected to the fractionation scheme described by Swain et
al. (11), as modified by Bell et al. (12). The scheme is
described in detail as applied to oils in Rubin et al. (13)
and in the first part of this presentation. As an example,
a summary of the results from a sample of Synfuel A-2 (9) is
given in Table 1. Subfractionation results are shown with
the neutral fraction chromatographed on a Florisil column.
The column was eluted with the solvents shown and, with this
sample, collected in one fraction. The data includes the
analytical weight analysis of the sample (column 1) along
with the specific mutagenic activity (slope of dose-response
curve) of each fraction (column 2). The product of these
(column 3) represents a weighted value of each fraction relative to the contribution to the starting test material.
Mutagenic activity is seen in both the acidic and basic
fractions along with the neutral subfractions. However, the
major contributors to the mutagenicity appear to occur in
the basic fractions, with activities also consistently present in the neutral materials.

A comparison of these activities and the total mutagenic
potential of the various oil and aqueous samples is given in
Table 2. Reasonable reproducibility is seen in similar

Table 1

Distribution of Mutagenic Activity
of Synthetic Oil[1] (Synfuel A-2)

Fraction[2]	Relative Weight (% of total)	Specific Activity[3] (rev/mg)	Weighted Activity[4] (rev/mg)
$NaOH_I$	20.9	1,700	356
WA_I	2.2	180	4
WA_E	4.9	1,260	62
SA_I	<0.1	30	0
SA_E	0.4	130	1
SA_W	0.4	120	1
B_{Ia}	6.8	38,700	2,633
B_{Ib}	0.1	1,270	1
B_E	2.0	36,200	725
B_W	0.6	570	3
Neutral	69.2	583 (570)[5]	403
TOTAL	107.6		4,189
Neutral Subfractions			
Hexane	72.7	340	244
Hexane/benzene	5.0	710	35
Benzene/ether	19.8	1,360	270
Methanol	2.3	1,460	34
Subtotal	99.8		583
Initial sample, g	26.166		
Chromatographed, g	10.664		

[1]All assays carried out in the presence of crude liver S-9 from rats induced with Aroclor 1254.

[2]I = insoluble (fractions a and b), E = ether soluble, W = water soluble, WA = weak acid, SA = strong acid, B = base.

[3]rev/mg = revertants/mg, the number of histidine revertants from Salmonella strain TA98 by use of the plate assay with 2×10^8 bacteria per plate. Values are derived from the slope of the induction curve extrapolated to a milligram value. NT = not tested.

[4]Weighted activity of each fraction relative to the starting material is the product of columns one and two. The sum of these products is given as a measure of the total mutagenic potential of each material.

[5]Comparable to "specific activity," but based on the activity of the total neutral fraction rather than the summation of the individual fraction.

samples, e.g., Synfuel A-1 and A-2, and Synfuel B-1 and B-2. Synfuel A-3 represents the same material without prior centrifugation of the solids. The consistency of activities seen in all oils considered is illustrated. On a relative scale, the synthetic fuels show more mutagenic activity than the natural crude "control" samples shown. Shale oil appears to be only slightly higher than the natural crudes. References are given to the complete published compilations on these samples. See Table 2.

Each determination represents the slope of the dose-response curve. All testing was carried out in the presence of the rat-liver microsomal activation system. Slight mutagenic activity without enzyme treatment was occasionally noted.

The routine screening employed strains TA100 (missense) and TA98 (frameshift); however, complete strain-specificity tests were carried out with selected materials. Fractions giving a positive response with straing TA98 were, in general, also positive with the other frameshift strains, TA1537 and TA1538. Additionally, positive results were routinely noted with the sensitive missense strain TA100; however, reversion of the missense strain TA1535 was rare. TA98 appeared to be the best general indicator of mutagenic activity of these materials. Furthermore, liver preparations

Table 2

Summary of Mutagenicity Testing Results with Synthetic Oils and Aqueous
Samples – Class Fractionation Scheme*

Sample	SA$_W$		B$_E$		Neutral		Total Wt. Act. (rev/mg)	Reference
	Rel. Wt. (%)	Spec. Act. (rev/mg)	Rel. Wt. (%)	Spec. Act. (rev/mg)	Rel. Wt. (%)	Spec. Act. (rev/mg)		
Composite Crude-1	0.1	400	0.2	150	84.2	277	241	(9)
Composite Crude-2	0.1	750	0.2	500	84.2	166	147	(10)
LA-MS Crude	0.1	240	0.2	180	80.7	90	76	(9)
Shale Oil	0.6	160	7.1	952	86.7	112	178	(10)
Synfuel A-1	0.3	240	2.0	28,900	73.6	517	4,032	(9)
Synfuel A-2	0.4	120	2.0	36,200	69.2	583	4,189	(9)
Synfuel A-3	0.3	1,010	3.1	43,300	56.4	1,094	7,308	(9)
Synfuel B-1	0.4	0	2.6	1,500	82.3	560	516	(9)
Synfuel B-2	1.6	0	1.8	3,800	89.3	465	484	(9)
Separator Liquor 1.3 w/v%	53.9	0	0.5	850	1.5	0	17	unpubl.
Gasifier Condensate 0.9 w/v%	30.8	0	0.5	4,000	1.9	100	211	(14)
Process Water (Shale Oil) 1.0 w/v%	65.0	0	2.7	1,575	2.4	52	68	(10)

*Strain TA98 with metabolic activation with Aroclor-induced preparation.

from rats induced with Aroclor 1254 (a gift from Monsanto) showed the best general applicability. However, individual differences in effectiveness do occur; for example, variously induced preparations show obvious differences between basic fractions and, e.g., the neutral/methanol fraction (9). An Aroclor-induced preparation reacts best with the neutral fraction (polynuclear aromatic hydrocarbons?), while a pheno-barbital-induced preparation works more efficiently with the basic fraction (heterocyclic nitrogen compounds?).

Primary candidates for the mutagens (and carcinogens?) responsible for activity in the basic fractions include quinoline, substituted quinolines, alkyl pyridines, acridine, naphthylamines, aza-arenes, benzacridines, and aromatic amines; in the neutral fractions, potential threats may consist of benzanthracenes, dibenzanthracenes, substituted anthracenes, benzopyrenes, benzofluorenes, pyrene, substituted pyrenes, and chrysenes (see Ho et al., 15). Thus, work with these pure compounds is being carried out concurrently.

Reproducibility of results was shown by comparison of data from similar samples. Although discrepancies exist from fraction to fraction, the general trend is apparent, and the sum of activities appears to be roughly reproducible. Again, when the major component, neutral fraction is assayable as with the Synfuel A, the summation of the subfraction values of the neutrals reflects the approximate additivity of the individual mutagenic determinations. For example, 570 revertants/mg with a direct assay of the neutrals from Synfuel A-2 compares with 583 revertants/mg based on the summation (Table 1).

An overview of the results points to a number of consistencies: (1) all crudes and synthetic fuels showed some mutagenic potential; (2) the neutral and basic fractions showed activities regardless of the source of the sample; and (3) the relative total mutagenic potentials varied over two orders of magnitude. Whether these results reflect a comparative biohazard of processes still under development is not the point in question here. The results simply show that biological testing--genetic reversion assays in this case--can be carried out with the newly developed tester systems, but only when coupled with the appropriate analytical separation schemes. Conceivably, the use of this approach could provide rapid information concerning health effects.

Aqueous Samples. Table 2 also lists sample results from a group of aqueous samples subjected to the class fractionation procedure (Stedman procedure). In general, greater activity is seen in the more polar, more water soluble fractions rather than in the nonpolar neutral materials. Caution has to be extended on work with any aqueous material because of the high potential for instability. Although we have used organic extraction here, techniques with resin concentration, e.g., XAD-2, may prove useful with aqueous samples (16,17). Only in exceptional cases is the mutagenic activity directly observable in an unconcentrated sample.

Liquid Chromatographic Fractionation

In the initial studies with coal liquefaction products, the crude oils were fractionated using the scheme originally developed for cigarette smoke condensates (Stedman procedure). The scheme yields class separations based on the relative acid-base properties of the components. The samples are partitioned between ethyl ether and 1 N NaOH in a single-stage, continuous procedure to yield an aqueous acid fraction and organic phase base and neutral fractions. The organic fraction is extracted with 1 N HCl to yield an aqueous basic fraction and an organic basic fraction and an organic neutral fraction. The neutral material is subsequently subfractionated on a Florisil column. These primary subfractions are then subjected to mutagenicity testing.

Realizing the potential for modification of the components within the procedure, we moved to consideration of a number of other methods. The fractionation procedure using Sephadex LH-20 can provide a gentle and large-scale class separation for (initially) crude oils from shale oil and coal liquefaction processes. The procedure involves three steps using the gel in different modes:

- Lipophilic-hydrophilic partitioning.

- Molecular size separation.

- Aliphatic-aromatic separation.

The procedure (18) was designed by Jones, Guerin, and Clark of the Analytical Chemistry Division. Using fractions prepared as above, we have started a comparison of this procedure and the Stedman procedure for usefulness in preparation for bioassay. The preliminary mutagenicity studies

confirm the suitability and utility of the method. Table 3
summarizes some of the results from shale oil. The method
appears to be generally applicable to complex organic mixtures and achieves the goal of presenting a gentle and rapid
separation scheme, useful with large-scale samples.

Subfractionation of Neutral Components. The polycyclic
aromatic hydrocarbons (PAH), presumably occurring in the
neutral fractions of the various schemes noted, have been
listed as major contributors to mutagenicity of the test
materials. With natural crudes, these components appear to
account for the bulk of the activity. With synthetic crudes,
the contributions of both the basic and neutral fractions
must be considered. Further work is needed to define (isolate and identify) the mutagenic components of these important classes.

We have carried out a preliminary study with synfuel
PAHs subfractionated and detected by the short-term mutagenicity assay. As shown in Table 4, shale oil (sample B) can
be separated into lipophilic and hydrophilic fractions with
the Sephadex LH-20 partition chromatography described previously (18,19). Further separation of the lipophilic
fraction is achieved by neutral alumina and LH-20 using
various solvents. The various subfractions can then be
assayed for mutagenicity with the Salmonella histidine reversion system. As seen in Table 4, activity seems to peak
in the 4- and 5-ring subfractions, those containing known
carcinogens/mutagens as benzo(a)pyrene, benzo(c)phenanthrene,
and 3-methylcholanthrene.

Subfraction of Basic Components. Again, considering the
results with class fractionation procedures, we developed a
procedure (20) specifically designed for subfractionation of
the basic materials, now realized to be a major contributor
to mutagenic activity. An elution sequence using alumina and
Sephadex LH-20 gel with a combination of solvents isolates 90%
of the mutagenic activity from basic compounds into 0.5 wt%
fraction of crude oil.

A basic alumina column eluted first with benzene and
then ethanol isolates the mutagenic components of the ether
soluble base fractions (ESB) of synthetic crude oils into a
fraction of about 25% of the ESB. A further separation is
achieved by eluting the ethanol isolate through a Sephadex
LH-20 gel column with isopropanol followed by acetone. About
90% of the basic mutagenic activity is recovered in the acetone subfraction which comprises ~0.5 wt% of the crude oil.

Table 3

Sephadex LH-20 Fractionation of Shale Oil Coupled With
Mutagenicity Testing

Test Material-Fraction	% of Total	Specific* Activity (rev/mg)
Crude Oil	100	233
Hydrophilic	6	1300
Lipophilic	93	196
● Polymer	5	54
● Hydrogen Bonding	5	1040
● Sieved	84	100
●● Polymer	1	0
●● Aliphatics	60	180
Aromatics		
● 1 and 2 Ring	14	24
● 3 and 4 Ring	5	132
● Polynuclear	4	1220

*Slope of dose-response curve with <u>Salmonella</u> strain TA98
plus rat-liver preparation induced with Aroclor 1254.

Development of this separation scheme was made possible using
the Ames microbial mutagenesis assay as the detector during
exploratory liquid chromatographic separations. Table 5
lists some of the preliminary data from these studies.

COMPARATIVE MUTAGENESIS

In order to validate and compare the results accumulated
in the Ames system with complex test materials from synthetic
fuel technologies, we selected specific fractions or subfractions on the basis of their activity in the histidine

Table 4

Subfractionation of Neutral Components From Shale Oil:
Distribution of Polycyclic Aromatic Hydrocarbons
and Mutagenic Activity

Subfraction	Weight %	Specific Activity* rev/mg of Fraction	
		Without S-9	With S-9
Aromatic Fraction	100	60	170
I (polymeric)	5.7	0	0
II (1-ring)	47.0	0	0
III (2-ring)	33.7	0	0
IV (3-ring)	8.0	0	1000
V (4-ring)	2.7	1600	4000
VI (5-ring)	0.6	2600	3800
VII (<5-ring)	0.5	600	1500
TOTAL	98.2	62	214

*Number of histidine revertants from Salmonella strain TA98
by use of plate assay with 2×10^8 bacteria per plate.
Values derived from slope of the induction curve. "With S-9"
indicates test carried out in presence of crude enzyme preparations from rats induced with Aroclor 1254.

reversion assay for further testing in the various other
tests designed to detect mutagenicity. Preliminary results
have been published in the Proceedings of the Second International Conference on Environmental Mutagens, Edinburgh,
1977 (21). For the purpose of a qualitative comparison,
the results are given in Table 6. The selected fractions
or subfractions used were basic and neutral isolates from
synthetic crude oils from coal liquefaction processes [Syn-
fuel A and B as described in Epler et al. (9)]. With
Drosophila (22) and in the mammalian cell gene mutation

Table 5

Subfractionation of Basic Fraction:
Distribution by Weight and Mutagenic Activity

	Shale oil		Synfuel A-3	
	wt%	rev/mg*	wt%	rev/mg*
Basic fraction (A)	100	2,500	100	30,000
Alumina				
Benzene (B)	78	600	76	0
Ethanol (C)	-	-	-	-
LH-20				
Isopropanol (D)	12	0	12	0
Acetone (E)	10	20,000	12	222,000

*Assayed with Strain TA98 with Aroclor-induced preparation.

assay (23), the detection has been a function of newly developed fractionation schemes (e.g., the use of LH-20) that result in higher specific activity (more highly purified) mutagenic subfractions. In general, the results validate the initial screening carried out in the Salmonella assay, but these other systems have not as yet been used to test exhaustively materials that are negative in the Ames system. Note also, however, that the preliminary results of Generoso (personal communication) show that the crude synthetic fuel does induce dominant lethals in mice although the basic fraction alone appears to be negative.

For the comparative studies with microbial systems given here, we selected four Synfuel fractions. The results with the frameshift strain TA98 with metabolic activation were considered. Fractions 6 [strong acid, water soluble (SA)]; 7 [base insoluble, fraction A (B)]; 9 [base, ether soluble (B)]; and 14 (neutrals/methanol) were selected on the basis of their ability to revert the frameshift alleles of the Ames system. In order to validate the mutagenicity results obtained from the Salmonella histidine-reversion system, we

Table 6

Comparative Mutagenesis of Fractions from
Synthetic Crude Oils[1]

Test System	Assay	Basic Fraction	Neutral Fraction	Crude[2] Synfuel
Salmonella	$his^- \rightarrow his^+$	+	+	+
E. coli	$arg^- \rightarrow arg^+$	+	+	NT
	$gal^- \rightarrow gal^+$	+	+	NT
Yeast	$his^- \rightarrow his^+$	+	+	NT
	$CAN^s \rightarrow can^r$	+	+	NT
Drosophila	SLRL	+	−	NT
CHO cells	6-thioguanine Resistance	+	NT	NT
Human leukocytes	Chromatid Aberrations	P	+?	NT
Mouse	Dominant Lethals	−	P	+
Carcinogenesis	Skin Painting[3]	P	P	P

[1]For references to published work or work in progress, see text. The fractions used were generally those from Synfuel A-3 or Synfuel B-2. + = mutagenic; − = nonmutagenic; NT = not tested; and P = in progress.

[2]Crude synfuels are generally too toxic to test in most systems.

[3]Work of J.M. Holland, Oak Ridge National Laboratory, in progress.

extended the treatment with the selected test fractions to the E. coli 343/113 system of Mohn (24). The results obtained in the forward (gal^+) and reverse-mutation (arg^+) assays with E. coli support the results obtained with

<u>Salmonella</u>. Both the basic fraction (#9) and the neutral subfraction (#14) are mutagenic upon metabolic activation with Aroclor-induced rat-liver homogenate (S-9).

Further validation of the bacterial results was obtained by assaying for both forward mutation and reverse mutation in the yeast system (21,25). The Synfuel A fractions tested were weakly mutagenic and were effective without metabolic activation. Some antagonistic effects were encountered when metabolic activation was incorporated. The most active fraction, the ether soluble bases (B_E), also reverted the putative frameshift marker, <u>hom3-10</u>. This fraction may contain acridines and other nitrogen heterocyclics. Unpublished results from our group have pointed to similar effectiveness without activation in the <u>Salmonella</u> system when suspension tests rather than plate assays are used with crude mixtures.

To ascertain the comparative effectiveness in the human leukocyte chromatid aberration assay, we treated selected test fractions from Synfuel B. The test materials used were the neutral subfractions and represent largely polycyclic aromatic hydrocarbons. The coal fractions were suspended in DMSO at a concentration of 20 mg/ml total solids. Two hours of control treatment with 5% DMSO and treatment with the four subfractions (neutrals as hexane, hexane/benzene, benzene/ether, and methanol subfractions) over a concentration range of 0.1-1.0% (20 µg-200 µg) were ineffective in producing chromatid aberrations (1600 cells scored). However, metabolic activation was not included with any exogenous enzyme source nor are the assumed constituents (PAH's) effective as chromosome breaking agents. Work with other fractions and the inclusion of metabolism is in progress.

Selected test fractions from Synfuel B were assayed in the <u>Drosophila</u> sex-linked recessive-lethal (SLRC). Fraction 13 (neutral/benzene/ether) is slightly effective as a mutagen for <u>Drosophila</u> at the higher concentrations fed.

Several other syncrude (similar crude) fractions which were scored as mutagenic in the <u>Salmonella</u> assays were tested in <u>Drosophila</u>. (All of the fractions require metabolic activation in the <u>Salmonella</u> assay.) Of the five fractions tested, only 12 and 13 gave any indication of an effect. Additionally, the highly active (in <u>Salmonella</u>) basic subfraction from the procedure previously described was tested. This basic material showed a significant dose-dependent response in the <u>Drosophila</u> SLRL assay. (See discussion by Nix and Brewen, this proceedings, reference 22.)

In conclusion, short-term tests with bacterial and fungal mutagenicity assays appear to detect effectively the mutagenic potential of complex environmental or industrial effluents; however, chemical fractionation is necessary to reduce toxicity and concentrate hazardous materials. Extension of the results to higher organisms, i.e., mammalian cells, Drosophila, and the mouse appears to be valid but needs more testing.

CONCLUSIONS

In these initial feasibility studies, the purpose has not been to reflect on whether a relative biohazard exists in comparison with other materials or processes. The results show that biological testing, within the limits of the specific system used, can be carried out with complex organic materials but perhaps only when coupled with the appropriate analytical separation schemes. An extrapolation to relative biohazard at this point would be, at least, premature. The primary use that such combined chemical and biological work may serve is to aid in isolating and identifying the specific classes or components involved. A number of precautions are listed below.

The detection or perhaps the generation of mutagenic activity may well be a function of the chemical fractionation scheme used. The inability to recover specific chemical classes or the formation of artifacts by the treatment could well corrupt the results obtained, in addition to the possibility of an inability to detect the specific biological endpoint chosen. Along with the obvious bias that could accompany the choice of samples and their solubility or the time and method of storage, a number of biological discrepancies can also enter into the determinations. For example, concomitant bacterial toxicity can nullify any genetic damage assay that might be carried out; the choice of inducer for the liver enzymes involved might be wrong for selected compounds; the choice of strain could be inappropriate for selected compounds; and additionally, the applicability of the generally used Salmonella test to other genetic endpoints and the validation of the apparent correlation between mutagenicity and carcinogenicity still remains a point of significant fundamental research. Furthermore, the short term assays chronically show negative results with certain substances, e.g., heavy metals and certain classes of organics. Similarly, compounds involved in or requiring cocarcinogenic phenomena would presumably go undetected.

However, as a prescreen to aid the investigators in ordering their priorities, the short-term testing appears to be a valid testing approach with complex mixtures. Overinterpretation at this stage of research especially with respect to relative hazard or negative results should be avoided.

REFERENCES

1. Ames BN, Lee FD, Durston WE: An improved bacterial test system for the detection and classification of mutagens and carcinogens. Proc Natl Acad Sci USA 71: 782-786, 1973

2. Klass DL: Synthetic crude oil from shale and coal. Chem Technol August:499-510, 1975

3. Kier LD, Yamasaki E, Ames BN: Detection of mutagenic activity in cigarette smoke condensates. Proc Natl Acad Sci USA 71:4159-4163, 1974

4. Nagao M, Yahagi T, Kawachi T, Seino Y, Honda M, Matsukura N, Sugimura T, Wakabayashi K, Tsuji K, Kosuge T: Mutagens in foods, and especially pyrolysis products of protein. In: Progress in Genetic Toxicology (Scott D, Bridges BA, Sobels FH, eds.). Elsevier/North-Holland Biomedical Press, 1977, pp 259-264

5. Hardigree AA, Epler JL: Comparative mutagenesis of plant flavonoids in microbial systems. Mutat Res, in press

6. Ames BN, Kammen HO, Yamasaki E: Hair dyes are mutagenic: Identification of a variety of mutagenic ingredients. Proc Natl Acad Sci USA 72:2423-2427, 1975

7. Ames BN, McCann J, Yamasaki E: Methods for detecting carcinogens and mutagens with the Salmonella/mammalian-microsome mutagenicity test. Mutat Res 31:347-364,1975

8. Chrisp CE, Fisher GL, Lammert JE: Mutagenicity of filtrates from respirable coal fly ash. Science 199:73-75, 1978

9. Epler JL, Young JA, Hardigree AA, Rao TK, Guerin MR,
    Rubin IB, Ho C-h, Clark BR: Analytical and biological
    analyses of test materials from the synthetic fuel
    technologies. I. Mutagenicity of crude oils determined
    by the Salmonella typhimurium/microsomal activation
    system. Mutat Res, in press

10. Epler JL, Rao TK, Guerin MR: Evaluation of feasibility
    of mutagenic testing of shale oil products and effluents.
    Environ Health Perspect, in press

11. Swain AP, Cooper JE, Stedman RL: Large scale fractiona-
    tion of cigarette smoke condensate for chemical and
    biologic investigations. Cancer Res 29:579-583, 1969

12. Bell JH, Ireland S, Spears AW: Identification of
    aromatic ketones in cigarette smoke condensate. Anal
    Chem 41:310-313, 1969

13. Rubin IB, Guerin MR, Hardigree AA, Epler JL: Fractiona-
    tion of synthetic crude oils from coal for biological
    testing. Environ Res 12:358-365, 1976

14. Epler JL, Larimer FW, Rao TK, Nix CE, Ho T: Energy-
    related pollutants in the environment: The use of
    short-term tests for mutagenicity in the isolation and
    identification of biohazards. Environ Health Perspect,
    in press

15. Ho C-h, Clark BR, Guerin MR: Direct analysis of organic
    compounds in aqueous byproducts from fossil fuel con-
    version processes: Oil shale retorting, synthane coal
    gasification and COED coal liquefaction. J Environ Sci
    Health All (7):481-489, 1976

16. Yamasaki E, Ames BN: Concentration of mutagens form
    urine by adsorption with the nonpolar resin XAD-2:
    Cigarette smokers have mutagenic urine. Proc Natl Acad
    Sci USA 71(8):3555-3559, 1977

17. Brown JP, Brown RJ, Roehm GW: The application of short-
    term microbial mutagenicity tests in the identification
    and development of nontoxic, nonadsorbable food addi-
    tives. In: Progress in Genetic Toxicology (Scott D,
    Bridges BA, Sobels FH, eds.). Elsevier/North-Holland
    Biomedical Press, 1977, pp 185-190

18. Jones AR, Guerin MR, Clark BR: Preparative-scale liquid chromatographic fractionation of crude oils derived from coal and shale. Anal Chem 49:1766-1771, 1977

19. Guerin MR, Epler JL, Griest WH, Clark BR, Rao TK: Polycyclic aromatic hydrocarbons from fossil fuel conversion processes. In: Carcinogenesis, Vol. 3 (Jones PW, Freudenthal RJ, eds.). New York, Raven Press, in press

20. Ho C-h, Clark BR, Guerin MR, Rao TK, Epler JL: Detection by bioassay in the liquid chromatographic isolation of mutagenic constituents of synthetic crude oils. Anal Chem, in press

21. Epler JL, Larimer FW, Nix CE, Ho T, Rao TK: Comparative mutagenesis of test material from the synthetic fuel technologies. In: Progress in Genetic Toxicology (Scott D, Bridges BA, Sobels FH, eds.). Elsevier/North-Holland, 1977, pp 275-284

22. Nix CE, Brewen BS: The role of Drosophila in chemical mutagenesis testing. Symposium on Application of Short-Term Bioassays in the Fractionation and Analysis of Complex Environmental Mixtures. Williamsburg, Virginia, 1978

23. Hsie AW, O'Neill JP, Sebastian JRS, Couch DB, Brimer PA, Sun WNC, Fuscoe JC, Forbes NL, Machanoff R, Riddle JC, Hsie MH: Mutagenicity of carcinogens: Study of 101 individual agents and 3 subfractions of a crude synthetic oil in a quantitative mammalian cell gene mutation system. Symposium on Application of Short-Term Bioassays in the Fractionation and Analysis of Complex Environmental Mixtures. Williamsburg, Virginia, 1978

24. Mohn GR, Ellenberger J: The use of Escherichia coli K12/343/113 (λ) as a multipurpose indicator strain in various mutagenicity testing procedures. Mutat Res, in press

25. Larimer FW, Ramey DW, Lijinsky W, Epler JL: Mutagenicity of methylated N-nitrosopiperidines in Saccharomyces cerevisiae. Mutat Res, in press

QUANTITATIVE MAMMALIAN CELL GENETIC TOXICOLOGY: STUDY OF THE CYTOTOXICITY AND MUTAGENICITY OF SEVENTY INDIVIDUAL ENVIRONMENTAL AGENTS RELATED TO ENERGY TECHNOLOGIES AND THREE SUBFRACTIONS OF A CRUDE SYNTHETIC OIL IN THE CHO/HGPRT SYSTEM

Abraham W. Hsie, J. Patrick O'Neill,
Juan R. San Sebastian, David B. Couch,
Patricia A. Brimer, William N.C. Sun,
James C. Fuscoe, Nancy L. Forbes,
Richard Machanoff, James C. Riddle,
and Mayphoon H. Hsie

Biology Division
Oak Ridge National Laboratory and
University of Tennessee-Oak Ridge
Graduate School of Biomedical Sciences
Oak Ridge, Tennessee

As science and technology advance, an extraordinary quantity of natural and synthetic chemicals is introduced continuously into our environment. Through conventional animal tests, some of these environmental chemicals have been found to be either highly toxic, mutagenic, carcinogenic, or teratogenic. Epidemiological studies have shown that among these harmful chemicals, a few also exhibit such detrimental effects in the human population. Because of the high cost and length of time required for the animal experiments, such tests have been confined to only a very small fraction of these environmental agents. Thus, the biological effects of the great majority of these chemicals, including ingredients of our daily foods and drugs, remain either incompletely tested or unknown.

During the past few years, evidence has accumulated that a high percentage (80-90%) of human cancer is linked to exposure to industrial and environmental chemicals identifiable as carcinogens (23,44). Since the expense of animal tests preclude their routine use to identify environmental carcinogens, many short-term assays have been developed as initial carcinogen screening tests. Studies of mutagenesis and DNA-repair in microorganisms, especially Salmonella typhimurium and Escherichia coli, have established that approximately 90% of chemical carcinogens cause mutation induction or DNA damage in these bacteria (2,3,26, 27,38,39,42,45,46). Such findings imply that the microbial tests are useful to identify not only potential mutagens but also carcinogens in the environment.

In view of the intrinsic limitation of the microbial
assay to respond to certain classes of chemicals, such as the
apparent failure of the Salmonella assay to demonstrate that
carcinogenic halogenated hydrocarbons and metallic compounds
are mutagenic (27), it appears that no single test system
will give 100% correlation between mutagenicity and carcinogenicity. The use of a battery of tests rather than any
single test in isolation has thus been proposed to reduce
the probability of false negatives (i.e., known carcinogens
are not mutagenic) and false positives (i.e., known noncarcinogens are mutagenic) (4,37).

It has been recognized that studies of mutagenesis in
prokaryotes may not reveal some fundamental mechanisms of
mutagenesis in mammals, because mammals differ from prokaryotes in their level of organization and repair of DNA,
mechanisms of metabolism of chemicals, and other related
functions. Some bacterial mutagens such as caffeine and
hydroxylamine do not appear to be mutagenic in mammalian
cells, while agents such as nickel and beryllium compounds
are mutagenic in mammalian cells but not in the Salmonella
system (Couch, San Sebastian, and Hsie, unpublished, 27).
In addition, it is well known that chromosomal abnormality
is a major cause of inheritable human diseases and is often
associated with the process of malignancy. The great majority of chemical carcinogens are known to induce chromosomal
aberrations (1,24) or sister-chromatid exchange (1). Dieth-
ylstilbestrol, a synthetic hormone associated with cancer in
women, causes chromosomal aberrations in cultured mammalian
cells (24), but does not cause mutation induction in Salmonella (27). Clearly, mammalian cell systems offer advantages over bacterial systems for studying genetic toxicity
at the chromosome and chromatid level.

Since the observation that treatment of mammalian somatic cells with conventional mutagens such as ethyl methanesulfonate (EMS) and N-methyl-N'-nitro-N-nitrosoguanidine
(MNNG) causes an increase in the number of cell variants that
differ from parental cells in either nutritional requirement
(5,36) or drug sensitivity (5), there has been much interest
in utilizing a quantitative mammalian cell mutation system
for studying mechanisms underlying the process of mammalian
mutation and, additionally, for assessing the genetic hazard
of environmental agents to the human population. Several
mammalian cell mutation systems, especially those utilizing
resistance to purine analogues such as 8-azaguanine (AG) and
6-thioguanine (TG) as a genetic marker (6), have been developed for such purposes. The selection for mutation induction

to purine analogue resistance is based on the fact that the wild-type cells containing hypoxanthine-guanine phosphoribosyl transferase (HGPRT) activity are capable of converting the analogue to toxic metabolites, leading to cell death; the presumptive mutants, by virtue of the loss of HGPRT activity, are incapable of catalyzing this detrimental metabolism and, hence, escape the lethal effect of the purine analogue (Table 1).

Table 1

CHO/HGPRT Mutation Assay [1]

---

(1) Enzyme system:

$$\begin{array}{ccc} H & & IMP, \\ G \xrightarrow{\quad HGPRT \quad} & GMP \\ (\text{or TG, AG}) & & (\text{or TG, AG)MP} \end{array}$$

(2) Mutation induction and selection for variants and revertants:

(a) wild type $\xrightarrow[\substack{(\text{induced by physical} \\ \text{or chemical agents})}]{\text{mutation}}$ variant cell

    genotype HGPRT$^+$ HGPRT$^-$

    phenotype TG^S,   TG^r,

    aminopterin positive   aminopterin
                                          negative

(b) Variant selection is based on resistance to TG

(c) Selection of revertants is based on growth in the presence of aminopterin.

(3) Characterization of TG^r variants:

(a) Direct enzyme assay for conversion of [^{3}H]hypoxanthine to [^{3}H]IMP.

(b) Cellular incorporation of [^{3}H]hypoxanthine into cellular macromolecules as revealed by either direct radioactivity measurement or autoradiographic determination.

(c) Sensitivity of clonal growth to aminopterin (10 μM) in medium F12FCM5 which contains hypoxanthine (30 μM), glycine (100 μM), and thymidine (3 μM).

---

[1]Table II of ref. 21.

The near-diploid Chinese hamster ovary (CHO) cell line
has been chosen for our study because a mutation assay,
referred to as the CHO/HGPRT system, has been well defined
(7-10,16-22,29-33). We have used CHO cells because these are
perhaps the best characterized mammalian cells genetically
(35,40). They exhibit high cloning efficiency, achieving
nearly 100% under normal growth conditions, and are capable
of growing in a relatively well-defined medium on a glass or
plastic substratum or in suspension with a population doubling time of 12-13 hr. In addition, the cells have a stable,
easily recognizable karyotype of 20 or 21 chromosomes (depending on the subclone) (11) and are suitable for studying
mutagen- or carcinogen-induced chromosome and chromatid aberrations (1) and sister-chromatid exchanges (1,24) (Table 2).

METHODS AND MATERIALS

Cell Culture

All studies to be described have employed a subclone of
CHO-K$_1$ cells (25), designated as CHO-K$_1$-BH$_4$ (16). It was
isolated following selection in F12 medium containing aminopterin (10 µM) (16). Cells are routinely cultured in Ham's
F12 medium (Pacific Biological Co.) containing 5% heat-
inactivated (56°C, 30 min), extensively dialyzed fetal calf
serum (Pacific Biological Co.) (medium F12FCM5) in plastic
tissue culture dishes (Falcon or Corning Glass Works) under
standard conditions of 5% CO_2 in air at 37°C in a 100% humidified incubator. These cells grow in medium which contains
aminopterin as well as in regular medium with 5 or 10% dialyzed fetal calf serum with a population doubling time of
12-13 hr. Cells are removed with 0.05% trypsin for subculture, and the number is determined with a Coulter counter
(model B, Coulter Electronics).

Treatment with Chemicals

We have standardized treatment procedures which are found
to be suitable for various chemicals (16,29). Briefly, CHO
cells are plated at 5 x 10^5 cells/25 cm^2 bottle in medium F12-
FCM5. After a 16- to 24-h growth period (cell number = 1.0-
1.5 x 10^6 cells/plate), the cells are washed once with saline
G, and sufficient serum-free F12 medium is added to bring the
final volume to 5 ml after the addition of various amounts of
microsome preparation (up to 1 ml) and 50 µl of chemical,
usually dissolved in dimethyl sulfoxide. Chemicals and/or

Table 2

Characteristics of CHO cells[1]

---

(1) Exhibit a stable karyotype over 20 years with a modal chromosome number of 20 which has a distinctly recognizable morphology.

(2) Have a colony-forming capacity of nearly 100% in a defined growth medium.

(3) Grow well in either monolayer or suspension with a relatively short population doubling time of 12-14 hr.

(4) Are genetically and biochemically well characterized, with many genetic markers available, including auxotrophy, drug resistance, temperature sensitivity, etc.

(5) Respond well to various synchronization methods, including the mitotic detachment procedure, which facilitate cell cycle study.

(6) Are useful in somatic cell hybridization experiments because they readily hybridize with different cell types, including human cells; when the CHO-human cell hybrid is formed there is subsequent rapid, preferential loss of human chromosome, which facilitates the assignment of marker genes to specific chromosomes or linkage groups in the human karyotype.

(7) Respond quantitatively to various physical and chemical mutagens and carcinogens with high sensitivity.

(8) Adapt to mutation induction either through coupling with a microsome activation system or through host (mouse) mediation.

(9) Are capable of monitoring induced mutation to multiple gene markers, chromosome aberration, and sister chromatid exchange in the same mutagen-treated cell culture.

---

[1]Table I of ref. 21.

microsomes are omitted from some plates to provide controls. The microsomal preparation has been prepared in this labora- tory according to the method of Ames et al. (3) from livers of Aroclor 1254-induced male Sprague-Dawley rats; the micro- some mix for biotransformation contains (per ml) 33 μmoles KCl, 8 μmoles $MgCl_2$, 4 μmoles NADP, 5 μmoles glucose-6-phos- phate, 100 μmoles phosphate buffer (pH 7.4), and 0.2 ml micro- some fraction. Cells are then incubated for 5 h and washed 3 times with saline G before 5 ml of F12FCM5 are added. Fol- lowing overnight incubation, cells are trypsinized and plated for cytotoxicity and specific gene mutagenesis to be described below. Treatment with physical agents has been described in detail elsewhere (17,19,29,30).

Cytotoxicity

The effect of chemicals on the cellular cloning effi- ciency is determined by use of the treated cells described above. For an expected cloning efficiency higher than 50%, 200 well-dispersed single cells are plated, and for an expec- ted survival lower than this, the number of cells plated is adjusted accordingly to yield 100-200 surviving colonies after standard incubation in medium F12FCM5 for 7 days. At the end of the incubation period, the plates are fixed with 3.7% formalin and stained with a dilute crystal violet solu- tion before the colonies are enumerated. A cluster of more than 50 cells growing within a confined area is considered to be a colony. Control cells, which do not receive treat- ment with mutagen, usually give 80% or higher plating effi- ciency under this condition. Neither the solvent-microsome mix nor these agents individually affect the cellular cloning efficiency. The effect of carcinogen on the cloning effi- ciency is expressed as percent survival relative to the untreated controls.

Specific Gene Mutagenesis

The CHO/HGPRT system has been defined in terms of medium, TG concentration, optimal cell density for selection (and, hence, recovery of the presumptive mutants), and expression time for the mutant phenotype (16,29). For the determination of mutation induction, the treated cells are allowed to ex- press the "mutant phenotype" in F12 medium for 7-9 days, at which time mutation induction reaches a maximum which is maintained thereafter (as long as 35 days examined) for several agents (EMS, MNNG, ICR-191, X ray, and UV) irrespec- tive of concentration or intensity of the mutagen (29-32).

Routine subculture is performed at 2-day intervals during
the expression period, and at the end of this time the cells
are plated for selection in hypoxanthine-free F12FCM5 containing 1.7 μg/ml (10 μM) of TG at a density of 2.0×10^5
cells/100 mm plastic dish (Corning or Falcon), which permits
100% mutant recovery in reconstruction experiments (29). We
find the use of dialyzed serum particularly important, presumably due to potential competition between hypoxanthine
and TG for transport into the cells and for catalysis by
HGPRT (29). After 7 to 8 days in the selective medium, the
drug-resistant colonies develop; they are then fixed, stained,
and counted. Such a protocol permits the maximum yield by
various physical and chemical agents of TG-resistant variants,
>98% of which have highly reduced HGPRT activity (7-10,16-22,
29-33). Mutation frequency is calculated based on the number
of drug-resistant colonies per survivor at the end of the
expression period.

RESULTS

Characteristics of the CHO/HGPRT System: Evidence of the
Genetic Basis of Mutation at a Specific Locus

     Conclusive, direct proof of the genetic origin of muta-
tions in somatic cells should theoretically rely on demonstration that the affected hereditary alteration has resulted in
a modified nucleotide sequence of the specific gene, causing
modified coding properties which result in the production of
altered protein with changes in the amino acid sequence. In
the absence of such proof, one must rely on indirect criteria
which are consistent with the concept that the observed phenotypic variations are genetic in nature. Such criteria include
stability of altered phenotype, mutagen-induced increase in
occurrence of stable variants, biochemical and physiological
identification of the variant phenotype, chromosomal locali-
zation of the affected gene, etc. (6,35,40,43).

     Over the past four years, we have used the assay proto-
col described (16,29) and have found in approximately 400
experiments that the spontaneous mutation frequency lies in
the range of $1-5 \times 10^{-6}$ mutant/cell. Various physical and
chemical agents are capable of inducing TG resistance.
Among all chemical mutagens examined, mutation induction
occurs as a linear function of the concentration (7-10,16-22,
29-33). For example, mutation frequency increases approximately linearly with EMS concentration in this near-diploid
cell line, conforming to the expectation that mutation induc-

tion occurs in the gene localized at the functionally mono-
somic X chromosome. However, in the tetraploid CHO cells,
EMS does not induce an appreciable number of mutations, even
at very high concentrations, as predicted theoretically (18).

We have been unable to detect any spontaneous reversion
with 13 TG-resistant mutants, all of which contain low, yet
detectable, HGPRT activity. More than 98% of the presumptive
mutants isolated either from spontaneous mutation or as a
result of mutation induction are sensitive to aminopterin,
incorporate hypoxanthine at reduced rates, and have less than
5% HGPRT activity (29). Studies in progress have also shown
that mutants containing temperature-sensitive HGPRT can be
selected, suggesting that mutation resides in the HGPRT
structural gene (O'Neill and Hsie, unpublished observations).

The CHO/HGPRT system appears to fulfill the criteria for
a specific gene locus mutational assay (Table 3) and should
be valuable in studying mechanisms of mammalian cell mutagenesis and as a system to determine the mutagenicity of
various physical and chemical agents.

Table 3

CHO/HGPRT Mutation Assay: Genetic Basis of Mutation at
HGPRT Locus in TG-Resistance Selection[1]

---

(1)   Spontaneous mutation frequency at 1-5 x 10^{-6} mutant/cell.

(2)   Mutation induction by physical and chemical agents with
      linear dose-response relationship.

(3)   Frequency of spontaneous reversion at less than 10^{-7}
      reversion/cell.

(4)   Failure to induce mutation in near-tetraploid cell lines.

(5)   Altered HGPRT activity in mutants.

    (a)   1179/1189 (98.4%) mutant colonies are aminopterin
        negative.

    (b)   121/122 (99.2%) mutant colonies show reduced hypox-
        anthine incorporation by autoradiography studies.

    (c)   81/83 (97.6%) isolated mutant clones show reduced
        HGPRT enzyme activity.

---

[1]Table III of ref. 21.

Mutagenicity of 70 Individual Energy-Technology-Related Environmental Agents

Polycyclic Hydrocarbons (Total of 27). Some of the most ubiquitous environmental organic pollutants in our environment are polycyclic hydrocarbons, many of which are carcinogenic. Coal- and synthetic-fuel-related energy technologies and gasoline-driven engines often generate high levels of polycyclic hydrocarbons which are detectable in urban air and water. We have studied the mutagenicity of benzo(a)pyrene [B(a)P] and its 19 metabolites, including 11 phenols, 3 epoxides, 3 diols, and 2 diolepoxides. For comparison, benzo(e)pyrene [B(e)P] and pyrene were added to this study. Also included were benz(a)anthracene (BA) and 4 related compounds (7,12-dimethyl BA, anthracene, and two phenolic derivatives of BA). The carcinogenic polycyclic hydrocarbons B(a)P, BA, and 7,12-dimethyl BA require metabolic activation to be mutagenic. The weak carcinogen B(e)P is less mutagenic than B(a)P. The noncarcinogenic polycyclic hydrocarbons, pyrene and anthracene, are nonmutagenic even with metabolic activation. B(a)P-4,5-epoxide and B(a)P-7,8-diol,9-10-epoxide(syn) are mutagenic. Since CHO cells cannot activate procarcinogens such as B(a)P, these cells appear to be most useful in screening for the mutagenicity of metabolites such as those of B(a)P (Hsie and Brimer, unpublished). Because of the limited availability of B(a)P derivatives, some of the experiments remain to be pursued in detail.

Metallic Compounds (Total of 15). The carcinogenic and mutagenic potential of certain toxic metallic compounds has become an environmental concern, especially with the increasing large-scale coal mining and coal firing of power plants. We found that $MnCl_2 \cdot 4H_2O$, $FeSO_4 \cdot 7H_2O$, $CoCl_2 \cdot 6H_2O$ and cis-$Pt(NH_3)_2Cl_2$ (an antitumor agent) are mutagenic, while $NiCl_2 \cdot 6H_2O$, $BeSO_4 \cdot 4H_2O$, and $CdCl_2$ are weakly mutagenic. Determination of metal mutagenicity is apparently complicated by the ionic composition of the medium. For example, we found that the mutagenicity and cytotoxicity of $MnCl_2$ were abolished by the excess of $MgCl_2$. The unusual environment required for demonstration of mutagenicity of $MnCl_2$ makes assessment of its biological hazard difficult. This too may account in part for varying results obtained in studying the mutagenicity of $AgNO_3$, $CaCl_2$, $Pb(CH_3COO)_2 \cdot 3H_2O$, $RbCl$, H_2SeO_3, $TiCl_4$, and $ZnSO_4 \cdot 7H_2O$ (10; Couch, San Sebastian, Forbes, and Hsie, unpublished).

Nitrosamines and Related Compounds (Total of 16). Nitrosamines are potent carcinogens for various animal species.
They are of environmental concern because it is known that
oxides of nitrogen produced at high temperature in internal
combustion engines and in coal-fired power plants can react
with atmospheric water to form nitrosamines. Nitrosamines
can also be formed in the human stomach by a reaction between
a common meat preservative, sodium nitrite, and various secondary and tertiary amines, many of which are often used as
counter or prescription drugs.

        Nitrosamines generally require metabolic activation to
be cytotoxic and/or mutagenic. In addition to investigating
two common aliphatic nitrosamines, dimethylnitrosamine (DMN)
and diethylnitrosamine (DEN), we have studied the mutagenicity
of 11 cyclic nitrosamines, including 3 nitrosopiperidines, 3
nitrosopyrrolidines, 3 nitrosopiperazines, and 2 nitrosomor-
pholines. Our studies also include the nitrosamine-related
chemicals dimethylamine, formaldehyde, and sodium nitrite.
We have found that all 9 carcinogenic nitrosamines (DMN, DEN,
2-methyl-1-nitrosopiperidine, 3,4-dichloro-1-nitrosopiperi-
dine, nitrosopyrrolidine, 3,4-dichloronitrosopyrrolidine,
1,4-dinitrosopiperazine, 1,5-dinitrosohomopiperazine, nitro-
somorpholine) are mutagenic and all 4 noncarcinogenic nitrosamines (2,5-dimethylnitrosopiperidine, 2,5-dimethylnitro-
sopyrrolidine, 1-nitrosopiperazine, nitrosophenmetrazine)
are nonmutagenic. Formaldehyde and sodium nitrite are nonmutagenic, and dimethylamine is mutagenic at high concentrations (San Sebastian, Couch, and Hsie, unpublished). Variable carcinogenicity data on the latter three chemicals exists
in the literature.

Quinoline Compounds (Total of 5). One class of potential
environmental contaminants from fossil-fuel energy is heterocyclic compounds such as quinolines. Quinoline, a known
carcinogen, is mutagenic with metabolic activation. Another
carcinogen, 4-nitroquinoline-1-oxide, is highly mutagenic;
its mutagenicity decreases when assayed in the presence of
the activation system. The carcinogenicity of 8-hydroxy-,
8-amino-, and 8-nitroquinoline is not known, but these compounds exhibit variably weak mutagenicity in preliminary
experiments (San Sebastian and Hsie, unpublished).

Physical Agents (Total of 7). The mutagenicity of both ionizing radiation such as X ray and nonionizing physical agents
such as UV light has been demonstrated. Fluorescent white,

black, and blue lights are slightly cytotoxic and mutagenic. Sunlamp light is highly cytotoxic and mutagenic, exhibiting the biological effects within 15 sec of exposure under conditions recommended by the manufacturer for human use. Cytotoxic and mutagenic effects are observed after five min of sunlight exposure; responses vary with hourly and daily variations in solar radiation. In view of man's constant exposure to various light sources, demonstration of their genetic toxicity suggests that daily exposure to these light sources, especially sunlight, should be minimized (17,19,30). The demonstration that the CHO/HGPRT system is capable of quantifying the cytotoxic and mutagenic effect of sunlight recommends it as a model mammalian cell system for studies of the genetic toxicology of sunlight per se and of the interactive effects between sunlight and other physical and chemical agents, leading ultimately to a better understanding of the effects of sunlight on humans and the environment.

Mutagenicity of 39 Other Chemicals

Direct-Acting Alkylating Agents and Related Compounds (Total of 11). Included are 10 alkylating agents {2 alkyl sulfates [dimethyl sulfate (DMS), diethyl sulfate (DES)], 3 alkyl alkanesulfonates [methyl methanesulfonate (MMS), EMS, and isopropyl methanesulfonate (iPMS)], 2 nitrosamidines [MNNG and N-ethyl-N'-nitrosoguanidine (ENNG)], 3 nitrosamides [N-methyl-N-nitrosourea (MNU), N-ethyl-N-nitrosourea (ENU), and N-butyl-N-nitrosourea (BNU)]} and a structural analogue of MNNG, N-methyl-N'-nitroguanidine (MNG). Among the alkyl sulfates and alkanesulfonates, cytotoxicity was found to decrease with the size of the alkyl group: DMS>DES; MMS>EMS >iPMS. The mutagenicity based on mutants induced per unit mutagen concentration was DMS>DES; MMS>EMS>iPMS. However, when comparisons were made at 10% survival, mutagenic potency was: DES>DMS; EMS>MMS>iPMS. Among the nitroso compounds, the order of the mutagenicity based on 10% survival was MNNG>ENNG>MNU>ENU>BNU. This is the same order of potency as observed for mutation induction per unit concentration of mutagen. MNG is not mutagenic (7-9; Couch, San Sebastian, and Hsie, unpublished).

Heterocyclic Nitrogen Mustards--ICR Compounds (Total of 10). A series of heterocyclic nitrogen half-mustards, the ICR-compounds, has been developed at the Institute for Cancer Research as antitumor agents. Apparently, the biological activities of these compounds are associated with their

ability to intercalate and covalently bind nucleic acid.
Ten ICR-compounds (ICR-191, -170, -292, -372, -340, -191-OH,
-170-OH, -292-OH, -372-OH, and -340-OH) have been studied.
The 2-chloroethyl side chain of the first 5 compounds (e.g.,
ICR-191, etc.) has been replaced by a hydroxy group in the
latter 5 (e.g., ICR-191-OH, etc.). The 10 compounds differ
in the heterocyclic nucleus (methoxyacridine for ICR-191 and
-170, benz(a)acridine for -292, and azaacridine for -372 and
-340) and the alkylating side chain (the same secondary amine
for ICR-191 and -372, and the same tertiary amine for -170,
-292, and -340). Those with 2-chloroethyl side chains are
highly mutagenic, with the tertiary amines 3 to 5 times more
mutagenic than the secondary amines. The 5 hydroxy derivatives are nonmutagenic, but remain highly toxic, indicating
that although the 2-chloroethyl group (nitrogen half-mustard)
is needed for mutagenicity, its replacement with a hydroxy
group does not alter cytotoxicity. Cytotoxicity and mutagenicity of ICR-compounds appear to be dissociable (32,33;
Fuscoe, O'Neill, and Hsie, unpublished).

Aromatic Amines (Total 5). Many aromatic amines are human
carcinogens. We have shown that the carcinogens 2-acetylami-
nofluorene and its N-hydroxy- and N-acetoxyl derivatives are
mutagenic, while fluorene, a noncarcinogenic analogue, is
nonmutagenic. 1-hydroxy-2-acetylaminofluorene appears to be
mutagenic at a very high concentration in one preliminary
experiment (Hsie, Sun, and Brimer, unpublished).

Miscellaneous Compounds (Total of 13). Three commonly used
organic solvents (acetone, dimethyl sulfoxide, and ethanol)
are noncarcinogenic and do not appear to be mutagenic. All
four metabolic inhibitors (cytosine arabinoside, hydroxyurea,
caffeine, and cycloheximide) are nonmutagenic in a preliminary
study without coupling with the metabolic activation system.
Hydrazine and hycanthone appear to be direct-acting mutagens.
N^6,O^2 -dibutyryl adenosine 3':5'-phosphate, an analogue of
adenosine 3':5'-phosphate and an important effector of growth
and differentiation in many biological systems, is not mutagenic. The pesticides captan and folpet are mutagenic. The
mutagenicity of an artificial sweetener, saccharin, appears to
be variable; its determination is complicated by the requirement of high concentrations to yield any biological effect
(O'Neill and Hsie, unpublished).

Correlation of Mutagenicity in the CHO/HGPRT Assay with Reported Carcinogenicity in Animal Tests

Among a total of 109 chemical and physical agents studied, at different stages of completion, 56 have been reported to be either carcinogenic or noncarcinogenic in animal studies. Mutagenicity in the CHO/HGPRT assay of 54 of these agents correlated with documented animal carcinogenicity. The concurrence (i.e., known carcinogens are mutagenic and noncarcinogens are nonmutagenic in CHO/HGPRT assays) of each class of agents so far tested is 100% except for nitrosamines and relatives (93%) and ICR compounds (83%) (Table 4). The existence of a high correlation [54/56 (96.43%)] between mutagenicity and carcinogenicity speaks favorably for the utility of this assay in prescreening the carcinogenicity of chemical and physical agents. However, this result should be viewed with caution, since so far only limited classes of chemicals have been tested and some of the preliminary results remained to be confirmed.

A possible false negative was formaldehyde, which has been shown to be either carcinogenic or noncarcinogenic depending on the way test animals are exposed to it. An apparent false positive was ICR-191, a potent mutagen for microorganisms and CHO and other mammalian cells, which has been shown to be noncarcinogenic in a recent study.

A Study of EMS Exposure Dose: Differential Effects on Cellular Lethality and Mutagenesis

Earlier, we found that EMS-induced mutation frequency to TG resistance in cells treated for a fixed period of 16 h is a linear function over a large range of mutagen concentrations (0.013-0.8 mg/ml), including both the shoulder region (0-0.1 mg/ml) and the exponentially killing portion (0.1-0.8 mg/ml). To investigate whether EMS-induced mutagenesis can be quantified further, cells were treated with several concentrations of EMS for intervals of 2-24 h. Mutation induction increased linearly with EMS concentrations of 0.05-0.4 mg/ml for incubation times of up to 12-14 h. However, cell survival decreased exponentially with time over the entire 24-hour period. This difference in the time course of cellular lethality vs. mutagenicity might be due to the formation of toxic, nonmutagenic breakdown products in the medium with longer incubation times, or it might reflect a difference in the mode of action of EMS in these two biological effects. Further studies using varying concentrations (0.05-3.2 mg/ml) of EMS for 2-12 h showed

Table 4

Correlation of Mutagenicity[1] in the CHO/HGPRT Assay with
Reported Carcinogenicity[2] in Animal Tests[3]

Agent[4]	Total No. Studied	Concurrence[5]	False Negatives[6]	False Positives[7]
Energy-technology-related substances				
Polycyclic hydrocarbons	27	6/6 (100%)	0	0
Metallic compounds	15	4/4 (100%)	0	0
Nitrosamines and relatives	16	14/15 (93.33%)	1/15 (6.67%)	0
Quinolines	5	2/2 (100%)	0	0
Physical agents	7	3/3 (100%)	0	0
Subtotal	70	29/30 (96.67%)	1/30 (3.33%)	0
Other chemicals				
Direct-acting alkylating agents and relatives	11	11/11(100%)	0	0
ICR compounds	10	5/6 (83.33%)	0	1/6 (16.67%)
Aromatic amines	5	4/4 (100%)	0	0
Miscellaneous compounds	13	5/5 (100%)	0	0
Subtotal	39	25/26 (96.15%)	0	1/26 (3.85%)
All agents	109	54/56 (96.43%)	1/56 (1.79%)	1/56 (1.79%)

[1]Agents studied are found to be either mutagenic regardless of "mutagenic potency") or nonmutagenic. The mutagenicity is assayed either directly or coupled with a metabolic activation system in vitro or in vivo. In the S-9 coupled assay the microsome used was prepared from livers of Aroclor 1254-induced male Sprague-Dawley rats. The effects of other inducers or of conditions of the activation system have not been investigated extensively and are under study.

Table 4 (continued)

[2]Agents studied are denoted as either carcinogenic, non-carcinogenic, or uncertain based primarily on published data from USPHS (44) and IARC (23), regardless of "carcinogenic potency." Carcinogenicity data on many compounds is not yet available. The search for such data is admittedly neither exhaustive nor updated.

[3]In part from Table VII of ref. 21.

[4]The data are compiled from all agents studied, excluding those whose carcinogenicity is either unknown or uncertain. Thus, only 56 out of 109 agents studied are compiled table.

[5]Known carcinogens are mutagenic in CHO/HGPRT assays, e.g., MNNG, ICR-292, Ni, B(a)P, hycanthone, UV.

[6]Known carcinogens are nonmutagenic in CHO/HGPRT assays, e.g., formaldehyde.

[7]Known noncarcinogens are mutagenic in CHO/HGPRT assays, e.g., ICR-191.

that the manifestation of cellular lethality and mutagenesis occurs as a function of EMS exposure dose in that the biological effect is the same for different combinations of concentration multiplied by duration of treatment which yield the same product. From these studies the mutagenic potential of EMS can be described as 310×10^{-6} mutants (cell mg ml^{-1} h)$^{-1}$. Thus, the CHO/HGPRT system appears to be suitable for dosimetry studies which are essential for our understanding of the molecular mechanisms involved in mammalian mutagenesis (31).

Screening for the Mutagenicity of Fractionated Synthetic Fuel

In addition to studying the mutagenicity of individual environmental agents such as polycyclic hydrocarbons, quinolines, nitrosamines, metallic compounds, etc., we have found that the CHO/HGPRT assay can detect the cytotoxicity and mutagenicity of a crude organic mixture, in this case three subfractions of a crude synthetic oil (fractionated by M.R. Guerin of the Analytical Chemistry Division, ORNL) supplied by the Pittsburgh Energy Research Center. The acetone effluent (which contains tentatively identifiable heterocyclic nitrogen compounds) derived from the basic fraction is most mutagenic in the presence of a metabolic activation system (Table 5) (Hsie and Brimer, unpublished). Earlier, it appeared that the extreme toxicity of the unfractionated crude fuel prevented meaningful mutagenicity studies in the CHO/HGPRT system (Hsie and Brimer, unpublished). The chemistry (14), mutagenicity in microbial systems (12), and environmental testing (13) of the Synfuel are presented elsewhere in this proceedings.

Preliminary Development and Validation of the CHO Genetic Toxicity Assay for the Simultaneous Determination of Cytotoxicity, Mutagenicity, Chromosome Aberrations, and Sister Chromatid Exchanges

We have so far shown that CHO cells are useful for studying the cytotoxicity and mutagenicity of various individual physical and chemical agents and a crude organic mixture. The CHO cells and other hamster cells in culture were also found to be suitable for studying carcinogen-induced chromosome and chromatid aberrations (1,15,14,18) and sister chromatid exchanges (1,34,41). In our preliminary studies, we have found that these assays are useful in evaluating the cytogenetic effects of B(a)P and DMN when CHO cells are coupled with the standard microsome preparation described earlier (San Sebastian and Hsie, unpublished).

The successful development and validation of the multiplex CHO cell genetic toxicity system will be extremely valuable from both the scientific and economic points of view in genetic toxicology, because this system will allow the simultaneous determination of four distinct biological effects: cytotoxicity or cloning efficiency measures the reproductive capacity of a single cell to develop into a colony; single gene mutagenesis involves changes in the nucleotide sequence of DNA of a specific gene resulting in the acquisition of a

Table 5

Cytotoxicity and Mutagenicity of Subfractions[1]
of Synfuel A Basic Fractions

Subfraction[2]	Concentration (µg/ml)	Relative Cloning Efficiency (%) Without S-9	With S-9	Observed Mutation Frequency (TG mutants/ 10^6 cells) Without S-9	With S-9
Benzene	0.25		92		4
	1		102		<1
	2.5		117		1
	10	109	91	<1	1
	25		90		16
	50		71		13
	100	<0.2	0.2	–	25
Isopropanol	0.25		95		1
	1		94		1
	2.5		103		6
	10	108	95	<1	4
	25		102		16
	50		82		7
	100	<0.2	58	–	2
Acetone	0.25		89		6
	1		101		5
	2.5		93		<1
	5		100		9
	10	58	96	13	22
	25	22	56	6	46
	50	<0.3	4	15	49
	100	<0.2	0.2	–	135
Controls					
EMS	200			279	–
B(a)P	8			–	557
Solvent		100	100	4	9

[1]Unpublished data of Hsie and Brimer.

[2]See ref. 14 for details about the chemical separation of Synfuel A.

novel or altered phenotype; <u>chromosome aberrations</u> involve
microscopically identifiable changes in the number and/or
structure of the chromosome; and <u>sister chromatid exchange</u>
measures the extent of double-strand exchange in the DNA
duplex after breaks and rejoining of subunits of chromatids
each of which consists of one DNA duplex.

SUMMARY AND CONCLUSIONS

Conditions necessary for quantifying mutation induction
to TG resistance, which selects for >98% mutants deficient in
the activity of HGPRT in a near-diploid CHO cell line, have
been defined. Employing this mutation assay, we have determined the mutagenicity of diversified agents, including 11
direct-acting alkylating agents, 16 nitrosamines, 10 heterocyclic nitrogen mustards, 15 metallic compounds, 5 quinolines,
5 aromatic amines, 27 polycyclic hydrocarbons, 12 miscellaneous compounds, and 7 ionizing and nonionizing physical
agents. The direct-acting carcinogen MNNG is mutagenic,
while its noncarcinogenic analogue N-methyl-N'-nitroguanidine
is not. Coupled with the rat liver S-9 activation system,
procarcinogens such as nitrosopyrrolidine, B(a)P and 2-
acetylaminofluorene are mutagenic while their analogues 2,5-
dimethylnitrosopyrrolidine, pyrene, and fluorene are not.
The mutagenicity of the 56 agents documented to be either
carcinogenic or noncarcinogenic correlated well [54/56
(96.43%)] with the reported animal carcinogenicity. A possible false negative was formaldehyde and a false positive
was ICR-191. Preliminary studies on a synthetic crude oil
show that the acetone effluent (tentatively identifiable as
heterocyclic nitrogen compounds) derived from the basic fraction of Synfuel A is the most mutagenic fraction. Thus the
assay appears to be applicable for monitoring the genetic
toxicity of crude organic mixtures in addition to diverse
individual chemical and physical agents. The quantitative
nature of the assay enables a study of EMS exposure dose:
the mutagenic potential of EMS can be described as 310×10^{-6}
mutants $(\text{cell mg ml}^{-1} \text{ h})^{-1}$. It is also feasible to expand
the CHO/HGPRT system for quantifying cytotoxicity and mutagenicity to determination of chromosomal aberrations and
sister-chromatid exchanges in cells treated under identical
conditions. Thus it is possible to study simultaneously
these four distinctive biological effects.

REFERENCES

1. Abe S, Sasaki M: Chromosome aberrations and sister chromatid exchanges in Chinese hamster cells exposed to various chemicals. J Natl Cancer Inst 58:1635-1641, 1977

2. Abrahamson S, Lewis EB: The detection of mutations in Drosophila melanogaster. In: Chemical Mutagens: Principles and Methods for Their Detection (Hollaender A, ed.). New York, Plenum Press, 1973, pp 461-488

3. Ames BN, McCann J, Yamasaki E: Methods for detecting carcinogens and mutagens with the Salmonella/mammalian-microsome mutagenicity test. Mutat Res 31:347-364, 1975

4. Bridges BA: Short term screening tests for carcinogens. Nature 261:195-200, 1976

5. Chu EHY, Malling HV: Chemical induction of specific locus mutations in Chinese hamster cells in vitro. Proc Natl Acad Sci USA 61:1306-1312, 1968

6. Chu EHY, Powell SS: Selective systems in somatic cell genetics. Adv Hum Genet 7:189-258, 1976

7. Couch DB, Hsie AW: Dose-response relationships of cytotoxicity and mutagenicity of monofunctional alkylating agents in Chinese hamster ovary cells. Mutat Res 38:399, 1976

8. Couch DB, Hsie AW: Mutagenicity and cytotoxicity of congeners of two classes of nitroso compounds in Chinese hamster ovary cells. Mutat Res, in press

9. Couch DB, Forbes NL, Hsie AW: Comparative mutagenicity of alkylsulfate and alkanesulfonate derivatives in Chinese hamster ovary cells. Mutat Res, in press

10. Couch DB, Hsie AW: Metal mutagenesis: Studies of the mutagenicity of manganous chloride and 14 other metallic compounds in the CHO/HGPRT assay. In: Proceedings of the Annual Meeting of the Environmental Mutagen Society. San Francisco, March 9-13, 1978, p 74

11. Deaven L, Peterson D: The chromosome of CHO, an aneuploid Chinese hamster cell line: G-band, C-band, and autoradiographic analysis. Chromosoma 41:129-144, 1973

12.   Epler JL: Short-term bioassay of complex organic mix-
      ture: Part II, Mutagenicity testing, this proceedings

13.   Gehrs CW: Short-term bioassay of complex organic mix-
      ture: Part I, Chemistry testing, this proceedings

14.   Guerin MR: Short-term bioassay of complex organic mix-
      ture: Part III, Environmental testing, this proceedings

15.   Heddle JA, Bodycote J: On the formation of chromosomal
      aberrations.   Mutat Res 9:117-126, 1970

16.   Hsie AW, Brimer PA, Mitchell TJ, Gosslee DG: The dose-
      response relationship for ethyl methanesulfonate-induced
      mutations at the hypoxanthine-guanine phosphoribosyl
      transferase locus in Chinese hamster ovary cells.   Somat
      Cell Genet 1:247-261, 1975

17.   Hsie AW, Brimer PA, Mitchell TJ, Gosslee DG: The dose-
      response relationship for ultraviolet light-induced
      mutations at the hypoxanthine-guanine phosphoribosyl
      transferase locus in Chinese hamster ovary cells.   Somat
      Cell Genet 1:383-389, 1975

18.   Hsie AW, Brimer PA, Machanoff R, Hsie MH: Further evi-
      dence for the genetic origin of mutations in mammalian
      somatic cells: The effects of ploidy level and selec-
      tion stringency on dose-dependent chemical mutagenesis
      to purine analogue resistance in Chinese hamster ovary
      cells.   Mutat Res 45:271-282, 1977

19.   Hsie AW,Li AP, Machanoff R: A fluence response study
      of lethality and mutagenicity of white, black, and blue
      fluorescent lights, sunlamp, and sunlight irradiation
      in Chinese hamster ovary cells.   Mutat Res 45:333-342,
      1977

20.   Hsie AW, Machanoff R, Couch DB, Holland JM: Mutagenicity
      of dimethylnitrosamine and ethyl methanesulfonate as
      determined by a quantitative host-mediated CHO/HGPRT
      assay.   Mutat Res, in press

21.   Hsie AW, Couch DB, O'Neill JP, San Sebastian JR, Brimer
      PA, Machanoff R, Riddle JC, Li AP, Fuscoe JC, Forbes NL,
      Hsie MH: Utilization of a quantitative mammalian cell
      mutation system, CHO/HGPRT, in experimental mutagenesis
      and genetic toxicology.   Chemical Industry Institute of
      Toxicology (CIIT) Workshop on "Strategies for Short-Term

Testing for Mutagens/Carcinogens," Research Triangle Park, North Carolina, August 11-12, 1977; also in CRC Uniscience Monographs in Toxicology (Butterworth BE, ed.), in press

22. Hsie AW, O'Neill JP, Couch DB, San Sebastian JR, Brimer PA, Machanoff R, Fuscoe JC, Riddle JC, Li AP, Forbes NL, Hsie MH: Quantitative analyses of radiation- and chemical-induced lethality and mutagenesis in Chinese hamster ovary cells. Radiat Res, in press

23. IARC, IARC Monograph on the Evaluation of Carcinogenic Risk of Chemicals to Man, Vols. 1-10. Lyons, IARC, 1972-76

24. Ishidate M Jr, Odashima S: Chromosome tests with 134 compounds on Chinese hamster cells in vitro: A screening for chemical carcinogens. Mutat Res 48:337-354, 1977

25. Kao FT, Puck TT: Induction and isolation of nutritional mutants in Chinese hamster ovary cells. Proc Natl Acad Sci USA 60:1275-1281, 1968

26. McCann J, Ames BN: Detection of carcinogens as mutagens in the Salmonella/microsome test: Assay of 300 chemicals: Discussion. Proc Natl Acad Sci USA 73:950-954, 1976

27. McCann J, Choi E, Yamasaki E, Ames BN: Detection of carcinogens as mutagens in the Salmonella/microsome test: Assay of 300 chemicals. Proc Natl Acad Sci USA 72:5135-5139, 1975

28. Natarajan AT, Tates AD, Van Buul PPW, Meijers M, De Vogel N: Cytogenetic effects of mutagens/carcinogens after activation in a microsomal systen in vitro. I. Induction of chromosome aberrations and sister chromatid exchanges by diethylnitrosamine (DEN) and dimethylnitrosamine (DMN) in CHO cells in the presence of rat liver microsomes. Mutat Res 37:83-90, 1976

29. O'Neill JP, Brimer PA, Machanoff R, Hirsch GP, Hsie AW: A quantitative assay of mutation induction at the hypoxanthine-guanine phosphoribosyl transferase locus in Chinese hamster ovary cells: Development and definition of the system. Mutat Res 45:91-101, 1977

30. O'Neill JP, Couch DB, Machanoff R, San Sebastian JR, Brimer PA, Hsie AW: A quantitative assay of mutation induction at the hypoxanthine-guanine phosphoribosyl transferase locus in Chinese hamster ovary cells (CHO/ HGPRT system): Utilization with a variety of mutagenic agents. Mutat Res 45:103-109, 1977

31. O'Neill JP, Hsie AW: Chemical mutagenesis of mammalian cells can be quantified. Nature 269:815-817, 1977

32. O'Neill JP, Fuscoe JC, Hsie AW: Mutagenicity of hetero-cyclic nitrogen mustards (ICR compounds) in cultured mammalian cells. Cancer Res, in press

33. O'Neill JP, Fuscoe JC, Hsie AW: Structure-activity relationship of antitumor agents in the CHO/HGPRT system: Cytotoxicity and mutagenicity of 8 ICR com-pounds. Proceeding of the Annual Meeting of the Ameri-can Society for Cancer Research, Washington, DC, April 5-8, 1978

34. Perry P, Evans HJ: Cytological detection of mutagen/ carcinogen exposure by sister chromatid exchange. Nature 258:121-125, 1975

35. Puck TT: The Mammalian Cell as a Microorganism. San Francisco, Holden-Day, Inc, 1972, p 219

36. Puck TT, Kao FT: Treatment of 5-bromodeoxyuridine and visible light for isolation of nutritionally deficient mutants. Proc Natl Acad Sci USA 58:1227-1234, 1967

37. Purchase IEH, Longstaff E, Ashby J, Styles JA, Anderson D, Lefevre PA, Westword FR: Evaluation of six short term tests for detecting organic chemical carcinogens and recommendations for their use. Nature 264:624-627, 1976

38. Rosenkranz HS, Gutter B, Speck WT: Mutagenicity and DNA-modifying activity: A comparison of two microbial assays. Mutat Res 41:61-70, 1976

39. San RHC, Stich HF: DNA repair synthesis of cultured human cells as a rapid bioassay for chemical carcinogens. Int J Cancer 16:284-291, 1975

40. Siminovitch L: On the nature of heritable variation in cultured somatic cells. Cell 7:1-11, 1976

41. Stetka DG, Wolff S: Sister chromatid exchange as an
     assay for genetic damage induced by mutagens/carcinogens.
     II. In vitro test for compounds requiring metabolic
     activation. Mutat Res 41:343-350, 1976

42. Sugimura T, Sato S, Nago M, Yahagi T, Matsushima T,
     Seino Y, Takeuchi M, Kawachi T: Overlapping of carcin-
     ogens and mutagens. In: Fundaments in Cancer Preven-
     tion (Magee PN et al., eds.). Tokyo, University of
     Tokyo Press/Baltimore, University Park Press, 1976,
     pp 191-215

43. Thompson LH, Baker RM: Isolation of mutants of cultured
     mammalian cells. Methods Cell Biol 6:209-281, 1973

44. USPHS, Survey of compounds which have been tested for
     carcinogenic activity. USPHS Publication No. 149,
     1972-73

45. Vogel E: The relation between mutational pattern and
     concentration by chemical mutagens in Drosophila. In:
     Screening Tests in Chemical Carcinogenesis (Montesano R,
     Bartsch H, Tomatis L, eds.). IARC Scientific Publica-
     tion No. 12, Lyons, IARC, 1976

46. Zimmerman SK: Procedures used in the induction of
     mitotic recombination and mutation in the yeast
     Saccharomyces cerevisiae. Mutat Res 31:71-86, 1975

ENVIRONMENTAL TESTING

C.W. Gehrs, B.R. Parkhurst, and D.S. Shriner
Environmental Sciences Division
Oak Ridge National Laboratory
Oak Ridge, Tennessee

INTRODUCTION

Environmental toxicology is a term that conveys different images to different people. In its broadest sense environmental toxicology encompasses all of the research necessary to evaluate the potential ecological effects, to determine the ultimate fate in the environment, and to identify critical pathways to man that might occur as the result of release of a particular material (Table 1). Evaluation of the potential toxicity of the material can be accomplished through three types of testing: short-term bioassays (environmental screening); subacute organismic, population, community, and ecosystem evaluation; and mechanistic studies. The latter two types are resource intense in that they generally require substantial manpower and time commitments. However, they are also essential for developing predictive capabilities regarding the potential for long-term chronic environmental effects resulting from the release of a complex effluent stream. Environmental screening, on the other hand, requires less manpower with results often obtained in less than a week's time.

Research sponsored by the Division of Biomedical and Environmental Research, U.S. Department of Energy, under contract W-7405-eng-26 with Union Carbide Corporation. Publication No. 1187, Environmental Sciences Division, ORNL.

Table 1

Description of Various Segments of
Environmental Toxicology Research

Environmental Toxicology

Effects	Transport
1) Short-term Bioassays $LC_{50}{}^{a}$, $GR_{50}{}^{b}$	1) Abiotic Processes hydrolysis, photolysis, sorption/sedimentation
2) Subacute Testing organism, population, ecosystem	2) Biotic Processes microbial degradation, uptake, bioaccumulation, bioconcentration, transformation, tissue distribution
3) Mechanistic Studies chemical structure/ biological activity, reproductive impairment, physiological responses	

---

[a] LC_{50},   the concentration of original test material that
will result in mortality of 50% of test organisms
in a certain time (usually 48 to 96 hours).

[b] GR_{50},   the concentration of original test material that
will reduce the growth of the test organisms during
a certain time (usually 48 hours).

Unfortunately, most environmental screening does not
provide the types of data necessary either for predicting
potential ecological effects or for estimating potential environmental risks. This paper is limited to a discussion of
short-term environmental bioassays and of potential uses of
information obtained from this type of testing. The integration of chemical characterization and fractionation into
the testing protocol is discussed, as well as the rationale
for selecting appropriate test systems. Two of the systems
currently employed are described.

Short-term bioassays have the potential for providing toxicity data for three sets of users. They can (1) provide guidance to control technologists and waste management personnel regarding which components of a complex mixture are biologically active and, hence, may require removal before effluent discharge; (2) provide guidance to the environmental scientist in determining which components of a complex mixture require further evaluation in subacute and mechanistic studies; and (3) provide semi-quantitative hazard assessment data when comparisons are made to standard reference compounds or toxicity data from other complex mixtures.

USE OF A BATTERY OF TESTS

A major consideration in the environmental testing of complex chemical mixtures is that such mixtures contain chemicals of a wide variety of classes and an equally wide range of concentrations potentially capable of causing additive, antagonistic, or synergistic responses in test organisms or at specific receptor sites. Such interactions may be a function of the combined dose of toxicants or of the inherent genetic susceptibility of a particular target organism. The genetic susceptibility of an organism may be due to (1) the presence or absence of effective barriers to absorption or translocation of a chemical; (2) the selective accumulation of the chemical in a bound or inactive form, or in tissues remote from a receptor site; (3) the presence or absence of the ability to detoxify the substance through biotransformation; (4) the existence or absence of specific target or receptor systems in exposed cell systems; or (5) some combination of the above factors (1). In addition, since a specific compound may influence different biologic receptor sites within an organism through a variety of pathways, the problem of environmental hazard assessment is extremely complex.

To establish with absolute certainty the degree of hazard posed by a particular material would require testing of all potential target organisms for their various responses. In the light of the large number of complex mixtures entering the environment each year which require evaluation for their potential hazard, it is readily apparent that the costs, both of time and money, for such an option prevent the making of assessments in a realistic time frame. However, at the other end of the spectrum, predictions concerning potential hazards based on the response of a single species to a chemical mixture have an unacceptably low level of certainty or dependability (because of differential species responses, etc.).

The task then is to arrive at some intermediate point in cost and effort that permits an acceptable estimate of hazard potential. One method of increasing the level of confidence is to use a series of test systems and develop a hazard estimate based on pooled results from these systems. This group of systems would ideally attempt to represent as broad a spectrum as possible of taxa of organisms, tissue ages, growth forms, routes of exposure, and environmental variables. In this manner, we anticipate being able to obtain estimates of potential hazard which will reflect better the range of response variability expected from a normal population of organisms.

CRITERIA FOR TEST SYSTEM SELECTION

For a specific system to be functional in environmental screening it must be of short duration, require minimal quantities of material, and be a standard test system, i.e., it must employ an organism for which a large base of toxicological data is available.

The test system must be of short duration, not only to minimize manpower costs, but also to prevent confounding interpretation of results that might occur from chemical changes in the aqueous media. Chemical separation to provide relatively discrete fractions is costly, time consuming, and produces only small amounts of material (2). It is impractical to use this approach, for example, for fish requiring several gallons of water per replicate, or plants requiring similar quantities of test media. The final criterion, using a test system for which a large data base already exists, is essential if even a semiquantitative assessment is to be made.

DESCRIPTION OF TEST SYSTEMS EMPLOYED

Bioassays of varying types have been used successfully over a wide range of applications. Prominent among these applications are tests with herbicides, insecticides, and plant hormones that have dealt essentially with effects of specific compounds. When employing a bioassay test in the screening of complex mixtures of chemicals, there are two basic assumptions that are made: (1) the species used will show an injury response in proportion to the concentration of the biologically active chemical species; and (2) the responses obtained are reproducible (3).

The two species used in the experiments discussed in this paper are the zooplankter, Daphnia magna, and the radish, Raphanus sativus. The parameter used in the zooplankton system is the 48-hr LC_{50}. Although these aquatic organisms are easily cultured in the laboratory, they have been found to be sensitive to aquatic pollutants. Four D. magna are placed in 80 ml of each toxicant test solution in 100-ml beakers covered with watch glasses. The small numbers of test organisms and the small volumes of media used are necessitated by the small quantities of test materials (often less than one gram) that are derived from the chemical fractionation procedure. Water temperature is maintained at 22 $\pm$ 0.5°C by placing the beakers in an environmental chamber. Photoperiod is maintained under a 12-hr light/dark regime. Toxicant solutions are prepared with filtered spring water (pH 7.8, alkalinity 119 mg/liter, hardness 140 mg/liter). All tests are run in triplicate. Serial dilutions with each concentration being 60% of the previous one are made for each test material. Controls of spring water without added toxicant are included. The range of dilutions are selected to bracket the 48-hr LC_{50}, which is obtained by computerized PROBIT analytical procedures (4). The presence of toxic interactions between fractions is determined using the additivity index of Marking and Dawson (5).

Radish seed responses to chemical mixtures are expressed in terms similar to the LC_{50} value used in the Daphnia studies. The value obtained represents the concentration of the original mixture (or concentration of specific components in the original mixture) which reduced the yield of a specific species by 50% (3,6,7), and is expressed as GR_{50} (concentration for 50% growth reduction). Measurement parameters used as estimates of yield include percent germination time to root emergence, fresh weight, and root elongation. A series of seed germination tests (according to the procedure discussed below) with water blank control solutions adjusted to a range of pH's are performed at different pH levels and serve as controls. The pH of each test material is determined when the material is received. Seed germination tests are then performed on each test material, which consists of seeds in petri dishes containing filter paper moistened with the toxicant. Germination percentages are determined, and length, fresh weight, and time to root emergence are measured and compared with control seeds. Toxicants showing inhibitory effects on germination percentages or root growth are diluted and tested as above to determine a GR_{50}. Toxicants showing no effects in the original test are not tested further. Specific fractions of material are also tested to determine GR_{50}.

ISOLATION OF TOXIC COMPONENTS

The sequential procedure employed to isolate and identify the toxic components of complex mixtures is shown in
Figure 1. In the first step the mixtures are screened for
acute toxicity (i.e., the Bodean test). Mixtures not found
to be acutely toxic do not undergo further testing. This
does not mean that potential problems do not necessarily
exist with those materials, but identification of such
potential problems would need to await chemical screening.
Those mixtures found to be acutely toxic are separated by
chemical extraction into their organic and inorganic components and tested for acute toxicity. If acute toxicity is
found in the organic components, the component is fractionated into acid, base, and neutral fractions. At this time,
the relative amount of each fraction in the total organic
component is also determined. The toxicity of each fraction
is determined and its contribution to the toxicity of the
original mixture calculated by relating the relative toxicity
of the fraction to its concentration in the original mixture.
If further isolation of the actual toxic compounds of the
organic fractions is desired, subfractionation and testing
can be performed. Up to 14 such subfractions have been separated from synthetic fuel process effluents for use in
mutagenesis testing, and these same subfractions could be
produced for acute toxicity testing (8). Ultimately, tests
with specific compounds can be used to evaluate the toxicity
of individual chemical components of the mixture.

A different approach is used to identify the toxic components of the inorganic fraction of the complex mixture.
If the inorganic fraction is found to be acutely toxic, it is
chemically characterized to determine its composition. The
toxicities of the individual components are ascertained and
the contribution of each component to the toxicity of the
original mixture assessed by the same method used for the
organic fraction (Figure 1).

In the last step of this environmental screening an effort is made to determine whether the testing procedure has
accounted for all of the toxicity of the original complex
mixture. The individually identified inorganic components
present in the original mixture are combined to produce a
"reconstituted" mixture. The acute toxicity of this mixture
is tested and compared to the toxicity of the original
mixture.

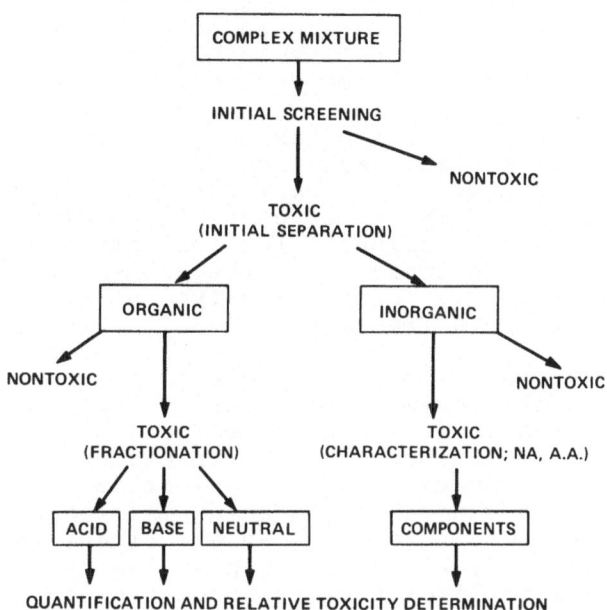

Figure 1.   Flow diagram that shows testing patterns used to identify toxic components.

The procedure outlined above has been used to identify the toxic components of several types of synthetic fuel aqueous effluents. The data from two are presented in this paper as examples. They include one in which inorganic components were the most toxic and one in which organic compounds were the most toxic. Only data from the Daphnia system are presented. Both of the effluents tested are complex mixtures of hundreds of individual organic and inorganic compounds which are byproducts of coal conversion processes. The first material tested was an untreated process-water effluent from the solvent refined coal (SRC) pilot plant in Ft. Lewis, Washington. The second material tested consisted of two effluents, untreated and biologically treated hydrocarbonization (HCZ) process-water from the process development unit (PDU) at the Oak Ridge National Laboratory.

The results of the application of the acute toxicity testing protocol to the SRC effluent revealed the total effluent to be acutely toxic to Daphnia, with a 48-hr LC_{50} estimated to be at a dilution of 15.7% (9). The inorganic portion of the effluents was found to be nontoxic; i.e., concentration was necessary before an LC_{50} could be produced. Further testing required chemical separation. Of the three fractions extracted from the organic portion of the effluent, the neutral fraction had the highest toxicity with a 48-hr LC_{50} of 9 mg/liter (Table 2). The 48-hr LC_{50}s of the acid and base fractions were 29.5 mg/liter and 45.8 mg/liter, respectively. Based on the concentrations of the individual fractions in the SRC effluent, the toxicity contribution of each fraction to the toxicity of the whole SRC effluent was calculated (Table 2). These toxicity contributions ranged from 53% for the acid fraction to 3.6% for the base fraction.

One of the values of being able to test the whole effluent as well as the various fractions is that toxic interactions between fractions can be detected and quantified. Testing of whole effluents (before fractionation) and reconstituted effluents enables gross evaluation of the role of the fractionation scheme in producing artifacts in the effluent components. The results of an experiment using effluents from an SRC facility (Table 3) suggest that neither toxic interactions or artifact formation occurred in the SRC effluents. The 48-hr LC_{50}s of the whole effluent and the reconstituted effluent were found to be 15.7% and 15.5%, respectively, calculated as a percent dilution of the effluent (Table 3). These values were not significantly different (t, $P = 0.05$), and indicated both that (1) the chemical fractionation procedure did not alter the toxicity of the fractions and (2) all of the toxicity of the effluents was accounted for

Table 2

Acute Toxicity of Organic Components
of Untreated SRC Effluents

Components	Acid	Neutral	Base
Concentration in effluent (mg/l)	99.5	24.8	10.6
Composition of total effluent (%)	28.5	7.1	3.0
Daphnia (48-hr LC_{50} in mg/l)	29.5	9.0	45.8
Relative contribution to effluent toxicity (%)	53.0	43.4	3.6

Table 3

Contributions of SRC Fractions to Total Toxicity of SRC Effluent

Fraction	48-hr LC_{50}	Fraction Concentration (mg/l) at 48-hr LC_{50} Concentration		Contribution to Whole Effluent Toxicity (%)[3]		Additive Index	
		Reconstituted Effluent	SRC Effluent	Reconstituted Effluent	SRC Effluent	Reconstituted Effluent	SRC Effluent
Acid	29.5	15.2	15.6	51.5	51.3		
Base	45.8	1.6	1.7	3.5	3.5		
Neutral	9.0	3.8	3.9	42.2	41.9		
Inorganic	1000[1]	32.7	33.6	3.0	3.3		
						0.00[2]	0.03[2]

[1] No mortality occurred at highest concentration in test series. 1000 mg/l is assumed to be a very conservative estimate of the 48-hr LC_{50}.

[2] Additive index > 0 = more than additive toxicity (synergism)
Additive index = 0 = simple additive toxicity
Additive index < 0 = less than additive toxicity (antagonism)

[3] The contribution of a fraction to the toxicity of the whole effluent is given by the expression:

$$\frac{Am}{Ai}$$

$$\frac{Am}{Ai} + \frac{Bm}{Bi} + \text{---} + \frac{Zm}{Zi}$$

where A, B --- Z, are the fractions, i the toxicities (LC_{50}'s) of the individual fractions, and m the fraction concentration at the LC_{50} concentration of the whole effluent.

in the four fractions. Using the method of Marking and Dawson
(5), the additivity index was 0.03 for the whole SRC effluent
and 0.00 for the reconstituted effluent (Table 3). These
values were not significantly different (t, P = 0.05), indicating that the toxicities of the individual fractions were
directly additive within the effluents. As a comparison, if
each fraction acted independently, an additivity index of
0.95 would be predicted.

The second example of the environmental screening approach deals with effluents from the Oak Ridge National Laboratory hydrocarbonization unit (HCZ). In the first step of
the toxicity screening of the HCZ effluent, the toxicities of
the untreated and treated effluents were compared. The treatment process was determined to have reduced the toxicity of
the effluent by 99% in the Daphnia system. Further testing
showed that the inorganic portion of the effluent contributed
99.5% of the toxicity (Table 4), with the remaining 0.5% contributed by phenols. Of the inorganic constituents, ammonia
was the principal toxic agent, contributing 96% of the toxicity.

The determination of interactions between the HCZ effluent components demonstrated a less than additive or antagonistic behavior. This indicated that the sum of the toxicities of the effluent components calculated individually was
greater than when they were present together in the effluent.

Table 4

Acute Toxicity Data for Hydrocarbonization

	Organics	Inorganics
Concentration in effluent (mg/l)	110	2846
Relative quantity (%)	3.7	96.3
Daphnia (48-hr LC_{50} in mg/l)	774	31.7
Relative contribution to effluent toxicity (%)	0.5	99.5

SUMMARY AND CONCLUSION

The previous discussion has shown how coupling of chemical separation and fractionation with environmental testing is able to identify those materials most biologically active, at least, in acute toxicity. In the case of the SRC effluent the primary activity was attributed to the acidic organic fraction, where the phenolics are located. Because of the relative removal efficiency of phenolics through chemical stripping and biological treatment (10), near field acute toxicity from effluent releases would not be expected from aqueous effluents released from a facility similar to the SRC pilot plant. The ultimate conclusion concerning potential effluents from the HCZ process development unit is the same, although the biologically active component was the inorganic NH_3 .

Environmental testing can help the control technologist and waste management engineer identify materials of potential environmental consequence. These screening activities are not intended to, and should not, replace subacute toxicity and mechanistic toxicity studies for developing data for predictive purposes with respect to chronic low level effects in the far field environment.

REFERENCES

1.   Loomis TA: Essentials of Toxicology.   Philadelphia, Lea and Febiger, 1970, p 162

2.   Guerin MR, Clark BR, Ho CH, Epler JL, Rao TI: Short-term bioassays of complex organic mixtures: Part 1. Chemistry.   Application of Short-term Bioassays in the Fractionation and Analysis of Complex Environmental Mixtures (1978) EPA-60019-78-027.

3.   Santelmann PW: Herbicide bioassay.   In: Research Methods in Weed Science.   Southern Weed Science Society, 177, pp 79-88

4.   Finney DJ: Statistical Methods in Biological Activity. London, Griffin Press, 1971, 2nd Edition

5.   Marking LL, Dawson VK: Method for assessment of toxic-
     ity of efficacy mixtures of chemicals.   Investigations
     in Fish Control No. 67, U.S. Department of Interior,
     Fish and Wildlife Services, Washington, D.C., 1975

6.   Sheets TJ: Effects of soil type and time on herbicidal
     activity of CDAA, CDEC, and EPTC.   Weeds 7:442-448,
     1959

7.   Upchurch RP: The influence of soil factors on the
     phytotoxicity and plant selectivity of diuron.   Weeds
     6:442-448, 1959

8.   Rubin IB, Guerin MR, Hardigree AA, Epler JL: Fraction-
     ation of synthetic crude oils from coal for biological
     testing.   Environmental Research 12:358-365, 1976

9.   Parkhurst BR, Gehrs CW, Rubin IB: The value of chemi-
     cal fractionation for identifying the toxic components
     of complex aqueous effluents.   Proceedings of ASTM 2nd
     Annual Symp. on Aquatic Toxicology.   Cleveland, in
     press

10. Herbes SE, Southworth GR, Gehrs CW: Organic contami-
     nants in aqueous coal conversion effluents: Environ-
     mental consequences and research priorities.   Proceed-
     ings of Trace Substances in Environmental Health-X,
     (Hemphill DD, ed.), University of Missouri, Columbia,
     1977, pp 295-303

INTEGRATING MICROBIOLOGICAL AND CHEMICAL TESTING INTO THE SCREENING OF AIR SAMPLES FOR POTENTIAL MUTAGENICITY

Edo D. Pellizzari, Linda W. Little,
Charles Sparacino, and Thomas J. Hughes
Research Triangle Institute
Research Triangle Park, North Carolina

Larry Claxton and Michael D. Waters
Health Effects Research Laboratory
U.S. Environmental Protection Agency
Research Triangle Park, North Carolina

A recent review of respiratory carcinogenesis (10) notes that presence of chemical carcinogens in the atmosphere, especially in certain urban and industrial environments, was described more than two centuries ago in Pott's studies in chimney sweeps and a century ago in the work of Harting and Hesse on uranium miners. In the last two decades, with the availability of sophisticated air sampling devices, analytical chemistry techniques, and bioassay procedures, the identity and carcinogenicity of many air pollutants, especially those organic compounds that can be extracted by solvents from particulates, has been documented (10).

Considerable work remains to be performed on those phases which are more difficult to collect, identify, and bioassay, including the insoluble portions of particulates (10) and the volatile organic components (17). A particular problem needing study, notes Van Duuren (17), is the role of aromatic hydrocarbon compounds, such as pyrene and fluoranthene, which are noncarcinogenic but act as potent cocarcinogens. In this regard, Weisburger (18) points out the necessity for studies to "delineate the carcinogenic risk of specific and rationally selected mixtures which may affect man."

The most direct measure of human risk would be provided by studies with human beings. Whereas in water pollution "the fish is the final arbiter of toxicity" (4), here the human being is the final arbiter of carcinogenicity. Information on human effects is usually obtained after the fact in case histories or epidemiological studies.

According to Kluyver's principle of the unity of bio-chemistry, a wide variety of organisms' effects at the sub-cellular level should show a great deal of similarity. Thus, other mammals, especially rodents, have been employed in assessment of carcinogenic risks to man. However, definitive bioassays are expensive because they require months to years for completion, large numbers of experimental animals, and large amounts of test sample.

Ames and associates (1,2,3,12,13) have developed and validated a bacterial mutagenesis assay which detects as mutagens approximately 85 percent of the known carcinogens that have been tested. The test is rapid and economical, requires little space, and is sensitive to nanogram or microgram levels of many mutagens. By addition of mammalian microsomal enzymes to the test system, the test can detect mutagens requiring metabolic activation.

Because of these advantages, the test has been recommended by EPA as one of bioassays to be used in "Level 1 Environmental Assessment" (6,8), designed by the EPA Industrial Environmental Research Laboratory to be a cost-effective approach in screening emissions to determine which "have a higher potential for causing measurable health or ecological effects." Thus, this test should receive priority for further assessment (8). The Ames test has been employed in detection of airborne mutagens by several investigators, including Talcott and Wei (16), Pitts et al. (15), and Flessel (7).

For the past eight months, an EPA-sponsored program has been underway at Research Technical Institute to develop a protocol that will define a minimal biological and chemical methodology to serve as a screen for the potential carcino-genicity of complex air pollutant mixtures. This protocol includes sampling, fractionation, chemical identification, and mutagenicity method development. Specifically, the program involves:

- Collection at selected locations of ambient air samples containing particulate and vapor phase material.

- Mutagenesis testing of crude particulate and vapor phase materials.

- Treatment of crude particulates by separating into major chemical classes through organic fractionation procedures.

- Chemical characterization of fractions showing mutagenic activity.

- Mutagenesis testing of vapor phase components.

Studies to date have focused on particulate phase components.

Sampling

Collection of a particulate sample is accomplished using the Battelle Maxi-sampler which partitions the sample into three particle size ranges, > 3.5 µm, 1.7–3.5 µm, and < 1.7 µm, the latter two ranges representing respirable particles. The Maxi-sampler can collect a relatively large amount of particulate, a gram or more per 24-hour period. This is important for ultimate bioassay and chemical identification procedures because of the sensitivity requirements associated with these processes.

FRACTIONATION

The well documented complexity of air particulate extracts, and the desirability of reducing the number of compounds that require identification, indicates the need for a suitable prefraction method. We have adopted a procedure that is reproducible, mild, and effective in which particulate extract is divided into a relatively small number of fractions, each with a similar chemical make-up.

The fractionation scheme proposed and used initially in this program involved sonication of the particulate with two solvents, cyclohexane and methanol. This approach has been shown to be effective in removing significant amounts of polar materials that are not removed by the more usual treatment with a single nonpolar solvent such as cyclohexane or benzene. It was hoped that the materials extracted by the two solvents could then be further treated separately to give two types of fractions for testing, namely, nonpolar and polar. The complete fractionation scheme yielded 13 separate samples for bioassays (Figure 1).

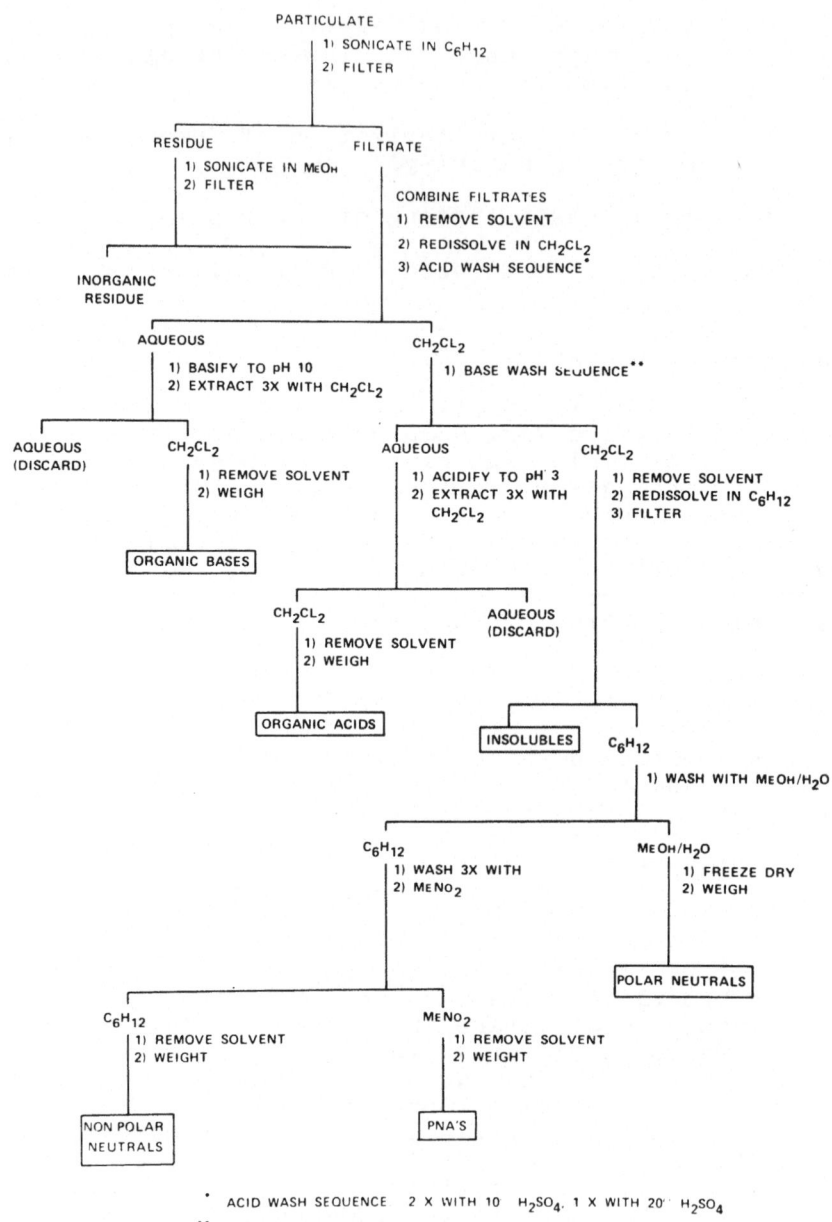

Figure 1.   Initial Fractionation Scheme.

It quickly became evident that the amounts of material available from the fractionation of the particulate extracts for bioassay were in some cases too low. It was also found through preliminary gc/ms results that the difference between the compositions of the polar and nonpolar fractions was small, i.e., essentially the same compounds were found in both sample types. For example, the nonpolar and polar organic base fractions were virtually identical in terms of components identified. Hill et al. (9) have shown that the composition of methanol vs cyclohexane extracts of air particulate samples are largely duplicative although there are real differences. We have thus modified the procedure by combining the cyclohexane and methanol extracts to produce a single crude particulate extract which is then fractionated through a solvent partition scheme to produce six fractions instead of 13 as before. This, of course, will provide more sample per fraction and thus reduce somewhat the burdens of high sensitivity on the bioassay.

The new partition scheme also incorporates other modifications which increase its usefulness and reliability. Attempts to validate the initial fractionation procedure with a known sample were disappointing. For example, a sample containing 53.5 mg of quinoline as the only organic base present was partitioned with 0.1 normal HCl. Recovery of the base from the acid phase produced only 6.4 mg (12%) of quinoline. Hence, a new procedure was adopted utilizing several sulphuric acid (10% and 20%) washes. When this procedure was applied to the 53.5 mg sample of quinoline, the recovery of base was quantitative. Other modifications to this scheme were based on published work. The entire scheme is depicted in Figure 2. The efficacy of the scheme was assessed by subjecting a mixture containing known amounts of compounds to the partition procedure. The mixture consisted of benzoic acid, phenol, quinoline, hexadecane, phenanthrene, and ethylene glycol. The compounds were chosen to represent the five classes of materials produced by the partition scheme. All these materials have been found in air particulate samples except ethylene glycol; no information on the composition of the polar neutral fraction is available. Ethylene glycol was included as a likely component of this fraction based on its known chemical properties. The experiment was conducted using both large and small mass samples. Recoveries were determined gravimetrically. The results are depicted in Table 1.

FRACTIONATION SCHEME

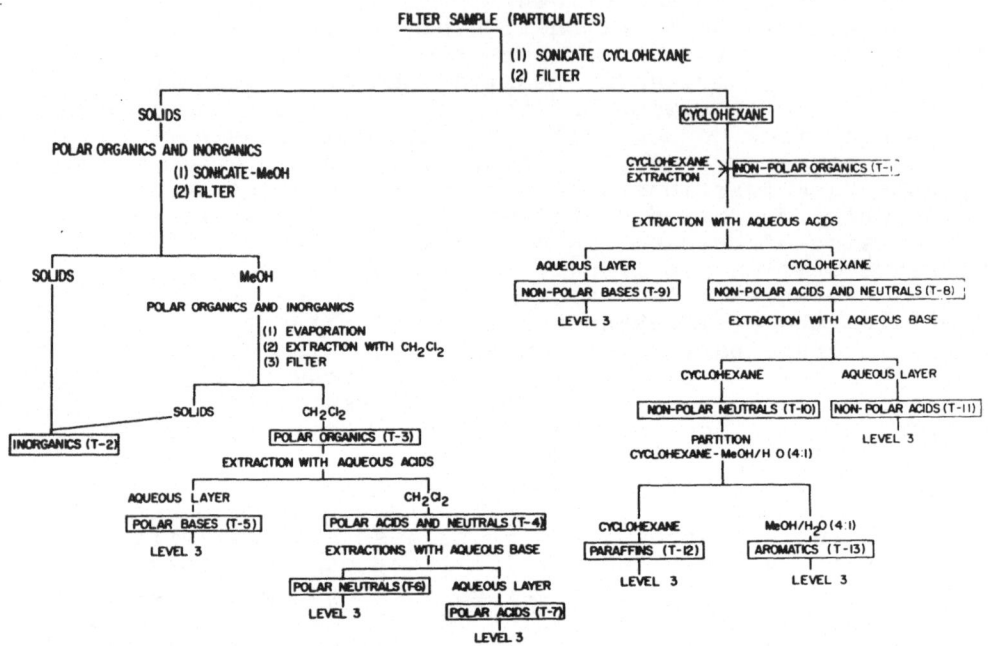

Figure 2.   Final fractionation scheme.

As a further check on the procedure, TLC scans were conducted on each fraction to ascertain the extent, if any, of spillover of one compound into other fraction(s). No such spillover was detected.

The first sample to be examined under this program was collected at South Charleston, West Virginia, during August, 1977, using the Maxi-sampler. Sampler plate scrapings were made at EPA and some four grams of material (less than 1.7 μm diameter) were received by Research Triangle Institute (RTI). This material was neither scraped nor stored by EPA under ideal conditions. However, the aim at this stage of the program was to develop the necessary extraction, partition, and identification methodology; the sample was thus considered acceptable for this use. Later samples have been received which have been properly handled both at EPA and RTI.

For identification of the fractional components, the following gc/ms run parameters were employed. The gc/ms

Table 1

Partition Scheme Validation Results

Compound	Class	Amount Added (mg)	Amount Recovered (mg)	Amount Recovered (mg)
Large Initial Mass				
Benzoic Acid/Phenol	Org. Acid	41.2/39.0	74.0	92.3
Quinoline	Org. Base	42.4	40.2	94.8
Hexadecane	Nonpolar Neutral	45.1	10.0	22.2
Phenanthrene	PNA	45.4	34.4	75.8
Ethylene Glycol	Polar Neutral	44.6	43.4	97.3
Small Initial Mass				
Benzoic Acid/Phenol	Org. Acid	0.8/6.4	7.0	97.2
Quinoline	Org. Base	5.5	5.3	96.4
Hexadecane	Nonpolar Neutral	8.6	6.9	80.2
Phenanthrene	PNA	4.3	4.3	100.0
Ethylene Glycol	Polar Neutral	5.0	3.6	72.0

studies were carried out using an LKB 2091 gas chromatograph/
mass spectrometer. The samples were chromatographed on an
OV-101 capillary column (25 m, SOCT, LKB) using a linear
temperature program. The column was held at 100°C for 2
minutes after injection, and then heated to 240°C at a rate
of 8°/min. Carrier gas flow rate was 2.0 ml/min with a split
ratio of 10:1. Injector temperature was 240°C. Mass spectral
scans were taken every 2 sec scanning from 50-492 mu. Total
ion current and mass plots were generated for interpretation.

Identifications were achieved by comparison of data from
generated mass plots with the Aldermaston 8 peak index of
mass spectra. Components which have been identified thus
far are shown in Table 2.

MUTAGENICITY TESTING

Materials and Methods

Bacterial Strains. Salmonella typhimurium strains used are
the histidine deficient mutants used to detect frameshift
reverse mutations (TA98, TA1537, and TA1538) and base pair
substitutions (TA100 and TA1535). All were obtained from
Dr. Bruce N. Ames, Biochemistry Department, University of
California at Berkeley.

Preparation of Liver Homogenate S-9 Fraction. Male Sprague-
Dawley or Craig-Dawley rats induced with Aroclor 1254 are
used in preparation of liver homogenates. For each preparation a minimum of 3 rats is used. Induction involves a single intraperitoneal injection of Aroclor 1254 in corn oil,
at a dose of 500 mg/kg, 5 days prior to sacrifice (3). The
S-9 fraction was prepared and stored in 2-5 ml aliquots at
-80°C for no longer than one month.

Presentation of Test Materials. Test materials are dissolved
in spectral grade dimethyl sulfoxide (DMSO), Schwartz-Mann or
Fisher brand. Other solvents are under investigation.

Test Procedure. For routine testing the plate incorporation
and spot test methods of Ames and associates are employed (3).
A well test procedure has been devised which allows detection
of toxicity, mutagenicity, and activation requirements, thus
decreasing the total amount of material required (11).

INTERFACING CHEMICAL AND BIOLOGICAL TESTING WITH PARTICULATES

Crude samples and fractions thereof, as described earlier
in the section on chemical fractionation procedures, must be
presented to the Salmonella assay system for determination of
mutagenicity.

For the Level 1 assessment mutagenicity test, the IERL-
RTP procedures manual (6) recommends that each sample undergo
a mutagenesis test, and if possible, a toxicity test, with the
mutagenesis test to include 4 tester strains; with and without

Table 2

Preliminary Identifications of Partition Fractions

Polar Bases

Nicotine
Di-butylphthalate
Butylbenzyl phthalate
Di-2-et-hexyl phthalate
Butyl benzoate
Butyl-α-methyl benzylamine
Di-i-bu phthalate
2-et-hexyl mercaptan
2-nitro-4,6-dichlorophenol
Phenyl benzoate

Polar Neutrals

2,6-di-t-butyl-p-cresol
n-pentylthiol-n-butyrate
2,5-dimethyl-4-isopropenyl-
    2,3-hexadien-5-ol
Butylbenzylphthalate
Di-n-octyl phthalate

Nonpolar Neutrals

Diphenyl diacetylene
Butryl benzyl phthalate
2,6-di-4-butyl-p-cresol
2,4-dimethylundecane
**Fluoranthene (17)
**Pyrene (17)
Triphenylene
Di-n-octylphthalate
Perylene
*Benzpyrene (5,17)
Methyl heptadecanoate
Benzanthrone
Naphtho-(2,1-b)thianaphthene
Triphenyl phosphate, 5-hydroxyimidazole
*Benz(a)anthracene (5), naphthacene
1,2-dichloro-3,3,4,4-tetrafluorocyclobutene
Di-2-ethylhexylphthalate

Polar Acids

p-fluoroacetophenone?
1-methoxycarbonyl pyrrolizine?
2-i-pr-1,3-dioxolane?
Nonyl-ß-naphthol?
Allopeucenin?
Methylene dichloride
3-cyclohexyleicosane
2,6-dimethyl-3-heptanol
2,4-dichlorobenzaldehyde
2-formyl-3,4-dihydropyran
2,6-dimethyl-2,5-heptadien-4-one
Methyl tridecanoate
Butyl benzyl phthalate
Di-2-et-hexyl phthalate
Diamylphthalate
Methyl-15-ethylheptadecanoate
Methyl-2,4-dichlorobenzoate
Dimethylphthalate
2,3-dihydro-2-methylbenzofuran
2,7-dimethylbenzo(b)thiophene
Methyl caproate?
Methyl-9-dodecenoate
Methyl palmitate
Methyl octadecanoate
Di-n-propyl phthalate
Di-n-octyl phthalate

---

* carcinogen
**cocarcinogen

the S-9 activation system; plate incorporation tests at concentrations of 0.01, 0.1, 1, and 10 mg/plate; all in duplicate. Sample size requirement for the initial mutagenesis
testing is thus 178 mg. Repeat studies over narrower concentration ranges, taking into account positive results in the
initial test, would require additional sample. Finally,
those compounds or samples producing positive results should
undergo testing to determine dose response curves.

    With the samples received to date, sample size has been
a critical limiting factor. Typically, one to four grams of
test material were obtained with 90-95% being inorganic, thus
only 50-400 mg of material have been available for further
fractionation, identification, and mutagenesis testing. The
small amounts available after fractionation are shown in the
first column of Tables 3-5.

    Tables 3-5 show the results obtained in spot and/or pour
plate tests of fractions obtained with the fractionation
scheme initially used (Figure 1), the amount tested being
dependent on the amount available. Positive results were
obtained in most Level 1 fractions (Table 3). On further
fractionation, activity was seen with both polar acids and
polar bases (Table 4). Some activity was seen with nonpolar
organics (Table 5) but interpretation is difficult due to
small test samples. In a few cases, dose response "curves"
were attempted. It must be stressed here that where sample
size was small, a negative result cannot be interpreted as
having much significance. There is no way to determine where
we are on a dose response curve, and if only spot tests can
be done we do not know if the samples have components which
cannot be evaluated in the spot test [for example, benzo-
(a)pyrene].

Approaches to Resolving the Problem of Sample Size

    To resolve the problem of limited sample size, a number
of options have been considered:

- Develop a tighter fractionation scheme with less
  fractions, as described above.

- Develop a modified system of assay that will give
  move information per unit of sample (11). (See
  Figure 3.)

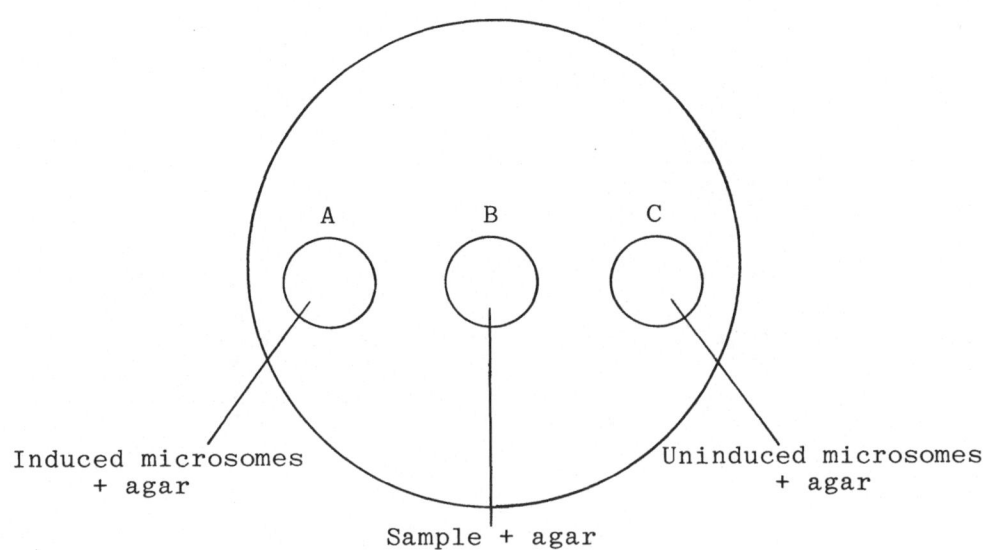

PROCEDURE:

1. Prepare base layer.
2. Cut wells into base layer.
3. Fill wells as indicated on diagram.
4. Overlay with agar containing microorganisms.

TEST RESULTS:

1. If toxic, clear zone around B. If not toxic, uniform lawn.
2. If sample requires activation, line of cells between A + B, weaker line between B + C.
3. If no activation is required, ring of cells around B.

ADVANTAGES:

1. Allows determination of induction requirement induced and uninduced, mutagenicity, and toxicity on one plate instead of separate plates for each case.
2. Do not have to worry about take-up of compound by filter or run-off of compound onto plate.

Figure 3. Use of wells instead of spot test to optimize test results.

Table 3

Ames Testing of Samples from Level 1 Fractionation
of Particulates (WV, Sample 2308)

Fraction	Description Chemical Type + Total Sample Amount	Dose µg/plate	Strain	Results, Avg. No. Colonies/Plate No Microsomes	Induced Microsomes	Mutagenic Ratio
A. SPOT TESTS						
T$_1$	Nonpolar Organics (23.6 mg)	790	98	19*	TNTC*	
		790	100	0	0	
		790	1535	2	2	
		790	1537	5	7	
		790	1538	3	6	
T$_2$	--NOT TESTED--					
T$_3$	Polar Organics (~0.78 mg)	26	98	0	0	
		26	100	0	0	
		26	1535	2	3	
		26	1537	6	8	
		26	1538	2	2	
T$_3$	Polar Organics (7.5 mg)	250	98	12*	6	
		250	100	0	0	
		250	1535	5	0	
		250	1537	2	3	
		250	1538	4	0	
T$_{3a}$	Polar Organics (12.0 mg) (portion insoluble in methanol)	400	98	0	0	
		400	100	TNTC**	0	
		400	1535	2	2	
		400	1537	5	7	
		400	1538	1	1	
T$_3$	Polar Organics (12.9 mg)	430	98	25*	15	
		430	100	0	0	
		430	1535	3	6	
		430	1537	9	6	
		430	1538	28*	16*	
B. POUR PLATE TESTS OF POSITIVE FRACTIONS						
T$_1$	Nonpolar Organics (23.6 mg)	790	98	89*		1.9
T$_3$	Polar Oragnics (7.5 mg)	250	98	63		1.3
T$_{3a}$	Polar Organics (12.0 mg) (portion insoluble in methanol)	400	1535	43		1.3
			100	46		0.9
T$_3$	Polar Organics (12.9 mg)	430	98	106*		2.3

* Postive Results.

**TNTC = Too numerous to count.

Table 4

Ames Testing of Samples from Level 2 Fractionation
(Polar Organics) of Particulates
(WV, Sample 2308)

Fraction	Description Chemical Type + Total Sample Amount	Dose µg/plate	Strain	Results, Avg. No. Colonies/Plate		Mutagenic Ratio
				No Microsomes	Induced Microsomes	
A. SPOT TESTS						
T_4	Polar acids and neutrals (1.0 mg)	33	98	0	0	
			100	0	0	
			1535	7	2	
			1537	5	4	
			1538	3	2	
T_{5a}**	Polar bases (0.9 mg)	30	98	0	0	
			100	0	0	
			1535	1	1	
			1537	7	3	
			1538	1	1	
T_{5b}**	Polar bases (43.2 mg)	1440	98	35*	28*	
			100	0	0	
			1535	2	5	
			1537	16*	7	
			1538	34*	5	
T_6	Polar neutrals	70	98	5	0	
			100	0	0	
			1535	3	2	
			1537	4	5	
			1538	2	3	
T_{7a}	Polar acids (1.5 mg)	50	98	0	0	
			100	0	0	
			1535	2	3	
			1537	5	6	
			1538	3	2	
T_{7b}	Polar acids (9.0 mg)	300	98	TNTC*	TNTC*	
			100	TNTC*	TNTC*	
			1535	2	TNTC*	
			1537	TNTC*	TNTC*	
			1538	TNTC*	TNTC*	
B. POUR PLATE TESTS OF POSITIVE FRACTIONS						
T_{5b}**	Polar bases (43.2 mg)	1440	98	154*		3.3
T_{7b}	Polar acids (9.0 mg)	300	1535	36		1.0
			100	47		0.7
			98	70		1.7

* Positive results.

  TNTC = Too numerous to count.

**5a and 5b are replicates, differing only in amount of starting material.

Table 5

Ames Testing of Samples from Level 3 Fractionation
(Nonpolar Organics) of Particulates
(WV, Sample 2308)

Fraction	Description Chemical Type + Total Sample Amount	Dose µg/plate	Strain	Results, Avg. No. Colonies/Plate No Microsomes	Induced Microsomes	Mutagenic Ratio
A.	**SPOT TESTS**					
T_8	Nonpolar acids and neutrals (in DMSO)		98	0	0	
			100	0	0	
			1535	2	TNTC*	
			1537	9	6	
			1538	0	0	
T_8	Nonpolar acids and neutrals (0.2 mg)	6	98	0	0	
			100	0	0	
			1535	3	3	
			1537	3	6	
			1538	1	2	
T_9	Nonpolar bases (0.6 mg)	20	98	0	0	
			100	0	0	
			1535	3	2	
			1537	3	4	
			1538	0	0	
T_{10}	Nonpolar neutrals (0.4 mg)	13	98	0	0	
			100	0	0	
			1535	3	5	
			1537	9	4	
			1538	3	2	
T_{11}	Nonpolar acids (17.4 mg)	580	98	0	0	
			100	0	0	
			1535	3	3	
			1537	4	4	
			1538	0	1	
T_{12}	Paraffins (1.7 mg)	56	98	0	0	
			100	0	6	
			1535	4	4	
			1537	5	5	
			1538	2	1	
T_{13}	Aromatics (1.3 mg)	43	98	0	0	
			100	0	TNTC*	
			1535	3	2	
			1537	4	6	
			1538	2	4	
B.	**POUR PLATE TESTS OF POSITIVE FRACTIONS**					
T_8	Nonpolar acids and neutrals (in DMSO)	--	1535		32	1.0
			100		47	0.9
T_{13}	Aromatics (1.3 mg)	43	1535		18	0.7
			100		35	0.5

* Positive results

**TNTC = Too numerous to count.

- Develop a scheme of priorities to be used with amounts of sample which do not allow complete testing, in order to maximize the amount of information that can be derived (Table 6).

- Request a larger sample.

Although the last option would appear to be the simplest and most obvious approach, technical difficulties are encountered in operation of the sampling system if longer sampling periods are used. Therefore, the other options have been adopted.

Interfacting Chemical and Biological Testing with Vapors and Gases

To date, only preliminary experiments have been conducted on approaches to quantitatively collect and present to the bioassay vapors and gases sorbed to Tenax and carbon. The collection system includes a Tenax GC cartridge backed up by a carbon cartridge to collect highly volatile materials that break through the Tenax. In the laboratory, Tenax-sorbed compounds may be transferred onto carbon with the assumption that the carbon can then be incorporated into the Ames assay system along with sufficient solvent to release the vapors into the plate.

Initial tests with model compounds, in which we are not limited by the availability of sample, indicate certain problems:

- Failure of the carbon to release some sorbed compounds at concentrations of solvent tolerated in the Ames test system.

- Where release is possible, assurance that the desorbed materials have the opportunity to contact the organisms before loss to the headspace in the plate.

Also, we must assure that with complex environmental samples, our method of presentation does not selectively release some of the sorbed materials, essentially "narrowing the window," by what in effect amounts to an additional fractioning or partitioning.

Table 6

Developing Priorities for Mutagenesis Procedures

To get the maximum information with limited quantities of test fractions, the following order of priorities is suggested, the most important being listed first:

(1)   Minimum of four strains with the well test and with induced microsomes.

(2)   Minimum of two doses (100 and 500 µg/plate), four strains, well test with induced microsomes.

(3)   As in 1, but with 5 strains.

(4)   As in 1, but with 5 strains and non-induced as well as induced microsomes.

(5)   As in 4, but with 2 doses.

(6)   Pour plate tests with >1 strain using induced microsomes. If plate is positive, repeat with no microsomes.

(7)   As in 6, but with 2 or more test sample concentrations.

Strains in order of decreasing importance are:

98, 1535, 1537, 1538, 100.

In all cases tests will be conducted in duplicate. Depending on the amount of sample available, as many as possible of the tests will be conducted.

Only one concentration of the sample fraction will be made for these tests. To obtain lower test concentrations, less sample volume per plate will be used. Pour plate tests will be conducted at concentrations twice those of the well tests, i.e., at 200 and 1000 µg/plate.

It might appear immediately obvious that we should consider suspension tests. Liquid suspension tests have been conventionally performed under conditions where sample size has not been limiting. Previous investigators have tested gaseous compounds for mutagenicity in both plate incorporation and liquid suspension tests, as recently reviewed by Malaveille and co-workers (14) in France. They examined the failure of liquid suspension tests to detect mutagenicity of certain compounds which were easily detected in soft agar plate incorporation tests. In tests with vinylidine chloride, they found that microsomes maintained their viability for up to 9 hr in soft agar, contrasted to about an hour in liquid suspension. Hence, for mass screening it would appear that a plate incorporation test would be preferable. Malaveille et al. (14) has devised a method for exposing soft agar plates to a test gas using a desiccator. However, this procedure involves a substantial quantity 2% (v/v), of the test material. Methods of this kind and novel techniques requiring small quantities of sample are currently under investigation.

SUMMARY

The goal of the research described in this paper is to adapt the Ames assay to mass screening, qualitatively and quantitatively, of both particulate and volatile components of complex air samples.

At present, problems are encountered due to

● Technical limitations on the amount of sample which can be obtained initially, thus limiting the amount of sample available for biotesting and pushing the assay to its limits of sensitivity.

● Difficulties in presenting the "entire sample" to the assay organisms without loss of some components or introduction of artifacts.

These problems have been addressed

● By development of a better chemical fractionation scheme, producing a smaller number of fractions with more material in each fraction.

- By developing a list of priorities in mutagenesis testing (strains, activation requirements, etc.) so that the maximum information can be obtained from a small sample.

- By developing a well test modification of the Ames test which allows testing of a number of parameters on a single plate.

The combined chemical fractionation/mutagenesis screening approach allows rapid identification of those fractions with mutagenic activity, i.e., those that should receive priority for further chemical identification.

REFERENCES

1. Ames BN: A bacterial system for detecting mutagens and carcinogens. In: Mutagenic Effects of Environmental Contaminants (Sutton HE, Harris JF, eds.). New York, Academic Press, 1972

2. Ames BN, Durston WE, Yamasaki E, Lee FD: Carcinogens are mutagens: A simple test system combining liver homogenates for activation and bacteria for detection. Proc Natl Acad Sci USA 70:2281-2285, 1973

3. Ames BN, McCann J, Yamasaki E: Methods for detecting carcinogens and mutagens with the Salmonella mammalian microsome mutagenicity test. Mutat Res 31:347-364, 1974

4. Brown VM: The prediction of the acute toxicity of river waters to fish. Proc Fourth Brit Coarse Fish Conf, Liverpool Univ, 1969

5. Dipple A: Polynuclear aromatic carcinogens. In: Chemical Carcinogenesis, ACS Monograph 173 (Searle CE, ed.). Washington, DC, American Chemical Society, 1976, pp 245-314.

6. Duke KM, Davis ME, Dennis AJ: IERL-RTP Procedures Manual: Level 1 Environmental Assessment Biological Tests for Pilot Studies. EPA-600/7-77-043. Washington, DC, US Environmental Protection Agency, 1977

7. Flessel CP: Mutagenic activity of particulate matter in California hi-vol samples. Berkeley, CA, Third International Symp Air Monitoring Quality Assurance, May 18-19, 1977

8. Hammersma JW, Reynolds SL, Maddalone RF: IERL-RTP Proce-
    dures Manual: Level 1 Environmental Assessment, EPA-600/
    2-76-160a. Washington, DC, US Govt Printing Office, 1976

9. Hill HH, Chan KW, Karasek FW: Extraction of organic
    compounds from airborne particulate matter. J Chromat
    131:245-252, 1977.

10. Hoffman D, Wynder EL: Environmental respiratory carcino-
    genesis. In: Chemical Carcinogenesis, ACS Monograph
    173 (Searle CE, ed.). Washington, DC, American Chemical
    Society, 1976, pp 324-365

11. Hughes TJ, Little L, Pellizzari E, Sparacino C, Claxton
    L, Waters M: Application of agar diffusion wells for
    microbial mutagenesis testing of air pollutants.
    Presented at Environmental Mutagen Society meeting,
    poster session, March 11, 1978

12. McCann J, Ames BN: Detection of carcinogens as mutagens
    in the Salmonella/microsome test: assay of 300 chemicals,
    part II. Proc Natl Acad Sci USA 73:950-954, 1976

13. McCann J, Choi E, Yamasaki E, Ames BN: Detection of
    carcinogens as mutagens in the Salmonella/microsome
    test: assay of 300 chemicals. Proc Natl Acad Sci USA
    72:5135-5139, 1975

14. Malaveille C, Planche G, Bartsch H: Factors for effi-
    ciency of the Salmonella/microsome mutagenicity assay.
    Chem-Biol Interactions 17:129-136, 1977

15. Pitts JN Jr, Grosjean D, Mischke TM, Simon VF, Poole D:
    Mutagenic activity of airborne particulate organic pol-
    lutants. Toxicol Letters 1:65-70, 1977

16. Talcott R, Wei E: Brief communication: airborne muta-
    gens bioassayed in Salmonella typhimurium. J Natl Can-
    cer Inst 58:499-451, 1977

17. Van Duuren BL: Tumor-promoting and cocarcinogenic
    agents in chemical carcinogenesis. In: Chemical Car-
    cinogenesis, ACS Monograph 173 (Searle CE, ed.). Wash-
    ington, DC, American Chemical Society, 1976, pp 24-51

18. Weisburger JH: Bioassays and tests for chemical carcin-
    ogens. In: Chemical Carcinogenesis, ACS Monograph 173
    (Searle CE, ed.). Washington, DC, American Chemical
    Society, 1976, pp 1-23

CHEMICAL AND MICROBIOLOGICAL STUDIES OF MUTAGENIC POLLUTANTS IN REAL AND SIMULATED ATMOSPHERES

James N. Pitts, Jr., Karel A. Van
Cauwenberghe, Daniel Grosjean,
Joachim P. Schmid, and Dennis R. Fitz
Department of Chemistry and Statewide
Air Pollution Research Center
University of California
Riverside, California

William L. Belser, Jr., Gregory B. Knudson,
and Paul M. Hynds
Department of Biology and Statewide
Air Pollution Research Center
University of California
Riverside, California

INTRODUCTION

In the early 1940's the organic extracts of ambient particulate matter (POM) collected from urban air in the United
States were found to be carcinogenic when administered subcutaneously to mice (1,2). Subsequently, this effect was
also observed in experimental animals injected with extracts
of ambient POM collected from Los Angeles photochemical smog
(3) and in seven other U.S. cities (4). Similar results now
have been found with ambient samples collected in various
urban centers throughout the world. This carcinogenicity is
customarily attributed to certain polycyclic aromatic hydrocarbons (PAH), e.g. benzo(a)pyrene (BaP), benz(a)anthracene
and aza-arenes such as benzocarbazoles in the neutral fraction of the organic particulates and benzacridines and
dibenzacridines in the basic fraction.

Several researchers, however, have found significant
discrepancies between the observed biological activity of
POM and the amounts of carcinogenic polycyclics determined
to be present. This is true not only of samples of ambient
particulate matter but also from the exhaust from spark-
ignition engines in light duty motor vehicles (3-10).

Thus, for example, Gordon and co-workers reported that, with
airborne particles collected in the Los Angeles area, the
benzene extract had 100 to 1000 times the cell transformation

activity that could be attributed to its known BaP content.
Furthermore, the methanol extract, while containing only
about 1/30 of the BaP in the total sample, showed an activity
comparable to the benzene extract (10).

Mohr et al. recently showed that auto exhaust had a pronounced carcinogenic effect on the lungs of Syrian golden
hamsters (100% rate of multiple pulmonary tumors). The
authors point out that "considering the relatively low total
dose of BaP contained in the condensate, this pronounced neoplastic response cannot be explained alone by the effects of
this well known carcinogenic hydrocarbon" (8).

We, among others, have felt for some time the need to
try to identify the chemical species responsible for this
"excess carcinogenicity" if one is to obtain a reliable estimate of the health impact on man of POM from whatever source
--ambient air, auto or diesel exhaust, or fly ash from coal
fired power plants. In short, one must fully characterize
the dose parameters in dose-response curves. The problem
is, for atmospheric particulates, characterization of the
"dose" requires a detailed knowledge of the physical and
chemical nature of the species present in POM at the site
of impact upon the biological target. Complications arise
because primary organic pollutants may, and do, undergo a
variety of chemical transformations in the presence of light,
oxygen, water, and a variety of copollutants. Thus, ambient
POM from polluted urban atmospheres is a highly complex mixture consisting of hundreds, probably thousands, of different
compounds.

Because of this complexity and also because of the costs
and time involved in animal tests for suspected carcinogens,
to date results from experiments directed to identify the
chemical structures of the compounds responsible for this
"excess carcinogenicity" have been relatively limited.
Therefore, we were most interested in applying to this problem the relatively inexpensive microbiological assay for
mutagenic activity developed by Ames and coworkers (11-13)
for fast screening of compounds for potential carcinogenic
activity. This assay is a reverse mutation system employing
histidine-requiring mutants of the bacterium Salmonella
typhimurium. It is now generally recognized as a useful,
though by no means exclusive, screening test for chemical
mutagens in complex environmental samples.

Specifically, we have been using the Ames test to screen
POM samples collected from real and simulated atmospheres for

mutagenic activity. Additionally, we have been using results
of this test to provide microbiological clues to the chemical
nature of the compounds responsible for the "excess" carcinogenicity, prior to their final characterization. Such clues
include:

- The type of mutation induced. Thus, many frameshift mutagens detected by strains TA1537, TA1538,
  and TA98 are planar molecules, capable of interca-
  lation between bases of the DNA strand (e.g., 9-
  aminoacridine); on the other hand, many base pair
  substitution mutagens, detected by strain TA1535,
  are alkylating agents (e.g., β-propiolactone).

- The distinction between compounds which are
  directly mutagenic and those which are promutagens
  requiring metabolic activation. Thus, BaP is a
  promutagen requiring treatment with S-9 liver
  homogenate, whereas its metabolite, 6-hydroxy-BaP,
  does not.

- The position of substitution in the compound will
  have a pronounced effect on the observed mutagenicity. For example, of the 12 hydroxy-isomers of
  BaP, only five phenols are directly acting frameshift mutagens. The 6- and 12-OH-BaP have strong
  activities; the 1-, 3-, and 7-OH-BaP are also mutagenic, but much weaker. The remaining seven isomers are nonmutagenic (14-16).

Soon after initiating our combined chemical-microbio-
logical experiments on POM (17,18), the need for standardization of the Ames test, including the number of cells per
plate, plate volume, concentrations of rat liver S-9 homogenate, etc., became increasingly apparent. Therefore, along
with the development of our HPLC separation and GC-MS identification procedures, we conducted a series of experiments
designed to give us a better understanding of the effects
of certain variables on its reproducibility. Thus, in this
paper we shall first deal with the chemical aspects of the
problem and then discuss some factors involved in the intra-
and interlaboratory standardization of this microbiological
assay.

MUTAGENIC ACTIVITY OF AMBIENT PARTICULATE MATTER

In 1975 we first reported the mutagenicity of the organic extracts from ambient particles collected at several sites in the Los Angeles Basin, as detected by the Ames assay system (19). This phenomenon has now been reported in studies at Ohmura and Fukuoka, Japan (20); Kobe, Japan (21); Buffalo, New York, and Berkeley, California (22); New York City, New York (23); and Chicago, Illinois (24).

More recently, we collected samples of airborne particulates at 11 urban sites in California's South Coast Air Basin (25,26). Using strains TA1537, TA1538, and TA98, all samples exhibited direct frameshift-type mutagenic activity, i.e., they did not require metabolic activation. Addition of the microsomal activation system (S-9 solution) did not significantly increase the activity of the majority of the samples tested. No activity was observed in any of the assays with strain TA1535, which, as noted earlier, is reverted by base pair substitution mutations.

Finally, in a size-resolved sample collected in downtown Los Angeles using a Sierra Hi-Vol cascade impactor, all mutagenic activity was found to be associated with the particles of diameter 1.1 micron or less. This is consistent with the well documented occurrence of PAH such as BaP in the respirable range of ambient particulates (27-30).

From these data we concluded that urban POM must contain direct mutagens in addition to carcinogenic PAH, such as BaP, which require metabolic activation. This is consistent with the numerous observations of "excess" carcinogenicity in animals or in cell transformation activity, as discussed above, and with the low average concentrations of BaP measured in the Los Angeles Basin (31).

We then formulated the hypothesis that some of these direct mutagens in ambient particulates might be formed in the atmospheric transformations of particulate BaP and other PAH by gaseous species such as ozone, nitrogen dioxide, peroxyacetyl nitrate (PAN), singlet molecular oxygen ($O_2 \, ^1\Delta g$), and free radicals present in photochemical smog such as OH and HO_2 (17,18,26).

Support for this idea can be found in several reports of the carcinogenic activity of polar fractions of organic particulates (4,5,10,32,33), of products of ozonized gasolines (34), of products of oxidation of aliphatic hydrocarbons (35),

and of the toxicity of the photooxidation products of a commercial fuel oil (36). Furthermore, it is known that some
of the oxygenated metabolites formed from BaP in mammalian
cells are directly mutagenic (14-16) and given the oxidizing
potential of photochemical smog, it seemed reasonable that
at least some analogous transformations might occur in polluted air.

     One complicating factor we faced initially was a direct
conflict in the literature over the chemical reactivity of
PAH in ambient particulates. Thus, two references stated
that "they are chemically inert and thus are removed from
the air only by rain or the slow sedimentation of the particulate" (37,38). However, we found this position difficult to reconcile with earlier literature data on the photochemical transformations of PAH adsorbed on a variety of
support materials such as filters, silica gel, and carbon
(soot) particles (3,39-43). These data show that certain
key PAH (e.g., BaP) can be quite reactive. As discussed
below, our experiments fully support the latter observations.

FORMATION OF MUTAGENIC POLLUTANTS FROM PAH IN REAL AND
SIMULATED ATMOSPHERES

     In order to test our hypothesis, experiments were
carried out in which several PAH, deposited on glass fiber
filters, were exposed to gaseous pollutants both in real and
in simulated atmospheres. We shall briefly summarize the
results; details are presented in two papers (17,18).

Exposure of Benzo(a)pyrene to Ambient Photochemical Smog

     In this series of experiments, BaP was exposed to the
gases present in ambient photochemical smog and the mutagenicity of the resulting products determined. The experimental set-up consisted of two conventional washed and fired
Gelman A/E glass fiber filters mounted in series in a high
volume sampler. The upstream filter, a "blank," collected
all ambient particulates and allowed only the gaseous pollutants present in photochemical smog to pass through to the
second filter. The latter was coated with BaP (~2 mg) that
could interact with the gaseous pollutants.

     After several days of exposure, the BaP coated filters
were extracted by ultrasonication and the concentrated
organic extracts tested with Salmonella strains TA98 and

TA100. They now showed direct mutagenicity indicating the
formation of products from the promutagen BaP.

In another experiment, the second BaP coated filter
was extracted and further separated into fractions by TLC on
silica gel plates. Each band was recovered in methanol and
analyzed by methane chemical ionization mass spectrometry.
Some of the tentatively identified products were, in order
of decreasing polarity: BaP-dihydrodiol(s) (mol. wt. 286),
BaP-diphenol(s) (mol. wt. 284), BaP-phenol(s) (mol. wt. 268),
and BaP-quinones (mol. wt. 282) (17,18).

Exposure of Benzo(a)pyrene to O_3 and Peroxyacetyl Nitrate

In order to obtain specific information about the reactions with BaP of certain key single pollutants present in
photochemical smog, another series of experiments was carried out under conditions similar to those used with ambient
smog. In these, glass fiber filters coated as above with
BaP were exposed in the dark to clean, particle-free air
containing 11 ppm O_3 (exposure time 24 hours at a flow rate
of 3 cfm), or 1.1 ppm PAN (16 hours, 3 cfm). Control runs
with BaP exposed to pure air (24 hours, 3 cfm) and with
blank filters exposed to O_3 or PAN were also included. No
mutagenicity was observed in any of the control runs, but
all the other exposures produced direct mutagenicity.

After exposure, the products and unreacted BaP were
separated by TLC, and the major bands analyzed by mass spectrometry. They were also tested separately for mutagenic
activity, both with and without metabolic activation. For
comparison purposes, a sample of BaP was also incubated for
30 minutes at 37°C with the liver S-9 homogenate solution
and the metabolites formed in this microsomal activation
system analyzed and tested for mutagenic activity.

We found that BaP reacted readily with these 1-10 ppm
levels of O_3 and PAN in air to form a variety of oxygenated
products. As expected, the TLC bands containing the unreacted BaP were not directly active and required metabolic activation. The band containing the BaP-quinones was complex
and contained, in addition to the inactive quinones (16), a
directly active compound of molecular weight 284, to date
still unidentified.

Products of the treatment of BaP with the S-9 mix
appeared as a series of TLC bands, one of which was complex.

This complex band was also seen with ambient smog but not with O_3 or PAN. It contained directly active mutagens. The R_f-values and the molecular weight of its components (mol. wt. 268) were consistent with those of isomers of hydroxy-benzo(a)pyrene. Positive identification will require comparison with the proper reference compounds.

Exposures of Benzo(a)pyrene and Perylene to NO_2

Exposure of BaP to 1.3 ppm of NO_2 (24 hours, 1 cfm), containing traces of nitric acid ($\sim10^2$ ppb) resulted in the appearance of only one major TLC band. This contained a directly active mutagen whose R_f and molecular weight of 297 were consistent with the nitrobenzo(a)pyrene (nitro-BaP) structure.

This band containing nitro-BaP was further resolved into two bands, one yellow and one orange, using TLC with toluene as the solvent. Comparison of the mass spectra and ultraviolet-visible spectra with those of authentic samples synthesized according to Dewar (44), allowed us to assign the structure 6-nitro-BaP to the component present in the yellow TLC band; the orange TLC band was a mixture of the 1-nitro and 3-nitro isomers.

In more recent studies, we have shown that the nitration of BaP by ppm levels of NO_2 in air is acid catalyzed by ppb levels of HNO_3. Furthermore, we obtained yields of $\sim18\%$ of nitro-derivatives from eight-hour exposures of BaP to only 0.25 ppm NO_2 (containing ~3 ppb HNO_3) in air. The value of 0.25 ppm is the air quality standard for NO_2 (one hour average) in California; during the late fall and winter months it is commonly exceeded in downtown Los Angeles, Pasadena and the coastal regions of the South Coast Air Basin.

Since exposure of BaP, a known carcinogen and activatable mutagen, to ppm and sub-ppm levels of NO_2 resulted in the formation of directly mutagenic nitro-derivatives (see discussion below), it seemed interesting to see if, under the same conditions, similar products could also be formed from a "noncarcinogenic" PAH, perylene (39,45). This isomer of BaP is also present in ambient POM and in POM emissions from a variety of combustion sources.

Thus, perylene, deposited on glass fiber filters as with with BaP, was exposed to 1 ppm of NO_2 for 24 hours at a flow

rate of 1 cfm. The major resulting TLC band (brick-red
color on silica gel) consisted of 3-nitro-perylene, identified by its mass spectrum and by comparison of its UV spectrum with literature data (46,47).

Figure 1 presents the mutagen test data on the parent
BaP, the filter derived nitro-BaP isomers and authentic
nitro-BaP isomers synthesized by the Dewar method. Panel A
shows a standard dose-response curve for BaP. Without activation we observed ~4 revertants/nmole; with S-9 activation
we observed ~120 revertants/nmole. The latter figure is in
good agreement with data of Ames for BaP (48).

Panel B shows results comparing the polluted air-filter
generated 6-nitro-BaP with the authentic laboratory prepared
6-nitro-BaP. In the absence of metabolic activation, this
isomer is more than six times as active as the parent BaP,
giving values of ~20 revertants/nmole for both the filter
generated and authentic 6-nitro-BaP. With metabolic activation, this isomer is more than three times as mutagenic as
the parent BaP, giving values of ~390 revertants/nmole and
~420 revertants/nmole for the filter generated and authentic
6-nitro-BaP, respectively.

Panel C shows the test data for the mixture of 1- and
3-nitro-BaP isomers. Note that the range of concentrations
used is lower than in the other tests, since we had only a
limited quantity of the filter generated material available.
Clearly this mixture of isomers is highly active. Thus, in
the absence of metabolic activation, the 1- and 3-isomer mix
yields approximately 40 times as many revertants as BaP
itself and approximately five times as many revertants as
the 6-nitro-BaP. The values are ~140 revertants/nmole and
~210 revertants/nmole for the filter generated and authentic
1- and 3-nitro-BaP mix, respectively. Activation of the 1-
and 3-isomers with S-9 gave values of ~3500 rev/nmole and
~5200 rev/nmole for the filter generated and authentic samples, respectively.

The shape of the dose response curve in this case is a
puzzle. Maximum mutagenesis occurs at very low doses, followed by some inhibition. The curve begins to rise again
at higher concentrations. At the moment we have no explanation for this, although our intuition is that this may be a
function of the relative abundance of the two isomers in the
mixture. Resolution of this issue will depend upon separating the 1- and 3-isomers and subjecting them to individual
tests.

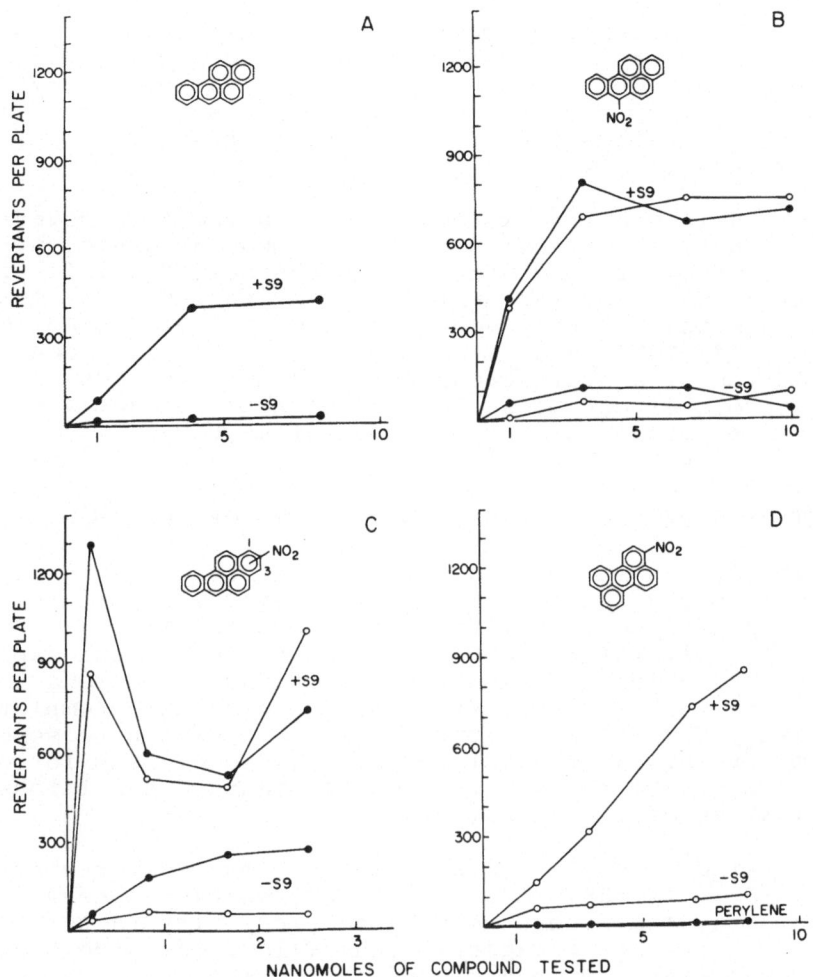

Figure 1. Dose-response curves obtained for BaP (A), 6-nitro-BaP (B), 1-nitro- and 3-nitro-BaP (C), and 3-nitro-perylene (D) in the Ames mutagenicity test with the most sensitive strain, TA100 (A,B), TA98 (C), and TA1538 (D) with (+S-9) and without (−S-9) metabolic activation. Each point represents the mean of at least triplicate plates. Subtracted spontaneous revertant backgrounds were: TA100:118–130 (−S-9), 130-152 (+S-9); TA98:24 (−S-9), 44 (+S-9); TA1538:12 (−S-9), 41 (+S-9). Also shown (full circles) are the corresponding curves for perylene (D) and for the nitro-BaP isomers (B,C) synthesized according to Dewar (44).

These data clearly demonstrate that nitration of the
promutagen BaP produces a direct mutagen and additionally
increases its mutagenic potency on S-9 activation. Furthermore, and not unexpectedly, the position of substitution in
the compound (6 versus 1 or 3) has a pronounced effect on
its mutagenic activity.

As shown in Panel D of Figure 1, with the Ames reversion assay, perylene itself proved to be nonmutagenic, at
low levels, either in the presence or absence of rat liver
homogenate.* However, the 3-nitro-isomer proved to be mutagenic in this test, giving ~40 rev/nmole without metabolic
activation and ~100 rev/nmole with metabolic activiation.
Thus, addition of a nitro-group in the 3-position converts
perylene into a direct mutagen. Furthermore, on activation
the 3-nitro-perylene is almost as potent as BaP.

SOME FACTORS AFFECTING THE REPRODUCIBILITY OF THE AMES TEST

Obviously, an increasing number of laboratories are
beginning to apply the Ames Salmonella reversion test to
mutagenicity studies of environmental samples (11-13). In
surveying the literature, and as a consequence of our experience in conducting the Ames test, we found there is a
pressing need for standardization of the procedures employed
in this assay for mutagenicity. Thus we present here some
data from our laboratory that may provide some basis for our
plea for the establishment of a set of standard conditions
for application of the Ames test.

In working with the standard set of Ames tester strains,
TA1535, TA1537, TA1538, TA98, and TA100, we have observed,
as have others, that there is strain specificity in the
response to different mutagens. Thus TA1537 may give a high
number of revertants with a given sample, while TA1538 and
TA98 are low. Hence, in working with environmentally derived
samples of unknown composition, we feel it is essential to
include all five tester strains in the preliminary screen.
In addition, although TA100 is widely recognized to be
strongly responsive to many frameshift mutagens, this tester
strain is often reported as specifically detecting base pair
substitution mutagens. Within its limitations TA100 is a
versatile strain, the fact of which we should not lose sight.

---

*Perylene has since been reported to be mutagenic in a for-
 ward mutation assay (49).

To satisfy ourselves on the reproducibility of the Ames test, we have examined the effect of growth media, cell density, agar plate volume, and S-9 concentration with the following results.

Effect of Growth

We have grown our tester cultures in a variety of media, including L-broth, nutrient broth, and Vogel and Bonner enriched medium with glucose as a carbon source. Except for slight variations in cell numbers, there is no observable effect in response to mutagens with different growth regimes.

Effect of Cell Density

We have conducted experiments on the effect of cell density on mutation frequency with interesting results. Using four of the Ames tester strains, we tested a single frame-shift mutagen (hycanthone), and for the fifth tester strain (TA1535) we used N-methyl-N'-nitro-N-nitrosoguanidine (NTG). We grew overnight cultures of each of the five strains to approximately the same turbidities. These were assayed by dilution and plate count. Each culture was tested undiluted and at 1:2, 1:4, 1:8, and 1:16 dilutions. The undiluted culture was tested at 0.2 and 0.1 ml per plate and the diluted culture was added at 0.1 ml per plate. Hycanthone was added at 15, 30, and 50 µg/plate.

The highest numbers of revertants were obtained at 2 x 10^8 cells per plate, with slightly fewer at 1 x 10^8 cells per plate. Over a range from 5 x 10^7 cells down to 6 x 10^6 cells per plate, the numbers of revertants were about 40% lower than the value at 1 x 10^8 cells per plate, but with no significant variations within this range.

Competition for the trace of histidine in the top agar is a prime factor in explaining these results. With only enough for two or three rounds of replication, adding too many cells will lead to the rapid exhaustion of histidine and hence less opportunity for mutation to occur. Similarly, lower numbers of cells allow for more background growth and can lead to an incorrect assessment of a compound's activity. Based upon these observations, it appears that 1 x 10^8 cells per plate is probably optimal. This is consistent with the observation of Rosenkranz (50).

Since the generation time of these organisms is less than 30 minutes in rich medium and inoculation sizes vary, we strongly recommend that the titer of overnight cultures be adjusted by optical density measurements to about 1 x 10^9 cells/ml, and that 0.1 ml inocula delivered to the test plates will then be close to the optimum cell density.

Effect of Agar Volume

Another variable in the experimental protocol is agar volume in the plates. This was also mentioned in Dr. Rosenkranz' presentation (50). In a sample of unknown composition, some compounds are certain to be water soluble. Diffusion caused by this solubility into different volumes of agar in the base layer could provide considerable variability in the dose of mutagen seen by the cells in the top layer.

Until recently we had been hand pouring our base agar layers. In preparation of large numbers of plates, considerable variation in agar volume occurs. Therefore, we conducted a controlled experiment in which we checked a random selection of hand-poured plates against plates poured by the Manostat automatic plate pourer. The machine was set to pour plates containing 15, 20, 25, or 30 ml agar per plate.

We used two mutagens, hycanthone and 2-aminofluorene, each at a single dose. For each mutagen we plated 30 replicates on the hand-poured plates and 30 replicates for each of the four volumes of the machine-poured plates. The tester strain was TA98, 2-aminofluorene was used at 0.5 µg/plate and hycanthone was used at 50 µg/plate. In Table 1 are presented the average number of revertants per plate.

From these data we conclude that the volume of the agar layer can affect the dose of the mutagen seen by the cells. At higher constant volumes, the number of revertants per plate falls off.

These data were also subjected to statistical analysis and the standard deviation and variance on the 30 replicates of each sample from the hycanthone data calculated. The variances of the constant volume samples were pooled and compared to the variance in the hand-poured plate sample using the F test. The variation among the hand-poured plates was significant at the $p = 0.05$ level. Thus, using variable volume hand-poured plates, in the smaller number of replicates

Table 1

Volume of Agar Per Plate

	15 ml	20 ml	25 ml	30 ml	Hand Poured
Hycanthone (Avg. rev/plate)*	270	223	189	188	167
2-aminofluorene (Avg. rev/plate)	666	687	531	490	504

*Average of 30 replicates in each case.

normally used (three to five plates per sample) in quanti-tative tests, could introduce substantial error into the results.

Based upon these studies, we recommend that when pos-sible constant volume plates containing 20 ml base agar be used. Although the average number of revertants observed on the 15 ml plates was higher in the hycanthone experiment, we recommend use of 20 ml plates for other reasons. Specif-ically, some of the 15 ml plates appeared to be drying out during the 48-hour incubation period, and we see this as introducing another potential problem in quantitation of results.

S-9 Suppression and Optimal Concentrations

In a number of our experiments, we observed suppression of reversion frequency when S-9 was added; therefore we set out to examine the effect of S-9 concentration on reversion frequency. Two different activatable mutagens were used, 2-aminofluorene and BaP. Each mutagen was tested at three concentrations and S-9 was tested at four concentrations.

In this experiment, we used TA100 as the tester strain. The S-9 liver homogenate was prepared according to Ames et al. (12). Sprague-Dawley rats were given a single i.p. injection of Aroclor 1254 at a dose of 500 mg/kg. The rats were starved for 12 hours and sacrificed on the fifth day past injection. The results of this experiment are shown in Figures 2 and 3.

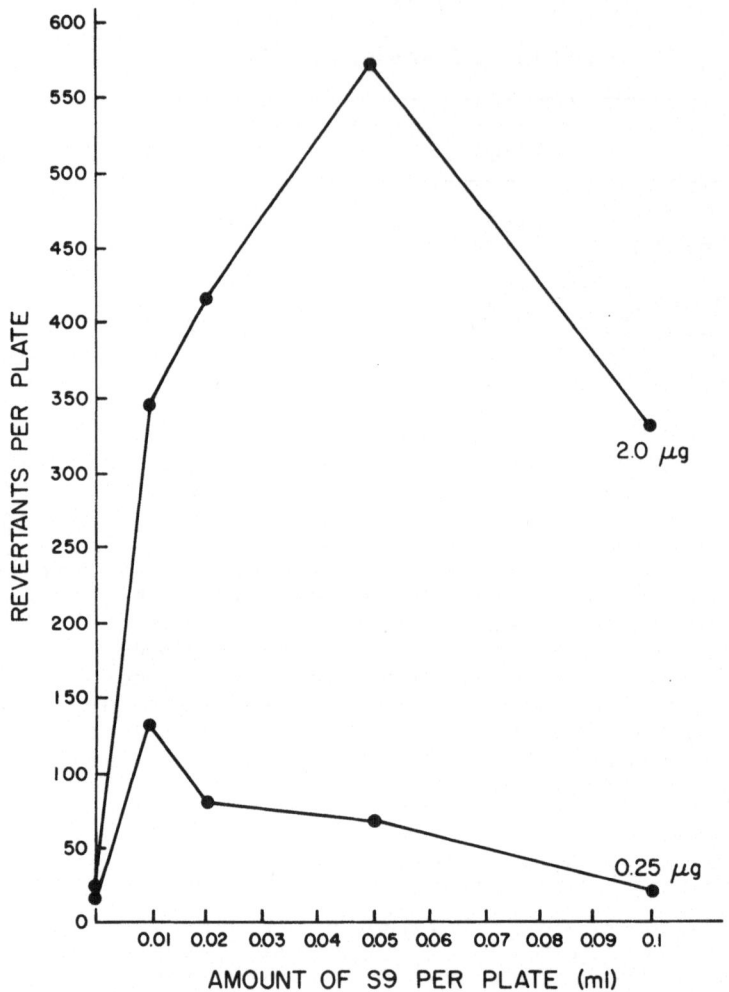

Figure 2. Dose response curve of S-9 for benzo(a)pyrene
with <u>Salmonella</u> <u>typhimurium</u> strain TA100.

Examination of the curve for the low concentration of
BaP shows that optimum activation occurs at 0.01 ml of the
homogenate, and that increasing concentrations of S-9 suppress the appearance of revertants. At the highest concentration of BaP optimum activation occurs at 0.05 ml S-9 per
plate with suppression at higher concentrations of S-9

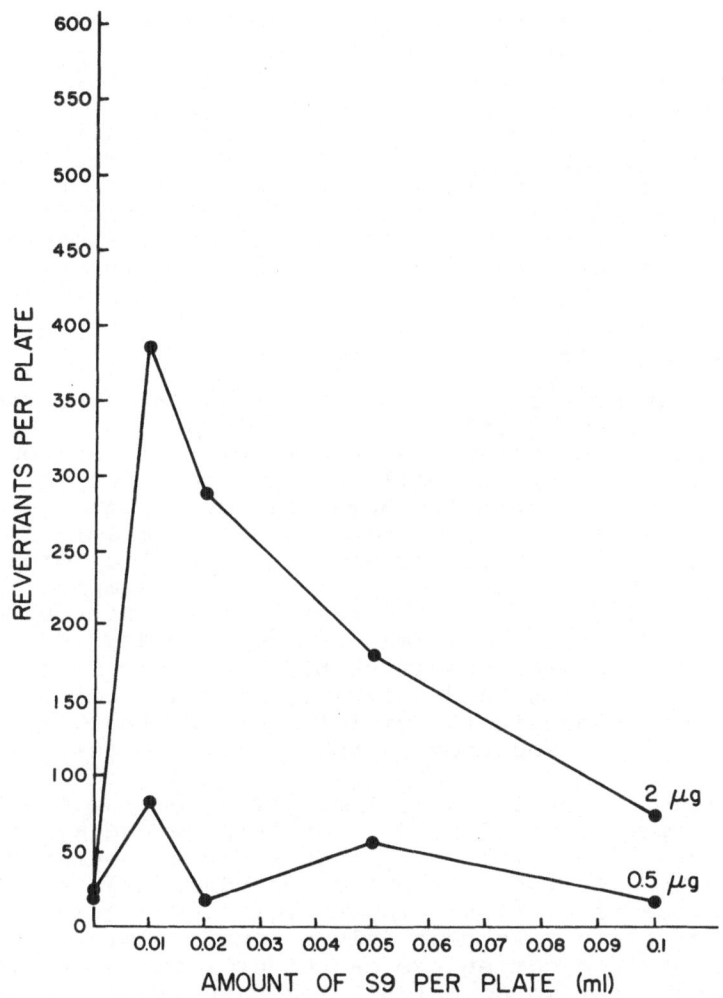

Figure 3. Dose response curve of S-9 for 2-aminofluorene
with Salmonella typhimurium strain TA100.

(Figure 2). The optimum S-9 concentration for the intermediate dose of benzo(a)pyrene (data not presented) was at
0.02 ml S-9 per plate with suppression at higher concentration of S-9. It is interesting that at the high and low
concentrations of 2-aminofluorene (Figure 3) the optimum
S-9 concentration for activation was identical and very low.

Thus, when screening unknown samples for mutagenicity, we recommend the use of two concentrations of S-9: a "low" (0.01 ml/plate) and a "high" (0.05 ml/plate) concentration. Thus, if a mutagen of the 2-aminofluorene type is present, it will not be suppressed by high S-9 to give a false negative, and if a mutagen such as BaP is present, a much better idea of the effective concentration range for further quantitation will be obtained. Before finally settling on the exact concentrations of S-9 to use, we plan to test several other activatable mutagens to confirm these findings.

Experimental Application

In our program, working with ambient air samples, we are faced with a problem in application of the test. Dr. Little (51), in her presentation, has emphasized this problem. The samples are often small and in low concentrations. Because of this, we essentially have "one shot" at a test effect. These samples are complex mixtures of chemicals and may comprise several mutagens as well as several cytotoxic compounds. Because of these properties of the samples, we quickly rejected use of a spot test as a preliminary screening procedure. This rejection was on two grounds. First, many of the mutagens are non-diffusable. Second, if a cytotoxic compound is present and diffuses, it would eliminate the positive reversions in the vicinity of the spot. This would lead to a high frequency of false negative test results.

This led us to develop a preliminary screening test based upon the agar layer method. For this we use all five tester strains. Each sample is tested at three concentrations over a thousand-fold concentration range. We use two levels of S-9 for activation: low = 0.01 ml and high = 0.05 ml S-9 per plate. We pour only a single plate for each test. From this screen, we cannot draw any quantitative conclusions, but we do establish a base line for the quantitative test. We learn whether a sample is mutagenic or not, what level of S-9 to use and what concentration range of sample to use for the quantitative test.

From this assay we select the most responsive strain or strains for the quantitative test.

For attempts to quantify mutagens, we use the most sensitive strains, select the high or low S-9 concentration, and test four concentrations of the sample based upon the highest number of revertants found in the preliminary test.

These concentrations are selected to cover one log of concentration surrounding the optimum.

Each day that samples are tested, it is essential that the full battery of controls be run. These provide the basis of reproducibility. It is necessary to constantly monitor the strains for presence of the plasmids and the mutations. In each experiment, spontaneous reversion frequencies must be tested. Furthermore, it is essential that the mutant strains be tested against known mutagen standards as an internal control on their response. It is desirable to quantify this response each time to control inherent biological variability.

In summary, our experiments support the following recommendations for standard application of the Ames test to samples of unknown chemical composition.

1. Preliminary test

    a. Grow overnight cultures of all strains.

    b. Adjust cultures to fixed optical density corresponding with 1×10^9 cells per ml.

    c. Use 0.1 ml of this cell suspension for assay plates.

    d. Use 20 ml constant volume agar plates where possible.

    e. Use the 2.0 ml soft agar overlay method to eliminate false negatives due to toxic chemicals.

    f. Test appropriate sample dilutions (we use a 3-log concentration range).

    g. Use two levels of S-9 (high and low) for metabolic activation, to avoid suppression.

    h. Include all five tester strains because of strain specificity in response to mutagens.

    i. Test five colonies from each "positive" sample for true reversion to exclude drug induced phenocopies.

    2.   Quantitative tests

        a.   Select the most responsive strain or strains.

        b.   Use three or four replicates plated on con-
             stant volume (20 ml) agar plates where pos-
             sible.

        c.   Use either high or low concentration of S-9
             as determined by the preliminary screen.

        d.   Hold cell densities constant at approximately
             1×10^8 cells per plate.

    As demonstrated in many of the papers of this proceed-
ings, assays of complex samples do not generate straight
line dose response curves. Dr. Commoner (52) has emphasized
the problem of making quantitative judgments from the complex curves. Our position is that a peak value derived from
such a nonlinear dose-response curve at least provides a
minimal estimate of the mutagenicity of the sample. True
quantitation depends upon subfractionation of these samples
to isolate the mutagenic agents.

CONCLUSIONS

    Directly active mutagens are formed upon exposure of
BaP of ambient photochemical smog as well as to sub-ppm levels
of several major gaseous components, NO_2, O_3, and PAN (17,18).

    However, we would like to emphasize that our studies
were conducted with PAH deposited on the surface of glass
fiber filters. Whether PAH adsorbed on the surface of airborne particles (soot, fly ash, etc.) will react in a similar fashion in the atmosphere is a complex problem. Thus,
the atmospheric reactions of PAH may be influenced by many
factors typical of surface chemistry as well as by pollutant
levels, particle size, sunlight intensity, atmospheric mixing, and transport time. Similarly, little is known about
the extent of possible reactions of PAH on glass fiber filters widely employed for decades to collect ambient particulates; our results suggest that they may indeed be significant. Therefore, the determination of possible filter
"artifacts" is of major importance since historically most
evaluations of the carcinogenic and mutagenic activity of
organic particulates have been based upon filter samples.

Finally, control experiments on the Ames <u>Salmonella</u> reversion test have resulted in a series of findings which support a standardized protocol for application of the test to ambient air samples and possibly to samples from other sources. These include recommendations for control of cell density, agar volume, S-9 concentration and strains used.

ACKNOWLEDGMENT

Much of this paper is based on "Atmospheric Reactions of Polycyclic Aromatic Hydrocarbons: Facile Formation of Mutagenic Nitro-Derivatives," <u>Science</u> (17), and "Photochemical and Biological Implications of the Atmospheric Reactions of Amines and Benzo(a)pyrene," <u>Philosophical Transactions of the Royal Society of London</u>, in press (18). These papers should be consulted for details.

We want to thank Dr. T.M. Mischke and Dr. T.L. Gibson of the Department of Chemistry, University of California, Riverside, who were involved with the chemical aspects of collection and analysis of the urban particulates, and Dr. V.F. Simmon and Mr. D. Poole, Stanford Research Institute, who kindly carried out the Ames tests during our initial screening program of urban aerosols collected in the Los Angeles Basin.

We also want to express our appreciation to the University of California and to the Federal agency who generously funded this research--the National Science Foundation-Research Applied to National Needs (Grant No. ENV73-02904-A04, Dr. R. Carrigan, Project Officer).

The contents do not necessarily reflect the views and/or policies of the NSF-RANN nor does mention of trade names or commercial products constitute endorsement or recommendation for use.

REFERENCES

1. Leiter J, Shimkin MB, Shear MJ: Production of subcutaneous sarcomas in mice with tars extracted from atmospheric dusts. J Natl Cancer Inst 3:155-165, 1942

2. Leiter J, Shear MJ: Quantitative experiments on the production of subcutaneous tumors in strain A mice with marginal doses of 3,4-benzpyrene. J Natl Cancer Inst 3:455-477, 1943

3. Kotin P, Falk HL, Mader P, Thomas M: Aromatic hydrocarbons. 1. Presence in the Los Angeles atmosphere and the carcinogenicity of atmosphereic extracts. Arch Indust Hyg 9:153-163, 1954

4. Hueper WC, Kotin P, Tabor EC, Payne WW, Falk HL, Sawicki E: Carcinogenic bioassays on air pollutants. Arch Pathol 74:89-116, 1962

5. Epstein SS, Mantel N, Stanley TW: Photo-dynamic assay of neutral subfractions of organic extracts of particulate atmospheric pollutants. Environ Sci Technol 2: 132-138, 1968

6. Rigdon RH, Neal J: Tumors in mice induced by air particulate matter from a petrochemical industrial area. Texas Reports on Biology and Medicine 29:110-123, 1971

7. Freeman AE, Price PJ, Bryan RJ, Gordon RJ, Gilden RV, Kelloff GJ, Huebner RJ: Transformation of rat and hamster embryo cells by extracts of city smog. Proc Nat Acad Sci 68:445-449, 1971

8. Mohr U, Reznik-Schuller H, Reznik G, Grimmer G, Misfeld J: Investigations on the carcinogenic burden by air pollution in man. XIV. Effects of automobile exhaust condensate on the Syrian golden hamster lung. Zbl Bakt Hyg, I Abt Orig B 163:425-432, 1976

9. Grimmer G: Analysis of automobile exhaust condensates. In: Air Pollution and Cancer in Man, IARC Scientific Publication #16 (Mohr U, Schmahl D, Tomatis L, eds.). Lyon, France, International Agency for Research on Cancer, 1977, pp 29-39

10. Gordon RJ, Bryan RJ, Rhim JS, Demoise C, Wolford RG, Freeman AE, Huebner RJ: Transformation of rat and mouse embryo cells by a new class of carcinogenic compounds isolated from city air. Int J Cancer 12:223-227, 1973

11. Ames BN, Durston WE, Yamasaki E, Lee FD: Carcinogens are mutagens: A simple test system combining liver homogenates for activation and bacteria for detection. Proc Nat Acad Sci 70:2281-2285, 1973

12. Ames BN, McCann J, Yamasaki E: Methods for detecting carcinogens and mutagens with the Salmonella/mammalian-microsome mutagenicity test. Mutat Res 31:347-364, 1975

13. Ames BN, McCann J: Carcinogens are mutagens: A simple test system. In: Screening tests in chemical carcinogenesis, IARC Scientific Publication #12 (Montesano R, Bartsch H, Tomatis L, eds.). Lyon, France, International Agency for Research on Cancer, 1976

14. Jerina DM, Lehr RE, Yagi H, Hernandez O, Dansette PM, Wislocki PG, Wood AW, Chang RL, Levin W, Conney AH: Mutagenicity of benzo(a)pyrene derivatives and the description of a quantum mechanical model which predicts the case of carbonium ion formation from diol epoxides. In: In Vitro Metabolic Activation in Mutagenesis Testing (de Serres FJ, Fonts JR, Bend JR, Philpot RM, eds.). Amsterdam, Elsevier, 1976

15. Jerina DM, Yagi H, Hernandez O, Dansette PM, Wood AW, Levin W, Chang RL, Wislocki PG, Conney AH: Synthesis and biological activity of potential benzo(a)pyrene metabolites. In: Polynuclear Aromatic Hydrocarbons: Chemistry, Metabolism and Carcinogenesis. (Freudenthal RI, Jones PW, eds.). New York, Raven Press, 1976, pp 91-113

16. Wislocki PG, Wood AW, Chang RL, Levin W, Yagi H, Hernandez O, Dansette PM, Jerina DM, Conney AH: Mutagenicity and cytotoxicity of benzo(a)pyrene, arene oxides, phenols, quinones and dihydrodiols in bacterial and mammalian cells. Cancer Res 36:3350-3357, 1976

17. Pitts JN Jr, Van Cauwenberghe KA, Grosjean D, Schmid JP, Fitz DR, Belser WL Jr, Knudson GB, Hynds PM: Atmospheric reactions of polycyclic aromatic hydrocarbons: facile formation of mutagenic nitro-derivatives. Science, Vol 202, 1978, pp 515-519

18. Pitts JN Jr: Photochemical and biological implications of the atmospheric reactions of amines and benzo(a)-pyrene. Philos Trans Royal Soc, in press, 1978

19. Pitts JN Jr, Doyle GJ, Lloyd AC, Winer AM: Chemical
     transformations in photochemical smog and their appli-
     cations to air pollution control strategies. Second
     Annual Report, National Science Foundation-Research
     Applied to National Needs, Grant No. AEN73-02904-A02,
     1975, p V-8

20. Tokiwa H, Tokeyoshi H, Morita K, Takahashi K, Soruta N,
     Ohnishi Y: Detection of mutagenic activity in urban
     air pollutants. Mutat Res 38:351-359, 1978

21. Teranishi K, Hamada K, Watanabe H: Mutagenicity in
     Salmonella typhimurium mutants of the benzene-soluble
     organic matter derived from airborne particulate matter
     and its five fractions. Mutat Res 56:273-280, 1978

22. Talcott R, Wei E: Brief communication: Airborne muta-
     gens bioassayed in Salmonella typhimurium. J Natl
     Cancer Inst 38:449-451, 1977

23. Daisey JM, Hawryluk I, Kneip TJ, Mukai FH: Mutagenic
     activity in organic fractions of airborne particulate
     matter. Presented at the Conference on Carbonaceous
     Particles in the Atmosphere, Berkeley, California,
     March 20-22, 1978

24. Commoner B, Madyastha P, Bronsdon A, Vithayathil AJ:
     Environmental mutagens in urban air particulates. J
     Toxicol Environ Health 4:59-77, 1978

25. Pitts JN Jr, Grosjean D, Mischke TM, Simmon VF, Poole
     D: Mutagenic activity of airborne particulate organic
     pollutants. Presented at: Symposium on mutagenic
     activity of airborne particulate organic pollutants,
     American Chemical Society, August 1977, and to be
     published in: Biological Effects of Environmental Pol-
     lutants (Lee SD, ed.). Ann Arbor, Michigan, 1978, in
     press

26. Pitts JN Jr, Grosjean D, Mischke TM, Simmon VF, Poole
     D: Mutagenic activity of airborne particulate organic
     pollutants. Toxicol Lett 1:65-70, 1977

27. Kertesz-Saringer M, Meszaros E, Varkonyi T: On the
     size distribution of benzo(a)pyrene containing particles
     in urban air. Atmos Environ 5:429-431, 1971

28. Natusch DFS, Wallace JR: Urban Aerosol Toxicity: The influence of particle size. Science 186:695-699, 1974

29. Pierce R, Katz M: Dependency of polynuclear aromatic hydrocarbons content on size distribution of atmospheric aerosols. Environ Sci Technol 9:347-353, 1975

30. Friedlander SK, Miguel A: Atmos Environ, in press, 1978

31. Gordon RJ, Bryan RJ: Patterns in airborne polynuclear hydrocarbon concentrations at four Los Angeles sites. Environ Sci Technol 7:1050-1053, 1973

32. Wynder EL, Hoffman D: Some laboratory and epidemiological aspects of air pollution carcinogenesis. J Air Pollut Control Assoc 15:155-158, 1965

33. Asahina S, Andrea J, Carmel A, Arnold E, Bishop Y, Joshi S, Coffin D, Epstein SS: Carcinogenicity of organic fractions of particulate pollutants collected in New York City and administered subcutaneously to infant mice. Cancer Res 32:2263-2268, 1972

34. Kotin P, Falk HL, McCammon CJ: The experimental induction of pulmonary tumors and changes in the respiratory epithelium in C57BL mice following their exposure to an atmosphere of ozonized gasoline. Cancer 11:473-481, 1958

35. Kotin P, Falk HL, Thomas M: Production of skin tumors in mice with oxidation products of aliphatic hydrocarbons. Cancer 9:905-909, 1956

36. Larson RA, Hunt LL, Blankenship DN: Formation of toxic products from a #2 fuel oil by photooxidation. Environ Sci Technol 11:492-496, 1977

37. Fishbein L: Atmospheric mutagens. In: Chemical Mutagens: Principles and Methods for Their Detection (Hollaender A, ed.). New York, Plenum Press, 1976, Volume 4, pp 219-339

38. Berry RS, Lehman PA: Aerochemistry of air pollution. Ann Rev Phys Chem 22:47-84, 1971

39. National Academy of Sciences: Particulate Polycyclic Organic Matter, Washington, DC, National Academy of Sciences, 1972

40.   Falk HL, Markul I, Kotin P: Aromatic hydrocarbons.
      IV. Their fate following emission into the atmosphere
      and experimental exposure to washed air and synthetic
      smog. AMA Arch Ind Health 13:13-17, 1956

41.   Tebbens BD, Thomas JF, Mukai M: Fate of arenes incor-
      porated with airborne soot. J Amer Ind Hyg Assoc 27:
      415-422, 1966

42.   Tebbens BD, Mukai M, Thomas JF: Fate of arenes incor-
      porated with airborne soot: Effect of irradiation. J
      Amer Ind Hyg Assoc 32:365-372, 1971

43.   Barofsky DF, Baum EJ: Exploratory field desorption
      mass analysis of the photoconversion of adsorbed poly-
      cyclic aromatic hydrocarbons. J Amer Chem Soc 98:8286-
      8287, 1976

44.   Dewar MJS, Mole T, Urch DS, Worford EWT: Electrophilic
      substitution. Part IV. The nitration of diphenyl,
      chrysene, benzo(a)pyrene and anthanthrene. J Chem Soc
      3573-3575, 1956

45.   Pfeiffer EH: Oncogenic interaction of carcinogenic
      and noncarcinogenic polycyclic aromatic hydrocarbons in
      mice. In: Air Pollution and Cancer in Man, IARC
      Scientific Publication #16 (Mohr M, Schmahl D, Tomatis
      L, eds.). Lyon, France, International Agency for
      Research on Cancer, 1977

46.   Dewar MJS, Mole T: Electrophilic substitution. Part
      II. The nitration of naphthalene and perylene. J Chem
      Soc 1441, 1956

47.   Hopff H, Schweizer HR: 251. Zur Kenntnis des Coronens
      Dienanlagerungen in der Perylen-und Benzperylenreihe.
      Helv Chim Acta 42:2315, 1957

48.   McCann J, Choi E, Yamasaki E, Ames BN: Detection of
      carcinogens as mutagens in the Salmonella/microsome
      test: Assay of 300 chemicals. Proc Nat Acad Sci 72:
      5135-5139, 1975

49.   Kaden DA, Thilly WG: Genetic toxicology of kerosene
      soot. Presented at: The Workshop on Unregulated
      Diesel Emissions and Their Potential Health Effects.
      Washington, DC, April 27-28, 1978

50. Rosenkranz HS: The use of microbial mutagenesis assay systems in the detection of environmental mutagens in complex mixtures. Presented at: Symposium on Application of Short-Term Bioassays in the Fractionation and Analysis of Complex Environmental Mixtures, Williamsburg, Virginia, February 21-22, 1978

51. Pellizzari ED, Little LW: Integrating of microbiological and chemical testing into the screening of air samples for potential mutagenicity. Presented at: Symposium on Application of Short-Term Bioassays in the Fractionation and Analysis of Complex Environmental Mixtures, Williamsburg, Virginia, February 21-22, 1978

52. Commoner B, Vithayathil AJ, Dolara P: Mutagenic analysis of complex samples of air particulates, aqueous effluents and foods. Presented at: Symposium on Application of Short-Term Bioassays in the Fractionation and Analysis of Complex Environmental Mixtures, Williamsburg, Virginia, February 21-22, 1978

Notes Added in Proof:

p. 360

We recently found that the half-life of BaP in air containing only 0.1 ppm ozone was less than one hour, and that certain of these products were direct mutagens. This ozone oxidation may be the most important fate of BaP on the surface of particulate matter.

Subsequent experiments using HPLC separation suggest that some quinones may have been formed on the TLC plate.

p. 364

Unpublished results from our own lab and that of Dickson and that of Eisenstadt (private communications) show that, at high S-9 levels (40% v/v) and higher concentrations of terylene than used above, is an activatable frameshift mutagen in the Ames test.

APPLICATION OF BIOASSAY TO THE CHARACTERIZATION OF DIESEL PARTICLE EMISSIONS

J. Huisingh, R. Bradow, R. Jungers,
L. Claxton, R. Zweidinger, S. Tejada,
J. Bumgarner, F. Duffield, and M. Waters
Health Effects Research Laboratory and
Environmental Sciences Research Laboratory
U.S. Environmental Protection Agency
Research Triangle Park, North Carolina

V.F. Simmon
SRI International
Menlo Park, California

C. Hare and C. Rodriguez
Southwest Research Institute
San Antonio, Texas

L. Snow
Northrop Services, Inc.
Research Triangle Park, North Carolina

PART I. CHARACTERIZATION OF HEAVY DUTY DIESEL PARTICLE EMISSIONS

INTRODUCTION

A wide variety of combustion sources produce soot, i.e., carbon aerosols containing variable quantities of organic matter. The most significant transportation-related sources of such materials are diesel engines. Diesel power has been used for railway locomotives, long haul trucks, and earthmoving equipment for many years. However, recently a strong trend has developed toward use of diesel engines in urban service vehicles and also taxicabs. In the near future substantial numbers of diesel-powered automobiles may be used by the general public.

These comparatively new developments not only increase the present rather small contribution from this source to ambient air particulate matter, but also shift the potential locale of the soot emission to more densely populated urban core areas. In fact, diesel engines have the greatest fuel economy advantages over gasoline engines in the low speed-light load situations characteristic of urban stop-and-go driving (1,2).

Some years ago the Environmental Protection Agency's Office of Research and Development recognized that this issue might come into prominence as petroleum-based fuels became scarce. Consequently, considerable efforts were made to develop procedures suitable for measuring diesel particle concentrations and composition (3,4). Subsequently, these methods have been used to describe the emission rates and general chemical character of the combustion products of a wide variety of small and large engines (2,4,5,6).

The most interesting aspect of these particles is the associated organic matter which varies widely in both emission rate and composition (3,7). Generally, the sources of these organic compounds appear to be unburned fuel and lubricant. However, there seems to be some partitioning of organic material between the gas phase and the particle-bound phase. Consequently, the soot-bound organic material is higher in average molecular weight than the fuel (3). The weight percentages of nitrogen and sulfur are also higher in soot organics than in the fuel. Further, there is substantial oxygen incorporation in the material, certainly as a result of partial combustion (3,4).

Diesel exhaust particulate, as well as other fossil fuel combustion products, are known to contain the carcinogenic and mutagenic chemical benzo(a)pyrene, among many other potentially hazardous and less well characterized components. Due to the potential proliferation of diesel powered vehicles, it is critical to identify those components which constitute a possible public health risk to facilitate their control.

To reduce the immensity of the organic analytical task, chemical fractionation and analysis were guided by short-term bioassays. In this way, crude fractions containing biological activity would be identified and prioritized for analytical efforts to characterize components in the most active fractions. This procedure also allows identification of relatively inactive materials and conserves resources which might otherwise be devoted to analysis of less important substances. The initial bioassays employed included cytotoxicity in mammalian cells and mutagenicity in bacteria. Since the fractions tested were not found to be highly toxic but were mutagenic, subsequent efforts concentrated on the use of bacterial mutagenesis bioassay in <u>Salmonella</u> <u>typhimurium</u> to guide fractionation.

This paper represents early, but very promising, results using such a procedure. Also described are engineering, chemical fractionation and analysis, and bioassay procedures currently being employed.

ENGINEERING PROCEDURES

Test methods for both heavy duty diesel truck engines and diesel passenger cars have been previously described in detail (3-5). The procedures used for the heavy duty engine experiments and the rationale for those procedures are outlined here.

Truck diesel engines, because of their very small speed range, tend to operate at or near constant speed much of the time. Consequently the current heavy-duty test procedure uses a series of 13 steady state operating speed-load conditions (modes) to simulate overall urban use. Independent gas analysis is made of each mode and weighing factors are used to arithmetically compose a cycle value. For the particulate sampling, however, it is more convenient to vary the time-in-mode to achieve a single physically composited filter sample. All heavy duty samples used in the present work were such time averaged 13 mode composites collected on glass fiber filters as previously described (4,8).

In order to obtain reasonable samples of particle-bound organics, it is important to consider the nature of the emission process. In the tailpipe of an operating diesel engine, the temperatures are sufficiently high (>200°C) that organic materials are generally in the gas phase. Thus, soot filtered at these temperatures contains very little extractable organic material; approximately one percent by weight can be extracted with methylene chloride, for example. However, when particles and gaseous exhaust enter the ambient air, as from automobiles and trucks, the mixture is quickly cooled and diluted. During this process, the overall temperature is reduced to the point that carbon particles begin to absorb organic material. Still further dilution may reduce the gas phase hydrocarbon concentrations to the point that further absorption ceases to occur. Thus, the particle composition may be stabilized at some point in the exhaust-air dilution process. This process has not been examined experimentally with real

vehicles, but considerable work has gone into laboratory
simulation of this process which is assumed to occur in the
ambient air.

A number of investigators have used air dilution tunnel
techniques to achieve this simulation (3,4,9,10), and a
wide variety of systems have been shown to be reasonably
effective in at least some sampling applications. Figure 1
presents the dilution-tunnel system used for these studies.
In the highest load modes of the 13-mode test procedure, exhaust volumes are so large that the capacity of a system
scaled to dilute the whole exhaust would have to be immense,
perhaps 500,000 to 1,000,00 liters/min. Consequently, heavy-
duty engine exhaust is first variably split, then diluted
10- to 15-fold in the dilution tunnel. The individual mode
dilution ratios are determined by the ratios of gas concentrations in the dilute and raw exhaust streams for CO_2 and
NO. Large samples of particulate material were obtained from
this apparatus by filtering the whole dilution tunnel contents at a flow rate of 12,000 liters/min.

Two test engines were chosen for this program, both
high production, naturally-aspirated, medium duty truck

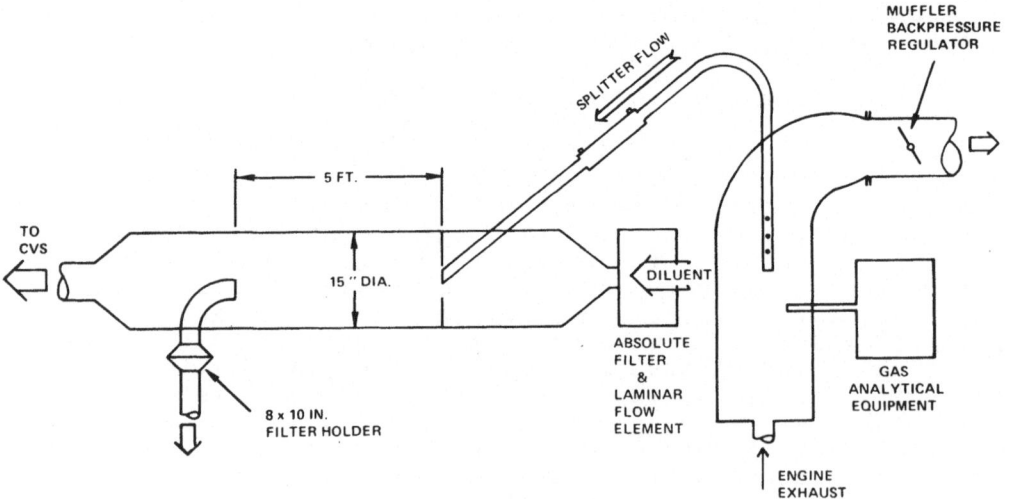

Figure 1. Dilution tunnel system for collection of heavy
duty diesel particulate.

power plants. The first, engine No. 1, was a typical city bus engine, a Detroit Diesel two-stroke-cycle 6V-71 in-line 6 cylinder engine. Engine No. 2 was a 4-stroke cycle, V-8, Caterpillar 3208, an engine now widely used in urban service vehicles.

CHEMICAL FRACTIONATION PROCEDURES

The procedure used for the extraction and separation of the organic components present in diesel exhaust particulate are outlined in Figures 2-5. Diesel exhaust particulate collected by filtration on glass fiber filters was extracted for six hour periods, first with dichloromethane (DCM) followed by acetonitrile (ACN). The majority of organic material was removed by the DCM extraction with some additional organic and inorganic material obtained by the subsequent ACN extraction (see Table 1). Initial characterization studies have dealt entirely with the DCM extracts. Fractionation of the DCM extracts for mutagenesis testing was carried out by procedures similar to those employed by Swain, et al. (11) for cigarette smoke condensate.

Table 1

Diesel Particulate Extracts

	Engine #1 2-Stroke 6V-71	Engine #2 4-Stroke 3208
Particulate emission rate	86.7 g/hr	42.5 g/hr
Total particulate collected	118.0 g	195.13 g
DCM extract	64.05 g	47.33 g
ACN extract	10.57 g	19.65 g

The solvent partitioning steps employed to obtain acid, basic, and neutral fractions are outlined in Figure 2. A sample of DCM extract was evaporated, weighed and reconstituted in ether. A small amount was ether insoluble and removed by filtration (INT fraction). The ether solution was extracted with 0.1N Na_2CO_3 to obtain the acid fraction (ACD) and then with 1N H_3PO_4 to obtain the basic fraction

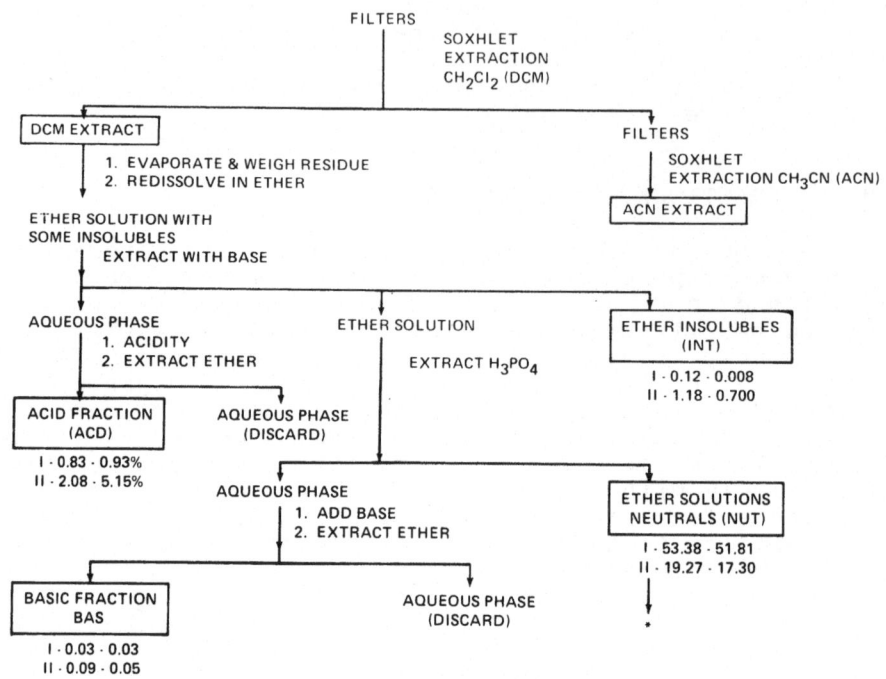

Figure 2.   Isolation and fractionation organics from diesel exhaust particulates.

(BAS). The remaining ether solution containing the neutral fraction (NUT) was then further fractionated by chromatography on silica gel (Bio-sil-A, 100-200 mesh) as shown in Figure 3. Elution was initiated with hexane, which removed the paraffins (PRF). When fluorescent materials, as observed under long wave-length UV, reached the bottom of the column, the aromatic fraction (ARM) was collected. The eluting solvent was then changed to 1% ether in hexane at which time a narrow yellow band moved down the column. This band did not fluoresce and quenched the bluish fluorescence ahead of it. The third or transitional fraction (TRN) was the yellow band of material which was also eluted with 1% ether in hexane. The remaining polar oxygenated compounds (OXY) were removed by elution with 50% acetone/methanol.

    The percentage that each fraction represented of the total exahust particulate originally collected is given in

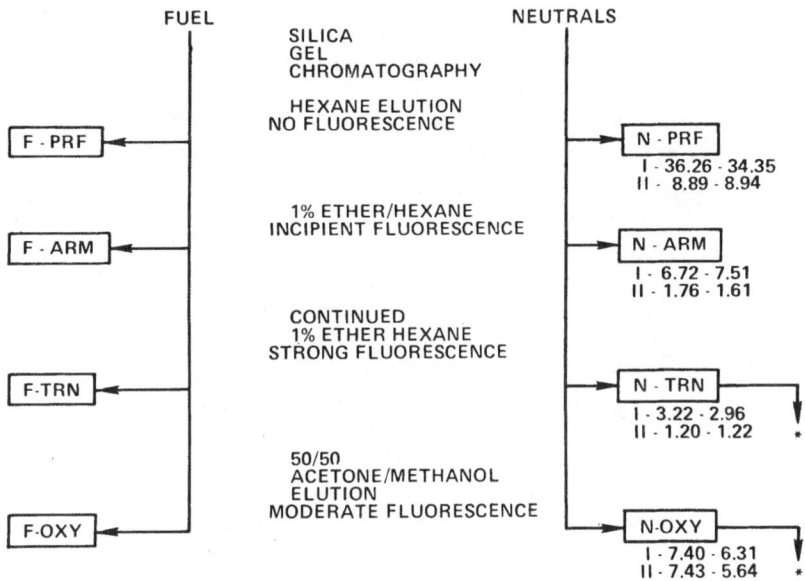

Figure 3. Silica gel chromatography fractionation of the neutral organics from both diesel exhaust particulate and uncombusted diesel fuel.

Table 2. The yield for each fraction was slightly affected by the base extraction employed in the initial solvent partitioning. The largest variations were seen in the ACD and OXY fractions and probably were due to incomplete extraction of phenols and other weak acids by 1 N Na_2CO_3. In addition to the DCM neutrals, a sample of diesel fuel was also chromatographed on silica gel in analogous fashion (Figure 3). The amounts obtained for each fraction are given in Table 3. Samples of all of the above fractions were prepared for mutagenesis bioassay by removing the solvents by evaporation and reconstituting in dimethylsulfoxide (DMSO).

On the basis of initial mutagenesis test results, further fractionation of the TRN and OXY fractions was accomplished by high pressure liquid chromatography (HPLC). The TRN fraction was chromatographed on a NH_2-bounded phase column

Table 2

Fractionation DMC Extracts
(% of Total Particulate)

Fraction	Engine #1 2-Stroke 6V-71		Engine #2 4-Stroke 3208	
	1.ON N_2CO_3	O.IN KOH	1.ON Na_2CO_3	O.IN KOH
ACD	0.83	1.93	2.08	5.15
BAS	0.03	0.03	0.09	0.05
INT	0.12	0.008	1.18	0.70
NUT	53.38	51.81	19.27	17.30
PRF	36.26	34.35	8.89	8.94
ARM	6.72	7.51	1.76	1.61
TRN	3.22	2.96	1.20	1.22
OXY	7.40	6.31	7.43	5.64

Table 3

Fraction of Diesel Fuel EM 239

Fraction	% of Fuel
PRF	74.04
ARM	21.51
TRN	0.69
OXY	0.33

TRANSITION SUBFRACTIONS:

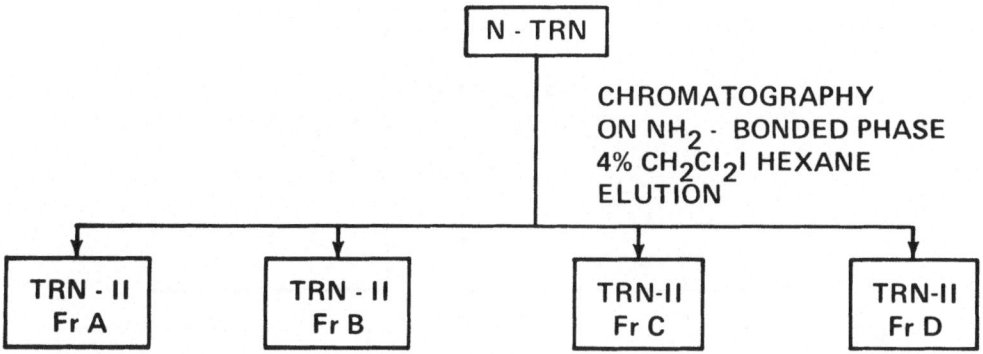

Figure 4. Chromatographic subfractionation of the neutral transitionals.

(Varian Associates) with 4% methylene chloride in hexane (Figure 4). Four complex fractions (A,B,C, and D) were collected for additional mutagenic testing.

The OXY fraction was further separated by gel permeation chromatography on 100 Å μ Styragel (Waters Associates) which has an exclusion limit of approximately 700 molecular weight. Two fractions arbitrarily divided into high (GPC-1) and low (GPC-2) molecular weight were collected using dichloromethane as eluent (Figure 5). These two fractions were of approximately equal mass and were submitted for mutagenesis testing along with the original OXY material (NEAT OXY).

Detailed characterization of the various fractions and subfractions aimed at identifying specific mutagens has been undertaken by gas chromatography on glass capillary columns with mass spectrometric detection. At present, some information is available concerning classes and types of compounds. The aromatic fraction (ARM) contains most of the PNA hydrocarbons such as benzo(a)pyrene. The TRN fraction contains substituted PNA's, phenols, ethers and ketones such as fluorenone and its methyl and dimethyl isomers. The known mutagen 2-aminofluorene has also been tentatively identified.

N - OXY SUBFRACTIONS:

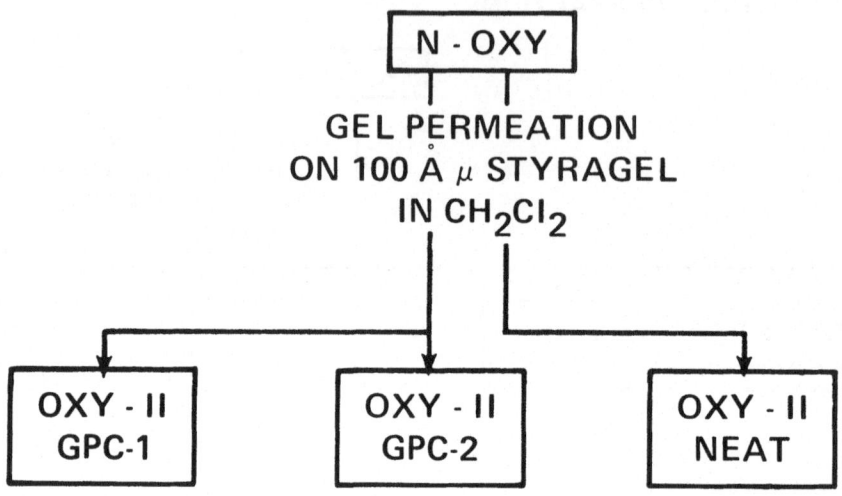

Figure 5. Gel chromatographic subfractionation of the neutral oxygenates.

It has been extremely difficult to work with the OXY fraction. Due to its very polar nature (up to 10% oxygen by elemental analysis), this material does not gas chromatograph well. Investigations are currently under way using HPLC fractionation schemes coupled with direct probe and field desorption mass spectrometry in both electron impact and chemical ionization modes.

BIOASSAY

The Ames Salmonella typhimurium/microsome mutagenesis bioassay was used to indicate which fractions of diesel exhaust were genetically active and to guide the fractionation of diesel exhaust so that biologically active components could be isolated and characterized. The plate incorporation procedure followed in these studies is described in detail by Ames, McCann, and Yamasaki (12). Histidine dependent strains of Salmonella typhimurium were obtained from Dr. Bruce Ames of the University of California at Berkeley. These strains were routinely checked for their genotypic characteristics and for the presence of the plasmid, as described by Ames et al. (12). Positive controls for each tester strain and the activation system as well as negative solvent controls were included with each experiment. In order to detect chemicals which are mutagenic only after metabolism by a mammalian enzyme system, a rat liver metabolic activation (MA) system is used in the bioassay. The Aroclor 1254-stimulated metabolic activation system was prepared as described by Ames et al. (12). Chemicals which are mutagenic without the metabolic activation system are referred to here as direct-acting mutagens.

Exhaust particulate samples from the heavy duty diesel engines were solvent extracted and fractionated as previously described. Bioassays were performed following removal of the solvent and addition of dimethyl sulfoxide (DMSO) to dissolve the mixture. After preliminary range-finding tests all samples were evaluated as described below except where sample size was limiting. The fractions were examined with five tester strains of Salmonella typhimurium (TA1535, TA1537, TA1538, TA98, TA100) with and without the liver metabolic activation (MA) system. The experiments were conducted in a dose response fashion (6-8 doses/fraction/tester strain) and each experiment was repeated where sample size permitted. Seven fractions from each engine were tested initially at SRI International. Subsequent subfractions and selected samples of the initial fractions were tested at Northrop Services Inc. and EPA (HERL/RTP) Laboratories. Identical samples tested in separate laboratories produced a similar mutagenic response.

In these initial investigations, extracted and fractionated samples were stored for several months prior to bioassay. Subsequent studies (reported in Part II) have shown that the mutagenicity of an unfractionated diesel extract was slightly reduced as a result of storage. No data are available on the effect of storage on the fractionated samples.

Figure 2 gives the original fractionation scheme used with heavy duty diesel exhausts. A summary of the bioassay results on these fractions is provided in Table 4. A fraction was considered positive if it gave a maximum response that was 2.5 times greater than the spontaneous rate for the particular strain used and if it gave a positive linear dose response in the major portion of the curve.

Table 4

Summary of Diesel Exhaust Fractions from Two Engines
in the Salmonella typhimurium Plate Incorporation Test

Engine I. Two Cycle Bus Diesel Engine

FRACTION	Bacteria:[1] Activation:	TA 1535 -MA[5]	TA 1535 +MA[6]	TA 1537 -MA	TA 1537 +MA	TA 1538 -MA	TA 1538 +MA	TA 98 -MA	TA 98 +MA	TA 100 -MA	TA 100 +MA
DCM (C)	S.A.[2]	15.9	12.28	18.08	22.44	66.61	122.5	93.17	132.85	370.50	329.89
	Slope (r)[3]	.68 (.4)	1.69 (.72)	11.95 (.98)	14.87 (.99)	50.51 (.99)	99.07 (.99)	67.25 (.99)	93.07 (.99)	255.34 (.99)	216.88 (.9)
	MR (dose)[4]	21 (.01)	18 (.50)	40 (3.0)*	53 (3.0)*	164 (3.0)*	324 (3.0)*	223 (3.0)*	327 (3.0)*	859 (3.0)*	760 (3.0)*
ACD (A)	S.A.	18.0	16.6	22.9	42.3	NT	NT	134.9	198	324	317
	Slope (r)	.7 (.86)	5.9 (.86)	20 (.45)	34 (.94)			96 (.94)	155 (.98)	195 (.99)	179 (.92)
	MR (dose)	20 (.05)	17 (1.0)	32 (.75)	45 (1.0)*			126 (1.0)*	188 (1.0)*	325 (1.0)*	309 (1.0)*
BAS (B)	S.A.	84	45.4	NT	NT			88.8	78.9	120.5	136.6
	Slope (r)	64 (.50)	32 (.50)					70 (.82)	45 (.57)	4.4 (.57)	49 (.57)
	MR (dose)	23 (.032)	15 (.016)					29 (.125)	47 (.25)	135 (1.25)	127 (.2)
INT (B)	S.A.	NT	NT	NT	NT	NT	NT	NT	NT	NT	216
	Slope (r)										119 (.53)
	MR (dose)										127 (.2)
PRF (A)	S.A.	18.5	10.5	3.4	5.5	10.3	20.3	18.6	24.9	110.38	122.1
	Slope (r)	1.0 (.26)	2.0 (.27)	0.6 (.11)	3.5 (0.4)	1.9 (.16)	9.9 (.96)	3.2 (.42)	15.8 (.70)	1.74 (.15)	30.9 (.72)
	MR (dose)	22 (.01)	12 (.33)	8 (.003)	10 (.33)	18 (.01)	22 (.1)	22 (.003)	30 (.33)	115 (.01)	128 (1.0)
ARM (A)	S.A.	15.1	13.7	6.0	10.4	13.8	24.3	22.4	41.4	89.0	144.0
	Slope (r)	1.1 (0.26)	1.96 (.27)	0.6 (.11)	3.5 (.43)	1.9 (.16)	9.96 (.96)	3.16 (.42)	15.8 (.70)	-10.9 (-.32)	30.9 (.72)
	MR (dose)	16 (.003)	16 (.003)	8 (.1)	11 (.1)	18 (.01)	24 (1.0)	24 (1.0)	39 (1.0)	101 (.003)	136 (.1)
TRN (A)	S.A.	26.2	50.2	37.2	61.8	NT	NT	291.6	333.3	789.5	558.2
	Slope (r)	12.2 (.98)	38.6 (.93)	29.7 (.86)	53.7 (.97)			258 (.99)	294 (.99)	599 (.95)	348 (.67)
	MR (dose)	23 (.75)	58 (1.0)*	31 (.5)*	55 (1.0)*			279 (1.0)*	328 (1.0)*	692 (1.0)*	760 (1.0)*
OXY (A)	S.A.	20.3	23.3	30.5	55.2	NT	NT	165.01	328.0	518.9	624.0
	Slope (r)	.77 (.07)	11.9 (.91)	20.6 (.79)	42.6 (.91)			126 (.98)	294 (.98)	380 (.99)	494 (.99)
	MR (dose)	23 (.25)	23 (1.0)	30 (.75)*	52 (.75)*			162 (1.0)*	322 (1.0)	507 (1.0)*	630 (1.0)*

[1] Column one gives the fraction tested and in which of three experiments (A,B,C) the fraction was tested.

[2] S.A.: Revertants per plate per mg fraction tested.

[3] Slope (r): The slope based on a linear regression and the correlation coefficient.

[4] MR (dose): The maximum response recordes as revertants per plate and the dose at which that response occurred.

[5] -MA: Without metabolic activation.

[6] +MA: With metabolic activation.

[7] *Indicates a positive response. The response was considered positive if it gave a maximum response that was 2.5 times greater than the spontaneous rate for the particular strain used and if it gave a positive linear dose response in the major portion of the curve.

Table 4 (continued)

Engine II. Four Cycle Diesel Engine

FRACTION	Bacteria: Activation:	TA 1535 MA	TA 1535 +MA	TA 1537 MA	TA 1537 +MA	TA 1538 -MA	TA 1538 +MA	TA 98 -MA	TA 98 +MA	TA 100 -MA	TA 100 +MA
DCM(C)	S.A.	12.2	34.7	544.5	344	1348	1715	2605	1667	4403	3986
	Slope (r)	5.8 (.03)	21.4 (.83)	534.9 (.99)	338 (.99)	1315 (.98)	1682 (.99)	2557 (.99)	1625 (.99)	4175 (.91)	3873 (.99)
	MR (dose)	21 (.03)	32 (1.0)*	166 (.30)	108 (.30)*	404 (.30)*	53 (.30)*	781 (.30)*	538 (.30)*	1333 (.30)	1256 (.30)
ACD(A)	S.A.	22.95	18.1	79	87.4	NT	NT	238.0	309.8	474.5	565.3
	Slope (r)	5.7 (.71)	3.9 (.39)	64.9 (.98)	74.0 (.97)			192.6 (.96)	248.3 (.94)	291.1 (.91)	402.1 (.98)
	MR (dose)	23 (.75)	22 (.50)	79 (1.0)*	84 (1.0)*			229 (1.0)*	266 (.75)	452 (1.0)*	547 (1.0)*
BAS(B)	S.A.	2.89	103.1	-1.38	39.5	NT	NT	65.9	162.5	89.9	180.5
	Slope (r)	-20.7(-.27)	91.6 (.80)	-8.0 (-0.12)	31.0 (.34)			43.8 (.91)	132.2 (.99)	-26.9 (.44)	74.3 (.78)
	MR (dose)	27 (.033)	22 (0.1)	10 (.033)	13 (.033)			38 (.33)	163 (1.0)*	136 (.33)	171 (1.0)
INT (A)	S.A.	17.9	16.0	27.3	21.6	69.8	121.7	83.3	106.9	159.6	200.9
	Slope (r)	1.0 (.13)	4.9 (.73)	21.0 (.96)	12.7 (.81)	52.5 (.92)	103.8 (.99)	53.9 (.91)	80.9 (.97)	53.0 (.87)	113.9 (.80)
	MR (dose)	20 (.10)	13 (.003)	27 (1.0)*	20 (1.0)*	65 (1.0)*	119 (1.0)*	81 (1.0)*	102 (1.0)*	157 (1.0)	195 (1.0)
PRP (B)	S.A.	17.3	11.8	2.7	5.8	14.8	19.8	21.7	32.4	118.0	134.6
	Slope (r)	-2.4(-.17)	.46 (.31)	-2.3(-.43)	0.52 (.11)	2.5 (.49)	.88 (.12)	-1.8(-.23)	3.9 (.41)	10.7 (.45)	32.4 (.86)
	MR (dose)	26 (.10)	12 (.003)	7 (.33)	9						
ARM (B)	S.A.	25.5	10.7	11.4	14.6	37.9	39.3	73.3	69.5	120.4	69.5
	Slope (r)	7.5 (.87)	-2.2 (.72)	6.3 (.83)	6.9 (0.58)	25.3 (.96)	19.0 (.95)	49.5 (.96)	30.1 (.87)	13.0 (.46)	(-49)(-.84)
	MR (dose)	26 (1.0)	14 (.01)	12 (1.0)	15 (1.0)	37 (1.0)*	38 (1.0)	73 (1.0)*	66 (1.0)	123 (1.0)	141 (.033)
TRN (B)	S.A.	60.9	374	634.7	1398	1999.9	5220	7593	2746	28730	6449
	Slope (r)	40.4 (.74)	354.9 (.94)	617.1 (.81)	1388 (.99)	1951.4(.85)	5171 (.98)	7521 (.97)	2617 (.90)	28553 (.97)	6149 (.72)
	MR (dose)	33 (.10)	127 (.33)*	68 (.10)*	144 (.10)	214 (.33)*	540 (.10)*	777 (.10)*	901 (.33)	1079 (.0333)*	898 (.033)
OXY (B)	S.A.	24.7	37.6	120	170.2	1279	2032	745.1	975	1896	1554
	Slope (r)	4.5 (.50)	24.5 (.89)	97.7 (.83)	145.2 (.88)	1225 (.94)	1984 (.98)	629.3 (.91)	798 (.79)	1693 (.87)	1317 (.80)
	MR (dose)	25 (.033)	36 (.10)*	205 (.33)*	150 (.33)*	430 (.33)*	680 (.33)*	677 (1.0)*	872 (.33)*	694 (.33)*	610 (.33)

CONTROL DATA

	TA 1535 MA	TA 1535 +MA	TA 1537 MA	TA 1537 +MA	TA 1538 -MA	TA 1538 +MA	TA 98 -MA	TA 98 +MA	TA 100 -MA	TA 100 +MA
Spontaneous A	14	13	5	3	10	13	22	25	124	108
Spontaneous B	21	13	8	10	16	16	21	28	123	98
Spontaneous C	16	8	5	8	11		22	36	93	105
Mean (S.D.)	17 (3.61)	11.33 (2.89)	6 (3.61)		12.33 (3.21)	16.66 (4.04)	21.66 (.58)	29.66 (5.69)	113.33 (17.62)	103.66 (5.13)

The total DCM extract and ACD, TRN, and OXY fractions from the 2-stroke cycle bus engine were positive, both with and without activation, in strains TA1537, TA98, and TA100. In addition the total DCM extract was positive with TA1538 both with and without activation. The TRN fraction in this engine was also positive with TA1538 when the activation system was added.

The total DCM extract and ACD, INT, TRN, and OXY fractions of the 4-stroke cycle truck engine were positive with and without activation, with strains TA1537, TA98, and TA100. TA1538 showed a positive response with and without activation to the DCM extract and INT, TRN and OXY fractions and to the ARM fraction without activation. When activation was used the DCM extract and TRN and OXY fractions also showed positive results with strain TA1535 in this engine. The BAS fraction of this engine was positive with TA98 with the added activation system.

A comparison of mutagenic response in TA1538 of the various fractions from the 4-stroke cycle truck engine is shown in Figure 6. In both engines the TRN and OXY subfractions of the neutral compounds are the most mutagenic when either the maximum fold increase (max. revertants/plate in sample minus solvent controls) or the specific activity (revertants/plate/µg sample) are compared. It can be noted that each positive fraction contained direct-acting mutagens. It also appears that the positive fractions contain compounds that need activation before being mutagenic. However, these fractions are complex mixtures and the metabolic activation system may also function to detoxify certain components thus allowing expression of the mutagenic potential of other components. Furthermore, as the concentration of some compounds is increased in the assay, metabolism of potentially active compounds may be altered such that a mutagenic metabolite is not formed.

The use of all five tester strains, as shown in Figure 7 for TRN II, yields information about the chemical structure and reactivity of the mutagens (13-17). Strain TA1535 is reverted to histidine independence by many mutagens which cause base-pair substitutions. Strains TA1537 and TA1538 are reverted by many frameshift mutagens. Strains TA98 and TA100 are more sensitive generally to mutagenic agents due to the addition of plasmids and may respond to mutagens which act either by base-pair substitution or frameshift mutation. Based upon the positive responses obtained in strains TA1537 and TA1538, it appears that the active components are mainly

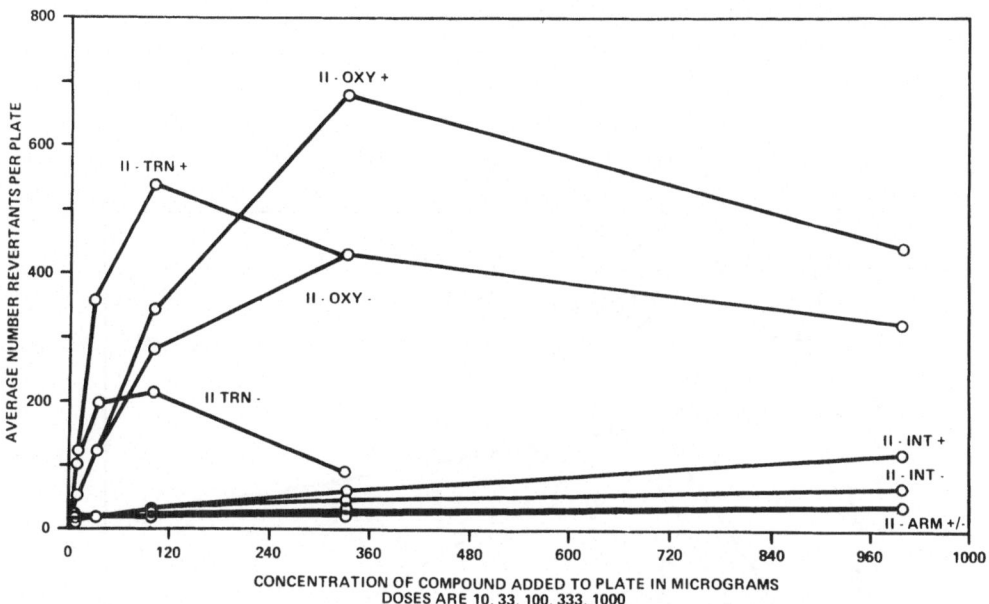

Figure 6. Comparison of the mutagenic response of various organic fractions from the 4-stroke cycle diesel truck exhuast particulate in <u>Salmonella</u> <u>typhimurium</u> strain TA1538.

frameshift mutagens. Furthermore, the activity within strains TA98 and TA100 appears to arise mainly from direct-acting mutagens. Since strain TA1538 showed activity with each of the positive fractions and also demonstrated a quantitative difference with and without activation, it was the strain of choice when only the use of one strain was possible due to sample size limitations.

Since most of the mutagenic activity was present in the neutral subfractions (TRN and OXY) it was decided to further separate these fractions. Chromatography of the TRN fraction yielded four subfractions (FrA, FrB, FrC, and FrD). Due to the lack of material each of these subfractions was assayed at only one dose in duplicate. FrA was negative, but the other three subfractions were positive. On a per weight basis, FrC was clearly the most mutagenic component (Table 5). The OXY fraction was subfractionated into two components, GPC-1 and GPC-2 by gel permeation on 100Å Styragel. An equal amount of components by weight went into GPC-1 and GPC-2.

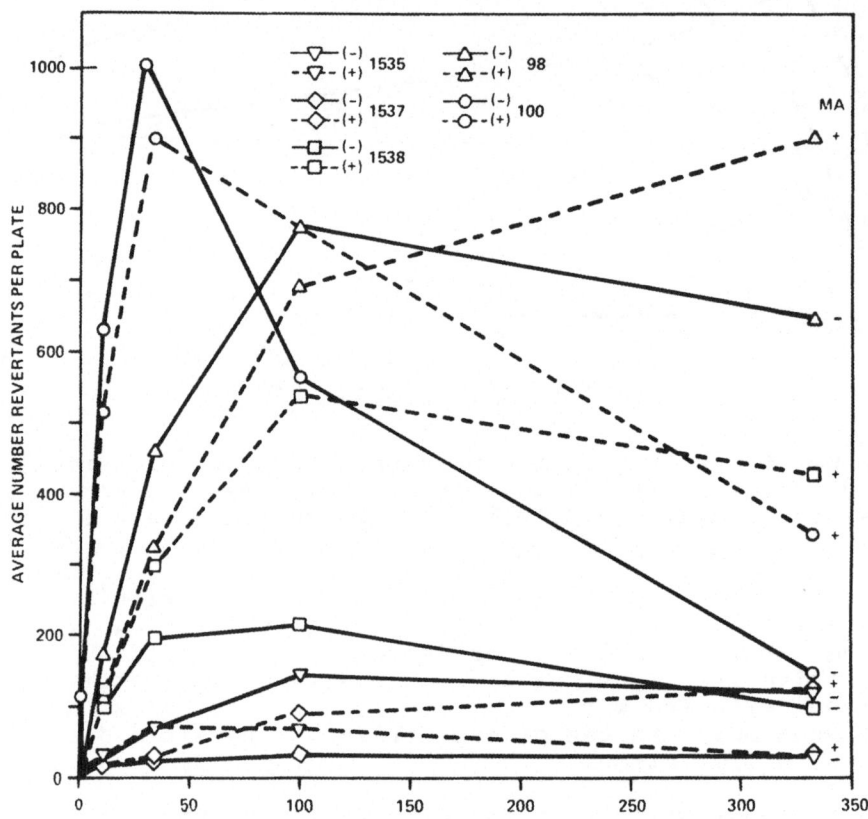

Figure 7. Comparison of the mutagenic response of all five
Salmonella typhimurium tester strains with (+) and without
(-) metabolic activation for the TRN II fraction.

Table 5

Mutagenicity of the TRN and OXY Subfractions in the
Salmonella typhimurium Plate Incorporation Test in TA1538

Subfraction	TRN:				OXY:		
	Fr	B	C	D	GPC-1	GPC-2	NEAT
µg/plate	184	370	218	410	360	360	333
Revertants/plate with activation	50	427	>1000	290	75	2333	1306

Again, due to small sample quantities only one dose (320 µg/
plate) could be tested. Nearly all of the activity was
recovered in GPC-2; therefore, there was a two-fold concentration of the active components into GPC-2. Fractionation
is continuing in order to identify the specific combustion
products that are mutagenic in this microbial system.

In order to investigate the possible source of mutagenic
compounds in the exhaust particulate, the unburned fuel was
fractionated and bioassayed. Neither the neat or fractionated fuel components were found to be mutagenic. The possibility that preparation and fractionation of exhaust particulate could convert otherwise inactive compounds into mutagens
seems very unlikely in view of these negative results with the
uncombusted fuel fractions. Blank filters, extracted and
carried through the fractionation and bioassayed were also
negative.

SUMMARY

Heavy-duty diesel particulate emissions from a 2-stroke
cycle and 4-stroke cycle engine were found to have 54 and 24%
organic extractable components. These organic extracts were
mutagenic in the Salmonella typhimurium/microsome bioassay.
Fractionation of these extracts yielded 53% (2-stroke cycle)
and 17% (4-stroke cycle) neutral components and substantially
smaller amounts of ether insoluble, acid and basic components.
All of these fractions showed some mutagenic activity. The
neutral components contained a major paraffinic fraction which
was not mutagenic. The other three fractions of the neutrals
were mutagenic, with the transitional and oxygenated fractions
being most mutagenic. Further fractionation and bioassay

suggest that these fractions contain a minimum of four sepa-
rable mutagenic components. These mutagenic fractions consist
of the more polar neutral compounds such as substituted polynuclear aromatics, phenols, ethers, and ketones.

The mutagenic activity compared among tester strains,
with and without metabolic activation, suggests that the mutagens are primarily direct-acting frameshift mutagens. Metabolic activation, in most cases, increases the mutagenic
response suggesting either the additional presence of promutagens or the detoxification of toxic components in the
mixture.

The mutagenic activity does not appear to result from
artifacts of extraction or fractionation of the samples.
Fractions of uncombusted fuel were not found to be mutagenic,
suggesting that the mutagens are products of the combustion
process.

PART II. APPLICATION OF A MUTAGENICITY BIOASSAY
MONITORING LIGHT DUTY DIESEL PARTICLE EMISSIONS

INTRODUCTION

The premise that diesel passenger cars will be used in
the future stimulated EPA to initiate the development of
methods for monitoring emissions products.

The predicted amount of particulate matter emitted from
both gasoline and diesel cars is illustrated in Table 6.* The
estimated particulate matter emitted from gasoline cars in
1990 assuming that zero diesel cars are sold is approximately
32,000 tons. This amount of particulate from gasoline cars
decreases insignificantly as the sales of diesels increase.
In contrast, the estimated amount of particulate matter emitted from the diesel cars ranges upwards dramatically, to
155,000 tons if the sales of diesels increase to the predicted
25%. Therefore, a 25% penetration of the market by diesel
vehicles could result in 181,000 tons of particulate emitted
per year by all vehicles with over 85% of the particulate
attributable to diesel cars. This estimate assumes no emission control on diesel passenger cars.

---

*This estimate does not reflect the influence of particle
 emission standards to be imposed as a result of the 1977
 Clean Air Act Amendments.

Table 6

Particulate Matter Emitted From Gasoline and Diesel Cars

% Diesel New Car Sales 1985 - 1990	Particulate Emissions in Tons by Gasoline Cars by 1990	Particulate Emissions in Tons by Light Diesel Cars by 1990
0	32,000	0
10	28,000	57,000
25	26,000	155,000

A means of enforceably, efficiently, and economically monitoring mobile source emissions is needed. This pilot study on particulate emissions from light-duty diesel vehicles employed several analytical tools including the Salmonella mutagenesis bioassay. The objective was to determine the feasibility of identifying factors which influence the mutagenicity of organic extracts of diesel particulate. Monitorable parameters currently used in particulate samples from a variety of sources include but are not limited to total suspended particulate, benzene soluble organics, and benzo-(a)-pyrene (BaP). Although BaP may not be the most biologically active component present in diesel exhaust emissions, an analytical scheme had been developed to measure BaP concentration rapidly and precisely in ambient air particulate (18-20) so this technique was used in this study.

ENGINEERING AND CHEMICAL PROCEDURES AND RESULTS

The particulate samples used in this study were collected using a dilution tunnel and sampling configuration described previously (21). In this case, all of the passenger car exhaust was diluted, but a fraction of the dilution tunnel contents was filtered. The filter samples were collected isokinetically on 20.32 by 25.4 cm (8"x10") Gelman type A glass fiber filters at a flow rate of 600 liters/min. The dilution tunnel flow was 10,000 liters/min. Each filter therefore represents 6 per cent of the total exhaust particulate mass.

The diesel vehicles used were a Volkswagen Diesel Rabbit, Mercedes 240D and Nissan - 4 cylinder. These diesel automobiles were operated on a chassis dynamometer using the

following standard driving cycles: hot start Federal Test
Procedure (FTP), cold start FTP, and 85 km/hr.

The fuels used in this study included those listed in
Table 7. These fuels were blended for this study (22) to
represent a cross section of diesel fuel available to the
public, including the Gulf National Average Fuel. EM 238-F
is a No. 2 diesel smoke test fuel with medium to high cetane
rating and medium-low aromatic content. EM 239-F is a No. 2
diesel Gulf National Average Fuel with medium to high cetane
rating and low aromatic content. EM 240-F is a No. 1 diesel
jet A fuel showing a high cetane rating and very low aromatic
content. This fuel was blended for aircraft use, but it can
be used in automotive power plants. EM 241-F is a No. 2
diesel minimum quality fuel having low cetane rating and a
high aromatic content. EM 242-F is a No. 2 diesel maximum
quality fuel having a high cetane rating and a low aromatic
content.

The particulate samples were treated as outlined in
Figure 8. A one by eight inch strip of the filter was cut
and processed. The samples were extracted for six hours in
a soxhlet extraction apparatus with 100 ml cyclohexane
(Burdick-Jackson) refluxing at a rate of eight times per
hour. The apparatus was allowed to cool to room temperature
and the extract transferred to a Kuderna-Danish concentrator,
which was placed in a water bath at a constant temperature of
50°C. To speed evaporation, the solvent surface was swept
with a stream of dry filtered nitrogen. The extract was
reduced to ten milliliters after two successive washes of
the container. Fifty microliters were removed and spotted on
a one centimeter channel of a 20x20 cm 20 per cent acetylated
cellulose TLC plate. The plate with samples, standards, and
blanks was developed to the 19 cm line in a solvent mixture
of 50 ml methylene chloride and 100 ml ethanol. The plates
were air dried and placed in a Perkin Elmer MPF-3 (or MPF
44A) fluorescence spectrometer. Each channel (18 total) was
scanned using an excitation wavelength of 388nm and read at
an emission wavelength of 430nm for BaP. All extraction and
fluorescence steps were carried out under filtered light
(Kodak Yellow Chrom II). A minimum detectable limit of
0.05 ng BaP per 50 microliter of extract was determined.
Extraction efficiencies for BaP from spiked blank filters
using cyclohexane were 98 $\pm$ 5% while recoveries from ambient
air filters were 93 $\pm$ 5% for cyclohexane, 94 $\pm$ 5% for benzene, and 88 $\pm$ 5% for methylene chloride.

Table 7

Diesel Fuels

Analysis	Couch #2	EM-238 F #2 Smoke Test	EM-239 F #2 Nat'l Avg.	EM-240 F #1 Jet A	EM-241 F #2 Min Qual	EM-242 F #2 Prem Qual
Gravity	34.6	35.9	35.9	43.9	32.4	38.7
Density	0.8515	0.8449	0.8449	0.8036	0.8618	0.8310
% Sulfur	0.17	0.27	0.17	0.011	0.30	0.27
Distillation						
IBP	376	394	369	344	378	377
10%	418	424	428	362	425	417
50%	498	498	496	397	483	488
90%	590	600	576	458	570	572
EP	642	658	624	506	610	610
% Carbon	84.62	83.64	83.12	83.98	84.36	84.00
% Hydrogen	14.80	14.72	14.96	14.86	13.95	14.98
FIA						
% Saturates	68.9	74.5	82.4	83.1	59.8	81.6
% Olefins	1.5	0.3	0.4	2.4	0.5	0.4
% Aromatics	29.6	25.2	17.2	14.5	39.7	18.0
Cetane Index	46.0	48.6	48.7	47.4	41.8	53.0

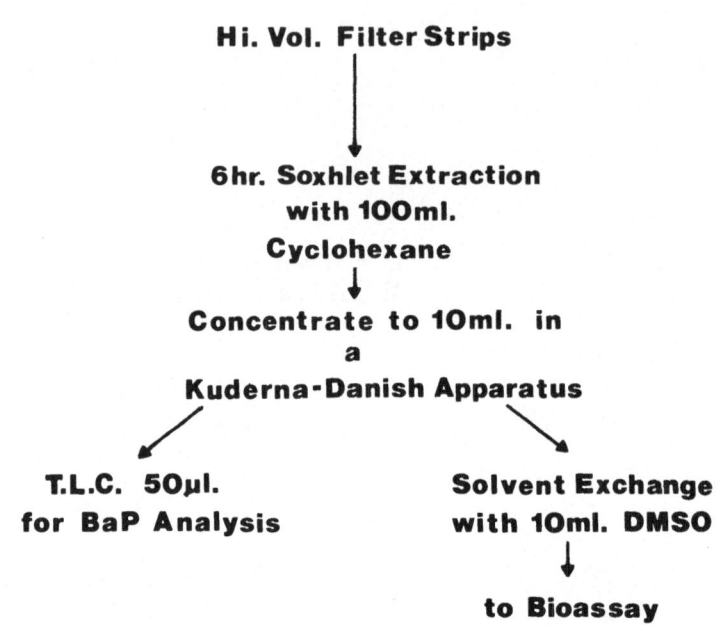

Figure 8. Extraction of organics from diesel exhaust particulates for benzo-a-pyrene analysis and bioassay.

The samples prepared for bioassay were solvent exchanged
with dimethysulfoxide (DMSO) in a Kuderna-Danish apparatus by
evaporating the cyclohexane to 3 ml and adding 10 ml DMSO.
The volume was then reduced to 7 ml under nitrogen and in a
50°C water bath. An additional quantity of DMSO was added to
bring the sample volume quantitatively to 10 ml. The samples
were placed in vials and frozen prior to bioassay.

BIOASSAY PROCEDURES AND RESULTS

The bacterial mutagenesis plate incorporation assay with
Salmonella typhimurium was performed according to the method
of Ames et al. (12) with the exception that the minimal histidine concentration was incorporated into the base layer of
the bacterial plates rather than into the overlay. None of
the light duty diesel exhaust samples have been chemically

fractionated. Instead, we examined either a dichloromethane (DCM) or a cyclohexane (CH) extract of the total exhaust.

The cyclohexane and dichloromethane extracted samples were solvent exchanged into DMSO as described above and either 0, 50, 100, 200 or 400 µl of the sample (except where indicated) were added to each plate. Strain 1538 was chosen for the experiments reported here due to limited sample size and the response of the total diesel extract observed previously with this strain as reported in Part I. All assays were performed in duplicate in the presence and absence of metabolic activation (MA). Average revertants per plate were calculated and adjusted by subtracting the spontaneous revertants from the control plates. The fold increase was calculated at each dose by dividing the number of revertants (rev./plate) in the treated plates by the control. The revertants/plate were plotted against the equivalent mg of particulate added and a linear regression was used to determine the slope. The specific activity in revertants/plate per 1 mg diesel particulate was calculated using this slope. The data acquired from the testing of light duty diesel are summarized in Table 8.

Variables inherent in the bioassay of organic extracts of diesel particulate were examined. These variables included the solvent systems and sample storage method and time. A comparison of two solvent systems was made using samples from two different engines each extracted with the different solvents, and tested in the bacterial plate incorporation test with and without activation. The dichloromethane (DCM) extracts gave consistently higher numbers of revertants per plate than did the cyclohexane (CH) extracts in either the presence or absence of metabolic activation, as shown in Figure 9. The DCM extracts were more mutagenic than the CH extracts when either fold increase or specific activity was compared (Table 8).

Since there is usually a time lapse of several days to several weeks before generated samples can be tested, storage may be an important factor. Equal portions of the same filters were taken for storage samples and fresh samples. Filter and extract samples were stored for eight weeks, refrigerated in sealed containers. The activities of the stored samples were then compared to fresh sample activity. Fresh samples were tested within 24 hours of collection. Results are illustrated in Figure 10. Whether or not metabolic activation is used, there seems to be some loss of mutagenic activity with storage. Direct acting components also seem to

Table 8

Response of Cyclohexane Extracts of Two Diesel Engines
Using Various Fuel Mixtures
Salmonella typhimurium – Plate Incorporation Test in Strain TA1538

SAMPLE:[1] Filter No. (Fuel-Solv)	Dose: mg Particulate	With M.A. Rev./Plate[2]	Adj. Rev./Plate[3]	Fold Increase[4]	Without M.A. Rev./Plate[2]	Adj. Rev./Plate[3]	Fold Increase	+MA Slope[5] & S.A.[6]	-MA Slope & S.A.	ngBaP/ mg Particulate
V.W. Rabbit										
8027 (238-CH)	2.48	170	129	4.14	65.5	40	2.57	50.3	17.5	6.97
	1.24	87	46	2.12	43.0	17.5	1.69	90.1	38.3	
	.62	79.5	38.5	1.94	25	-0.5	0.98			
	0.00	41		1.0	25.5		1.0			
8060 (239-CH)	2.0	73	32	1.78	29	3.5	1.14	14.1	1.45	14.52
	1.0	63	22	1.54	27	1.5	1.06	61.5	27.6	
	0.5	62	21	1.51	28	2.5	1.09			
8081 (240-CH)	2.08	209	168	5.10	74.5	49	2.92	80.6	24.3	10.31
	1.04	104	63	2.54	62.5	37	2.45	115.0	52.2	
	.52	77	36	1.88	37.5	12	1.47			
	0.00	41		1.0	25.5		1.0			
8125 (242-CH)	3.32	252	211	6.15	89.3	63.8	3.50	65.4	18.1	0.77
	1.66	132.5	91.5	3.23	60.5	35	2.37	94.7	49.0	
	0.83	71.5	30.5	1.74	53.5	28	2.10			
	0.00	41		1.0	25.5		1.0			
8140 (242-CH)	1.84	163	122	3.98	72	46.5	2.82	65.9	23.2	10.52
	.92	105.5	64.5	2.57	45.5	20	1.78	109.0	52.0	
	.46	75	34	1.83	47	21.5	1.84			
	0.00	41		1.0	25.5		1.0			
8096 (241-CH)	2.88	705	673	22.03	213.5	196.2	12.3	230.4	64.5	18.83
	1.44	485	453	15.16	208	190.7	12	315.2	127	
	.72	296.5	264.5	9.27	148.5	131.2	8.6			
	.36	150	118	4.69	73.5	56.2	4.2			
	0.00	32		1.0	17.3		1.0			

[1]Column 1 describes the sample used by engine type, the solvent extraction system used, and age of the sample.
[2]Rev./Plate: Average number of revertants per plate.
[3]Adj. Rev./Plate: Adjusted number revertants per plate (revertants/plate minus spontaneous revertants per plate).
[4]Fold Increase: (revertants/plate)/(spontaneous revertant/plate).
[5]Slope: Slope based on linear regression.
[6]S.A.: Specific Activity: Revertants/plates per one mg particulate.

Table 8 (continued)

SAMPLE: Filter No. (Fuel-Solv)	Dose: mg Particulate	With M.A. Rev./ Plate	Adj. Rev./ Plate	Fold Increase	Without M.A. Rev./ Plate	Adj. Rev./ Plate	Fold Increase	+MA Slope & S.A.	-MA Slope & S.A.	ngBaP/ mg Particulate
V.W. Rabbit										
8096 (241-DCM)	.467	971	928.5	22.59	507.5	486	23.6	1897.2	1000.7	----
	.234	730	687	16.98	392	370.5	18.23	2063.9	1086.5	
	.117	474.5	431.5	11.03	240.5	219	11.19			
	0.0	43		1.0	21.5		1.0			
8116 (241-CH)	3.52	1009.5	977.5	31.6	311	293.7	17.9	262.2	75.9	16.05
	1.76	876.0	844	27.4	310	292.7	17.9	482.4	173.2	
	.88	614.5	582.5	19.2	223	205.7	12.9			
	.44	299.5	267.5	9.4	126.5	109.2	7.3			
	0.0	32		1.0	17.3		1.0			
8116 (241-DCM)	.531	668.5	625.5	15.6	466	444.5	21.7	1068.6	762	----
	.256	753	710	17.5	356.5	335	16.6	1308.3	867	
	.128	472	429	11.0	275	253	12.8			
	0.0	43		1.0	22		1.0			
8104 (241-CH)	3.92	1083	1051	33.8	393.5	376.2	22.7	245.6 (476.2)*	88.7 (196.8)*	26.49
	1.96	973	941	30.4	399.5	382.2	23.1	524.2 (596.6)*	208.8 (242.7)*	
	.98	741	709	23.2	306.5	289.2	17.7			
	.49	369	337	11.5	135.5	118.2	7.8			
	0.0	32		1.0	17.3		1.0			
Mercedes										
8024-1 (238-CH)	3.16	67.5	26.5	1.7	23	-2.5	.71	8.12	0.43	0.76
	1.58	51	10	1.24	37.5	12	1.47	49.0	25.9	
	.79	49	8	1.20	18	-7.5	.90			
	0.0	41		1.0	25.5		1.0			
8024-2 (238-CH)	3.16	81	43	2.13	29	9.5	1.49	12.8	3.2	----
	1.58	51.5	13.5	1.36	25	5.5	1.28	51.1	22.3	
	.79	53.5	15.5	1.41	20.5	1.0	1.05			
	0.0	38		1.0	19.5		1.0			
8053-2 (239-CH)	3.12	138.5	100.5	3.64	76	56.5	3.90	31.1	18.4	1.87
	1.56	79	41	2.08	51.5	32	2.64	70.0	38.0	
	.78	70	32	1.84	32	12.5	1.64			
	0.0	38		1.0	19.5		1.0			

Table 8 (continued)

SAMPLE: Filter No. (Fuel-Solv)	Dose: mg Particulate	With M.A. Rev./ Plate	With M.A. Adj. Rev./ Plate	With M.A. Fold Increase	Without M.A. Rev./ Plate	Without M.A. Adj. Rev./ Plate	Without M.A. Fold Increase	+MA Slope & S.A.	-MA Slope & S.A.	ngBaP/ mg Particulate
Mercedes										
8053-1 (239-CH)	3.12	177	136	4.32	64.5	39	2.53	43.9	13.2	----
	1.56	88	47	2.15	43	17.5	1.68	76.7	35.4	
	.78	65	24	1.59	28	2.5	1.10			
	0.0	41		1.0	25.5		1.0			
8078-1 (240-CH)	2.40	70	29	1.71	33.5	8	1.31	11.9	2.2	1.47
	1.20	57	16	1.39	27.5	2	1.08	53.9	30.1	
	.60	50	9	1.22	34.5	9	1.35			
8078-2 (240-CH)	2.4	93	55	2.45	36.5	17	1.87	23.5	6.9	----
	1.20	48.5	10.5	1.28	29.5	10	1.51	54.2	27.3	
	0.60	42	4	1.11	25	5.5	1.28			
	0.0	38		1.0	19.5		1.0			
8088- (241-CH)	3.48	296	258	7.80	157.5	138	8.08	75.8	39.6	2.18
	1.74	197.5	159.5	5.20	91.5	72	4.69	117.1	60.5	
	.87	95.5	57.5	2.51	56	36.5	2.87			
	0.0	38		1.0	19.5		1.0			
8111- (241-CH)	3.68	716	678	18.84	354	334	18.2	180.9	91.6	
	1.84	454	416	11.95	261.5	242	13.4	258.9	133.2	
	.92	269	231	7.08	121	101.5	6.2			
		38		1.0	19.5		1.0			
8114 (241-CH)	4.16	261.5	223.5	6.88	94	74.5	4.82	54.9	17.8	
	2.08	142	104	3.74	82.5	63	4.23	84.6	46.4	
	1.04	77	39	2.03	48	28.5	2.46			
	0.0	38		1.0	19.5		1.0			
8133-1 (242-CH)	3.84	165	124	4.02	172	146.5	6.75	32.9	38.2	
	1.92	82	41	2.00	79	53.5	3.10	64.3	56.9 \	
	.96	58.5	17.5	1.43	55	29.5	2.16			
	0.0	41		1.0	25.5		1.0			
8133-2 (242-CH)	3.84	247	209	6.5	144	124.5	7.38	56.5	31.7	
	1.92	124	86	3.26	86	66.5	4.41	79.1	55.8	
	.96	61	23	1.61	60	40.5	3.08			
	0.0	38		1.0	19.5		1.0			

Table 8 (continued)

SAMPLE: Filter No. (Fuel-Solv)	Dose: mg Particulate	---With M.A.--- Rev./ Plate	Adj.Rev./ Plate	Fold Increase	---Without M.A.--- Rev./ Plate	Adj.Rev./ Plate	Fold Increase	+MA Slope & S.A.	-MA Slope & S.A.	ngBaP/ mg Particulate
Mercedes										
8142-1 (242-CH)	2.68	93.5	52.5	2.28	50	24.4	1.96	19.8	9.7	
	1.34	62	21	1.51	40	14.5	1.57	58.5	34.2	
	.67	51	10	1.24	28	2.5	1.09			
	0.0	41		1.0	25.5	1.0				
8142-2 (242-CH)	2.68	81	43	2.13	43	23.5	2.21	16.1	9.0	
	1.34	47	9	1.23	31	11.5	1.59	49.9	27.6	
	.67	44.5	6.5	1.17	23.3	3.8	1.20			
	0.0	38		1.0	19.5	1.0	1.0			

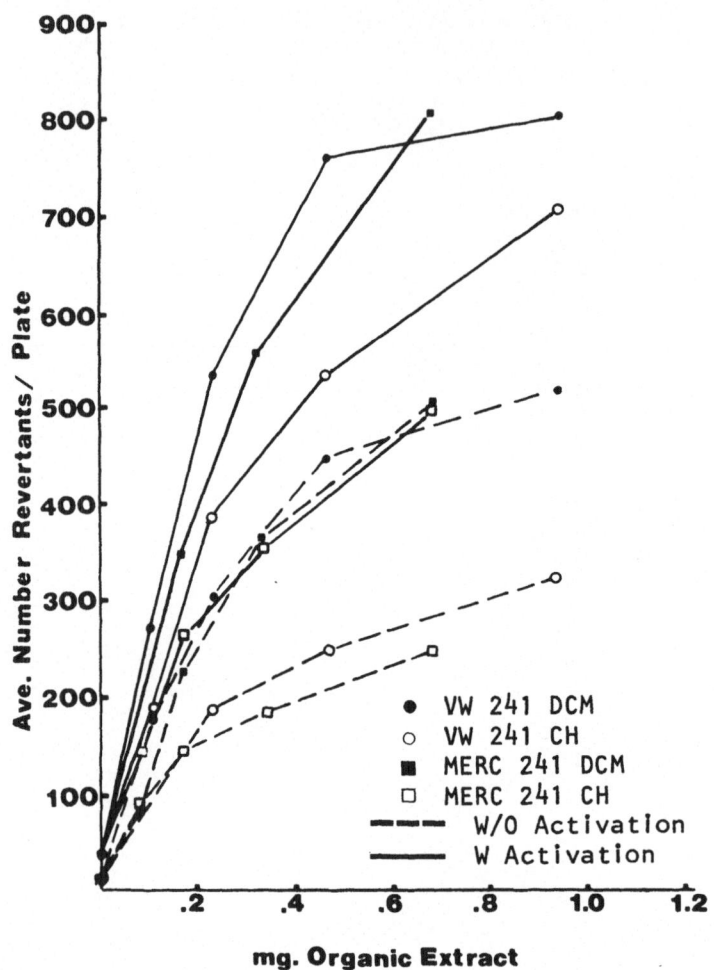

Figure 9.   Comparison of the mutagenic response of organics
extracted from diesel particulate with cyclohexane (CH) and
dichloromethane (DCM) in Salmonella typhimurium strain TA1538.

increase in toxicity with storage. If the linear portions of
the dose response curves are compared, the effect of storage
is minor. The major differences are apparent at higher sample
concentrations where the differences may be due to toxicity
factors. In no case did mutagenicity increase with storage.

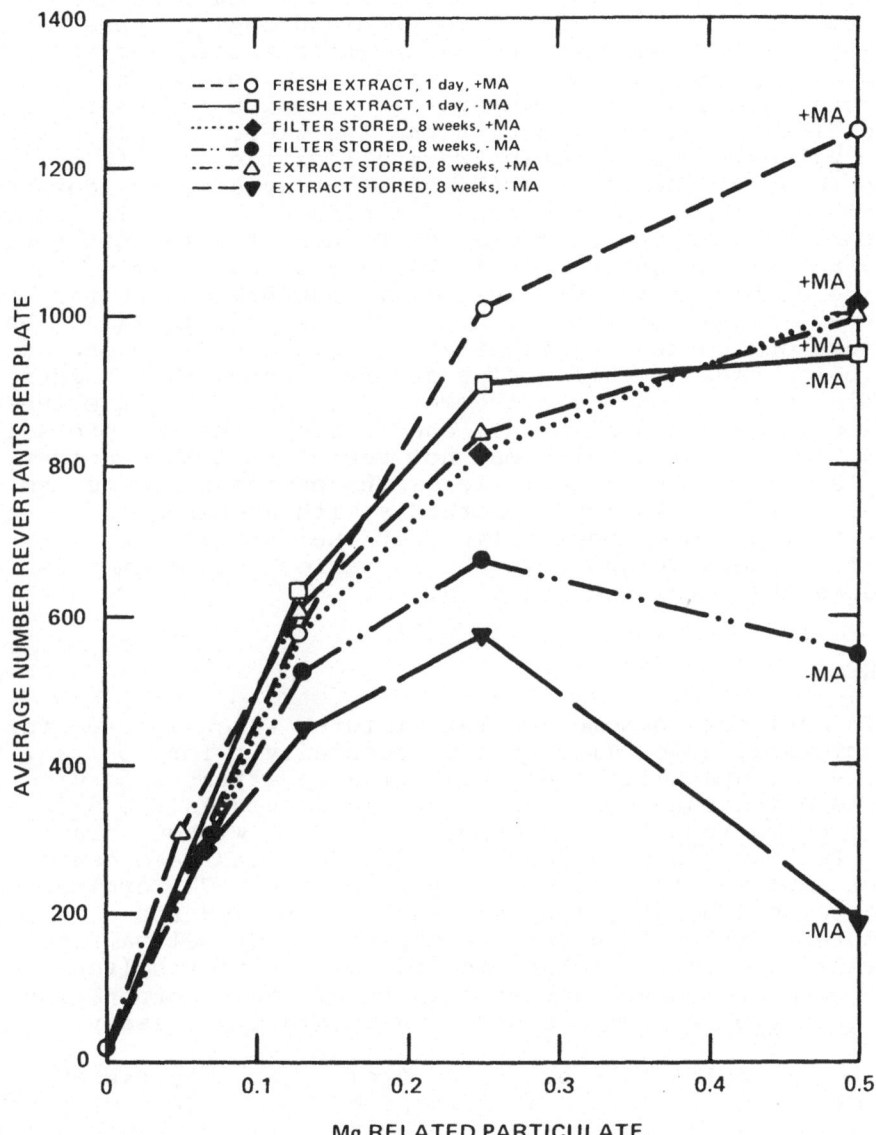

Figure 10. Effect of storage on the mutagenicity of organic extracts from diesel particulate when bioassayed in <u>Salmonella typhimurium</u> strain TA1538.

Under specific testing conditions a comparison can be
made between the different fuel types and engine types.
Diesel particle samples which had been extracted and analyzed
for BaP as part of a fuel study with diesel passenger cars
(22) were selected for bioassay. The samples chosen represented five fuels, two vehicles and the widest possible range
of BaP values. These samples were all extracted with cyclohexane (CH), solvent exchanged and bioassayed under identical
conditions. The data from these samples are shown in Table 9
and summarized in the histogram in Figure 11. In both vehicles, the minimum quality fuel (241) resulted in emissions
with the highest mutagenic activity. The BaP concentration
of the emissions was also highest with this fuel (241). Under these engine testing modes and with the cyclohexane extract being used, the VW engine generally created a higher
specific activity than did the Mercedes engine. Since these
vehicles were tested with only one testing mode and one solvent system, these results may not represent a true comparison of emissions characteristic of the engine being tested.
This histogram could change markedly with a change in solvent,
engine testing mode, engine type, or fuel characteristics;
however, it does demonstrate that a variety of parameters
influence the mutagenic activity.

SUMMARY

In contrast to some complex mixtures which are too toxic
to be bioassayed for microbial mutagenicity prior to fractionation, e.g., synthetic fuel (23), organic extracts of diesel
particle emissions were found to be mutagenic in the Salmonella typhimurium plate incorporation tests without fractionation. The mutagenic response is dependent on the organic
solvent employed to extract the particulate. The cyclohexane
extraction and benzo-(a)-pyrene analysis procedures developed
for ambient air particulate was applied to diesel particulate
emissions. Selected cyclohexane extracts, after solvent exchange were bioassayed directly in the plate incorporation
Salmonella typhimurium/microsome mutagenicity bioassay.

Diesel particulate emissions from light duty passenger
cars were found to have a wide range of both benzo-(a)-pyrene
content and mutagenic activity. The results from this pilot
study indicate that both methodologies are applicable to
evaluation of diesel particulate emissions. Variables which
affect these determinations, such as the extraction solvents
employed and the method of storage, need to be optimized and
subsequently standardized. This is particularly important

Table 9

Effect of Storage of Diesel Exhaust Samples from a Nissan Engine
on the Salmonella typhimurium Plate Incorporation Test in Strain TA1538

SAMPLE: Filter No. (Fuel-Solv.) [1]	Dose [1]	Organics mg.	------ With M.A. ------			------ Without M.A. ------			+MA Slope[5] & S.A.[6]	-MA Slope & S.A.
			Rev./ Plate[2]	Adj. Rev./ Plate[3]	Fold Increase[4]	Rev./ Plate	Adj. Rev./ Plate	Fold Increase		
Nissan:										
DCM Fresh Extract One Day	400	1.0	850	825	34.00	---	---	---	2361	1695
	200	0.5	1246	1221	49.84	946	929	55.65	2569	1951
	100	0.25	1043	1018	40.72	914	897	52.76		
	50	0.125	583	558	22.32	631	614	36.12		
	0	0.0	25	0	1.0	17	0	1.0		
DCM Filter Stored Eight Weeks	200	0.5	1045	1012	31.67	546	526	27.30	1922	891
	100	0.25	828	795	25.09	678	658	33.90	2116	1137
	50	0.125	583	550	16.67	525	505	26.25		
	25	0.063	283	250	8.58	298	278	14.90		
	0	0.0	33	0	1.0	20	0	1.0		
DCM Extract Stored Eight Weeks	200	0.5	1022	989	30.97	191	171	9.55	1849	2086
	100	0.25	846	813	24.64	572	552	28.60	2068	2190
	50	0.125	612	579	17.55	442	422	22.10		
	25	0.063	311	278	8.42	297	277	14.85		
	0	0.0	33	0	1.0	20	0	1.0		

[1] Column 1 describes the sample used by engine type, the solvent extraction system used, and age of the sample.

[2] Rev./Plate: Average number of revertants per plate.

[3] Adj. Rev./Plate: Adjusted number revertants per plate (revertants/plate minus spontaneous revertants per plate).

[4] Fold Increase: (Revertants/plate)/(spontaneous revertant/plate).

[5] Slope: Slope based on linear regression.

[6] S.A.: Specific Activity: Revertants/plates per one mg particulate.

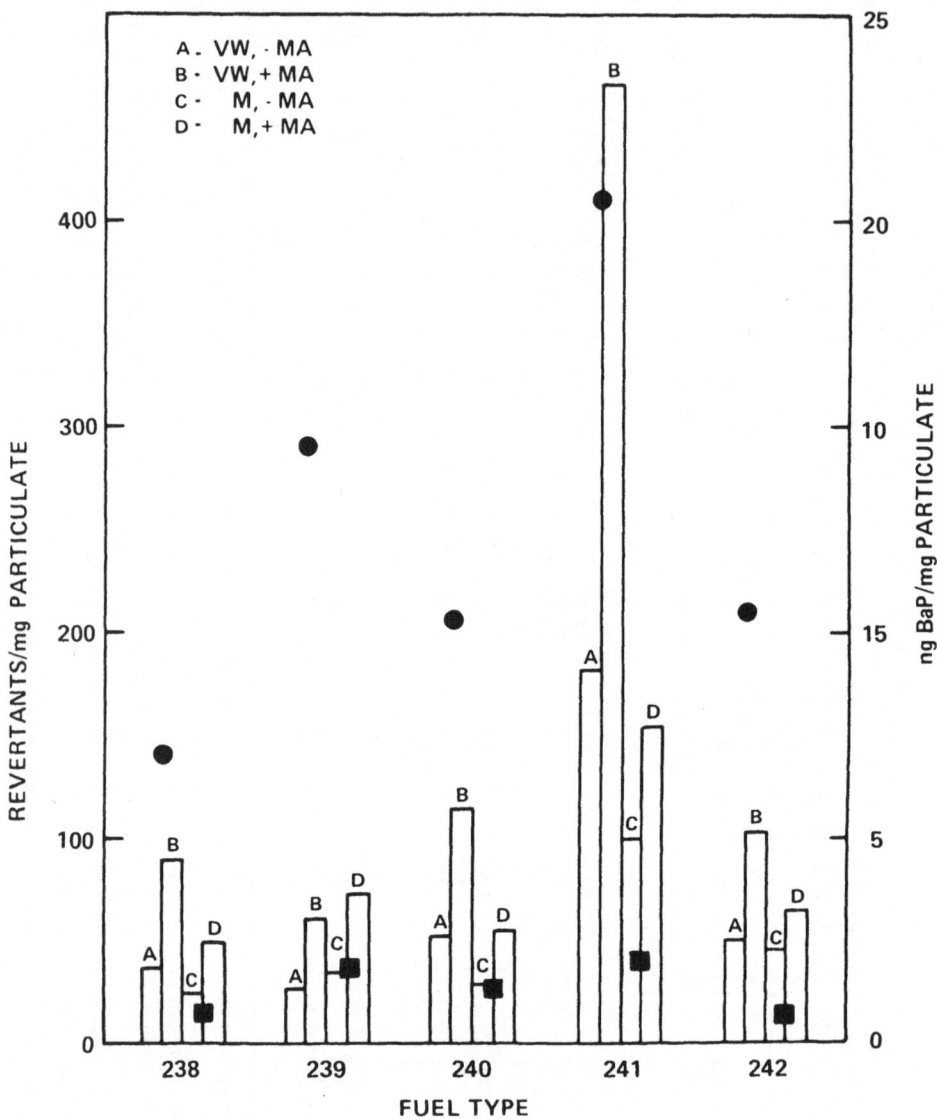

Figure 11. Comparison of the mutagenic activity in <u>Salmo-nella typhimurium</u> strain TA1538 and benzo-a-pyrene content of organic extracts from diesel particulate collected from two diesel passenger cars and five fuels.

with the solvent employed since both the absolute mutagenic response and relative response in comparing samples was found to change when dichloromethane extracts from the same filter were compared to cyclohexane extracts.

Storage of samples, either on filters or as extracts resulted in slight decreases in specific mutagenic activity when the linear portions of the dose response curves were compared. The absolute number of revertants per plate at higher, non-linear concentrations was more clearly reduced after storage. The loss in mutagenic activity was more pronounced in samples tested without metabolic activation and may have been due to an increase in direct acting toxic compounds as a result of sample storage.

Although sample replication was limited in this pilot study, it appears that the mutagenicity of the particulate emissions is influenced by the fuel and to a lesser extent by the vehicle. Both the BaP content and the mutagenic activity of the emissions were the highest when the minimum quality fuel (241) was used. This fuel has the lowest cetane value, highest aromatic content, and highest nitrogen content of the five fuels compared. The relationship of mutagenic activity to other fuel variables is being explored.

Sample 1804 with the highest BaP content (26.5 ng/mg particulate) provided only 0.1 µg BaP per plate; therefore this sample would not contain sufficient concentrations of BaP to be detectable in the mutagenesis bioassay (the minimum detectable limit in the plate incorporation assay is approximately 1 µg per plate). As described in Part I of this paper, the most mutagenic fractions were not the fraction (ARM) in which BaP would be found. Nevertheless, the sample containing the highest concentration of BaP was found to have the highest mutagenic specific activity, suggesting that BaP may be a useful indicator chemical.

ACKNOWLEDGMENTS

The authors wish to acknowledge the following technical support: D. Swanson, R. Hedgecoke, and C. Morris for the assistance in benzo-a-pyrene analysis; J. Hein for the analysis of physical properties of fuel; P. McBride and H.G. Shan for technical assistance in microbial mutagenesis; T. Baines and SWRI personnel for the light duty diesel particulate samples from which the exhaust particulate samples were obtained for the light duty pilot study Part II.

REFERENCES

1.   Springer KJ, Asby HA: The low emission car for 1975--
     enter the diesel.   SAE Paper No. 739133, Philadelphia,
     PA, August 1973

2.   Springer KJ, Stahman RC: Emissions and economy of four
     diesel cars.   SAE Paper No. 750332, Detroit, MI, Feb 1975

3.   Braddock JN, Bradow RL: Emissions patterns of diesel-
     powered passenger cars.   SAE Paper No. 750682, Houston,
     TX, June 1975

4.   Hare CT, Springer KJ, Bradow RL: Fuel and additive
     effects on diesel particulate-development and demonstra-
     tion of methodology.

5.   Braddock JN, Gabele PA: Emissions patterns of diesel-
     powered passenger cars--Part II.   SAE Paper No. 770168,
     Detroit, MI, Feb 1977

6.   Springer KJ, Baines TM: Emissions from diesel versions
     of production passenger cars.   SAE Paper No. 770818,
     Detroit, MI, Sept 1977

7.   Springer KJ: Investigation of diesel-powered vehicle
     emissions VII.   EPA Report No. EPA-460/3-76-034, Feb 1977

8.   Hare CT: Characterization of diesel gaseous and particu-
     late emissions.   Final Report on EPA Contract No. 68-02-
     1777, Sept 1977

9.   Beltzer M, Compion RJ, Petersen WL: Measurement of
     vehicle particulate emissions.   SAE Paper 740286, 1974

10. Begeman CR, Jackson IW, Nebel GJ: Sulfate emissions
     from catalyst-equipped automobiles.   SAE Paper 741060,
     1974

11. Swain AP, Cooper JE, Stedman RL: Large scale fraction-
     ation of cigarette smoke condensate for chemical and
     biological investigations.   Cancer Res 29:579-583, 1969

12. Ames BN, McCann J, Yamasaki E: Methods for detecting
     carcinogens and mutagens with the Salmonella/mammalian
     microsome mutagenicity tests.   Mutat Res 31:347-364, 1975

13. McCann J, Choi E, Yamasaki E, Ames BN: Detection of carcinogens as mutagens in the Salmonella microsome test: Assay of 300 chemicals. Proc Nat Acad Sci USA 72:5135-5139, 1975

14. Ames BN, Gurney EG, Miller JA, Bartsch H: Carcinogens as frameshift mutagens: Metabolites and derivatives of 2-acetylaminofluorene and other aromatic amine carcinogens. Proc Nat Acad Sci USA 69:3128-3132, 1972

15. Ames BN, Lee FD, Durston WE: An improved bacterial test system for the detection and classification of mutagens and carcinogens. Proc Nat Acad Sci USA 70:782-786, 1973

16. Ames BN, Durston WE, Yamasaki E, Lee FD: Carcinogens and mutagens: A simple test system combining liver homogenates for activation and bacteria for detection. Proc Nat Acad Sci USA 70:2281-2285, 1973

17. McCann J, Spingar NE, Kobori J, Ames BN: Detection of carcinogens as mutagens: Bacterial tester strains with R factor plasmids. Proc Nat Acad Sci USA 72:979-983, 1975

18. Swanson D, Morris C, Hedgecoke R, Bumgarner J, Jungers, R: A rapid analytical procedure for the analysis of benzo(a)pyrene in environmental samples, in press. EMSL, MD-78, Research Triangle Park, North Carolina

19. Human population exposure to coke oven atmospheric emissions, pp 64-67, EPA draft report, OAQPS (J Manning, MD-12), US Environmental Protection Agency, Research Triangle Park, North Carolina

20. Human population exposure to coke oven atmospheric emissions, p 48, EPA draft report, OAQPS (J Manning, MD-12), US Environmental Protection Agency, Research Triangle Park, North Carolina

21. Bradow RL, Moran JB: Sulfate emissions from catalysts cars--A review. SAE Paper No. 750090, 1975

22. EPA Contractor 68-02-2417 with Southwest Research Institute

23.   Epler JL, Young JA, Hardingree AA, Rao TK, Guerin MR,
      Rubin IB, Ho CH, Clark BR: Analytical and biological
      analysis of test materials from the synthetic fuel
      technologies. I. Mutagenicity of crude oils determined
      by the Salmonella typhimurium/microsomal activation sys-
      tem. Mutat Res, in press

MEASUREMENT OF BIOLOGICAL ACTIVITY OF AMBIENT AIR MIXTURES USING A MOBILE LABORATORY FOR *IN SITU* EXPOSURES: PRELIMINARY RESULTS FROM THE *TRADESCANTIA* PLANT TEST SYSTEM

L.A. Schairer and J. Van't Hof
Biology Department
Brookhaven National Laboratory
Upton, New York

C.G. Hayes and R.M. Burton
Health Effects Research Laboratory
U.S. Environmental Protection Agency
Research Triangle Park, North Carolina

Frederick J. de Serres
National Institute of
Environmental Health Sciences
Research Triangle Park, North Carolina

A variety of short-term bioassays has been developed to
assess the mutagenicity of industrial chemicals. Many of
these assays work well when used under laboratory conditions
but are not suitable for monitoring ambient air under field
conditions. To facilitate exposures of biological systems
to ambient air pollution in natural or industrial sites a
plan was implemented to design, assemble, and test a mobile
laboratory. The _Tradescantia_ plant test system was chosen
for these initial field studies because of its high sensitivity to both physical and chemical mutagens and its versatility and adaptability to monitoring the mutagenicity of
gaseous pollutants. Positive results to date support the
further development of the mobile laboratory and _Tradescantia_
system as a useful method for monitoring biological activity
of complex environmental mixtures _in situ_.

Several species of the family _Commelinaceae_, of which
Tradescantia is a member, have features particularly well
suited for certain radiation and chemical mutagen studies.
The effects of chemicals and/or ionizing radiation that are
easily measured include the following:

- Chromosome aberrations in microspores, root tips,
  and stamen hairs

- Somatic mutations in petals and stamen hairs in
  clones heterozygous for flower color

• Pollen abortion

• Cell sterility in stamen hairs

Of the four features mentioned, somatic mutation in stamen hairs is the most versatile as it requires the least complicated techniques and is more sensitive than the other endpoints to both physical and chemical mutagens. The pattern and magnitude of response of phenotypic changes in pigmentation in stamen hair cells have been studied after treatment with X rays (10), gamma rays (4,8), ^{3}H-β rays (Schairer LA, unpublished data), nitrogen ions (12), monoenergetic neutrons (13), and low gravity of space flight (7). X-ray and neutron dose-response curves as well as those for chronic gamma exposures show straight-line relationships over wide dose ranges with no evidence of a threshold dose even at levels as low as 250 mrad of X rays, 10 mrad of 0.43 MeV neutrons and 33 mR/h of ^{137}cesium gamma (8,10).

The significant mutagenic response to an accidental exposure to a gaseous chemical (5) as well as the high radiosensitivity were factors that prompted the use of Tradescantia as a test system to assay for the mutagenicity of various chemicals and air pollutants (9,11). Newly developed chemical exposure and dosimetric techniques verified the high sensitivity of the Tradescantia stamen hair system to gaseous chemical mutagens and these demonstrated its potential for monitoring ambient air pollution for mutagenicity (2,6,9).

Individual compounds or air pollutants can best be studied in the laboratory, but the mutagenicity of unusual and even unique ambient mixtures in urban or industrial sites must be assayed in the field. Perhaps the greatest advantage the stamen hair system affords over other test organisms is its versatility and adaptability to field studies.

THE TRADESCANTIA STAMEN HAIR SYSTEM

The stamen hair system has been described in detail elsewhere (1,11) so only certain features will be reviewed here. The plant used exclusively in the field studies to be described here is clone 4430, an interspecific hybrid (T. subacaulis x T. hirsutiflora) produced at Brookhaven (Figure 1a). This clone is a hybrid between pink- and blue-flowering parents with blue being dominant over pink. The

Figure 1(a). Normal stock plant of Tradescantia clone 4430
showing several mature inflorescences.

visible marker used in this test system is the phenotypic
change in pigmentation from blue to pink in mature flowers.
The pigmentation change (hereafter called mutational or pink
events) is induced in young developing floral tissue and is
expressed 5 to 18 days later as isolated pink cells or groups
of pink cells in the stamen hairs of mature flowers (Figure
1b, c). The pink events are essentially nonlethal so large
mutant sectors indicate genetic injury early in the development of that tissue.

The stock plants are easily maintained by vegetative
propagation and flower continuously throughout the year in
controlled-environment growth chambers. The material treated
consists of a group of unrooted, fresh cuttings containing
young inflorescences which contain flower buds in a range of
developmental stages as shown in Figures 1b and 2. Following
exposure to either chemical or physical mutagens, the cuttings
are grown in aerated Hoagland's nutrient solution under standard conditions and the flowers are analyzed each day as they
bloom for approximately three weeks after treatment. Induced
pink-event rates are expressed as the mean of the rates for
several consecutive peak response days, usually days 11 to 15
for acute X rays and 7 to 12 for acute chemical exposures
(Figure 2). Detailed descriptions of laboratory techniques
for radiation and chemical exposures and calculating mutation
rates are given elsewhere (7,9,11). The only modification
that has been adapted in the scoring method is that any inter-
rupted series of pink cells within one hair is considered to
be the result of a single mutational event (1). This conservative approach has only a slight effect upon the mutational
frequency at the levels described in this paper. The techniques for field exposures are new and, although described
briefly by Schairer et al. (3), they are reviewed below.

THE MOBILE MONITORING VEHICLE

The vehicle selected for the mobile monitoring project
was a 24-foot Clark mini-van trailer. The trailer shown in
Figure 3 was insulated and air conditioned to permit year-
round operation of the laboratory. In order to maintain a
semiclean environment for these studies, the trailer air was
recirculated through activated charcoal and HEPA particulate
filters. Three Model M-13 growth chambers (Environmental
Growth Chambers, Chagrin Falls, Ohio) were installed. One of
the chambers serves as a clean air control, the second is used
for ambient air exposure and the third is used as a backup
unit for either control or ambient air exposures (Figure 4).

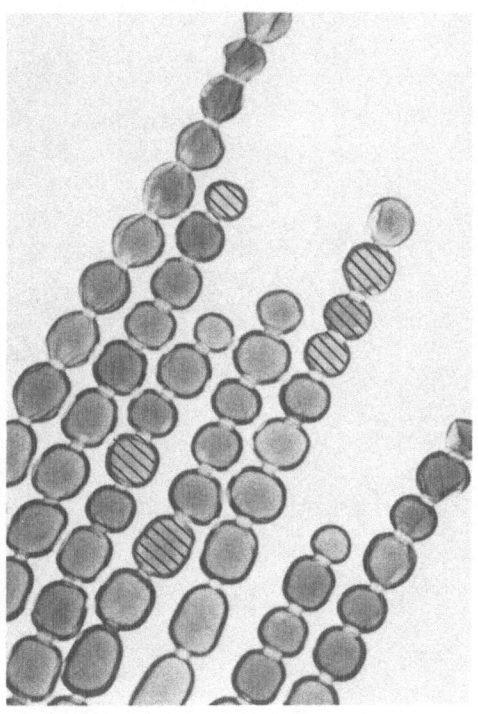

Figure 1(b). Enlarged view of single inflorescence show-ing range in bud size from meiotic stage to mature flower.

Figure 1(c). Enlargement of stamen hairs with pink mutant events indicated by shading. Mutant events in the flower color locus are not usually lethal; chains of pink cells represent daughter cells of the initial mutated cell.

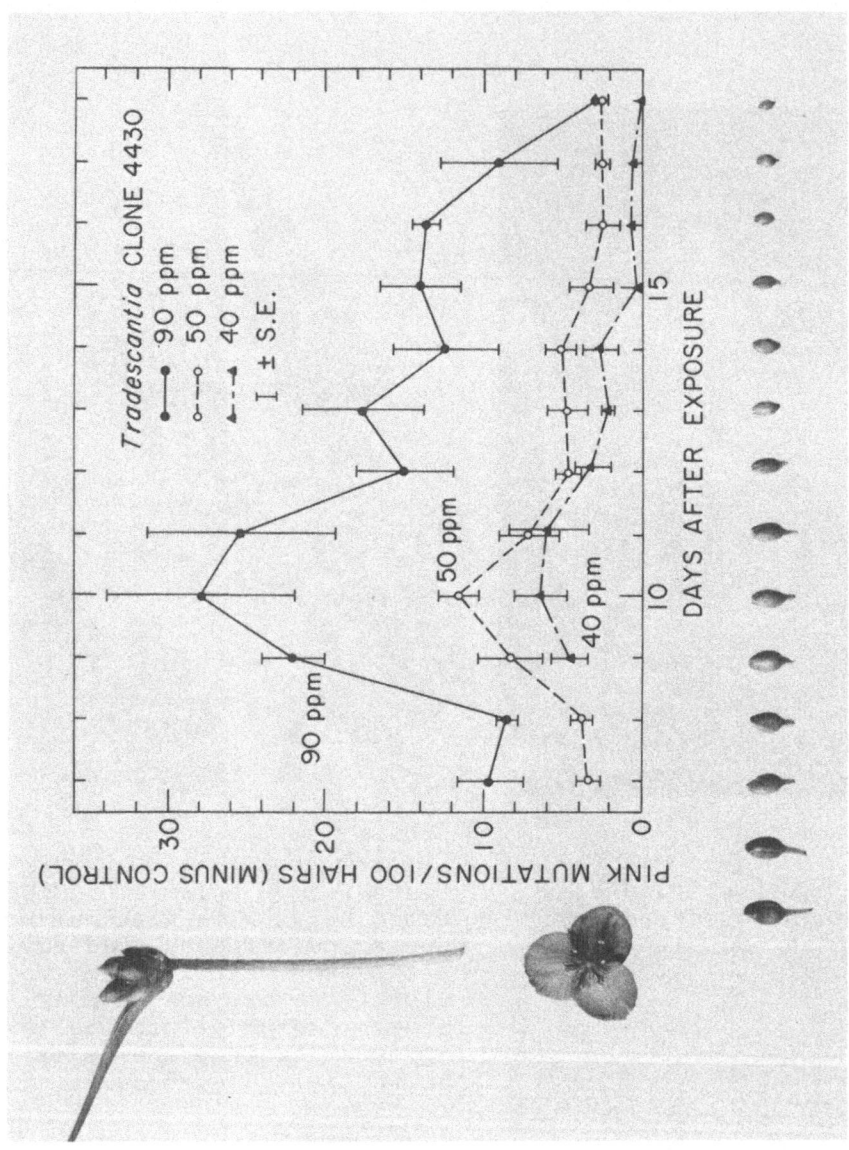

Figure 2. Pink mutation rates in Tradescantia stamen hairs following 6-hour exposure to gaseous EMS reach peak values 7-12 days after exposure. Pink events are analyzed in mature flowers; the corresponding bud sizes at time of treatment are shown along the abscissa.

Figure 3. View of the front of the Mobile Monitoring Vehicle (MMV) showing remote mounting of the air conditioning and filter train for the trailer and the three heat exchangers for the growth chambers.

Figure 4. Interior of mobile monitoring vehicle (MMV) showing rear exposure chamber (with cuttings) and a control chamber on the right. Round air filter cannisters are mounted on
brackets above chamber door.

Ambient air is drawn into the fumigation chamber through a four-inch glass duct at continuous flow rates up to about 18 cubic feet per minute, a maximum of one air change every two minutes. Each chamber is equipped with an air filter train composed of activated charcoal and HEPA particulate filters. This filter train is used to scrub the air continually in the chamber serving as the concurrent control. The total external electrical power requirement for the trailer air conditioning and chamber operation is a 100 amp, 220 volt service.

FIELD EXPOSURE TECHNIQUE

Field exposures were accomplished in the following manner: fresh cuttings of Tradescantia clone 4430 were made from stock plants grown in controlled environment chambers at Brookhaven National Laboratory; they were hand-carried to the test site by car or airplane; cuttings were placed in the chambers in glass containers filled with Hoagland's nutrient solution, and exposures were made for a ten-day period. At the end of the exposure the cuttings were taken back to Brookhaven National Laboratory for posttreatment analysis of the flowers as they bloomed each day. The peak mutation response period following a ten-day exposure is 11 to 17 days after the start of the exposure. The mean of the mutation rates for the seven-day scoring period resulted in an observed rate for a given test site based on an average stamen hair population between 300,000 and 400,000. A population of 300 cuttings in each ambient air and control chamber will yield enough data to resolve as small as a 10% increase in pink events over the background frequency.

CHEMICAL EXPOSURES UNDER LABORATORY CONDITIONS

Exposures to a standard chemical mutagen, the alkylating agent 1,2-dibromoethane (DBE), in the gaseous state, showed that the number of mutational events increased linearly with the product of concentration and hours of exposure to DBE, at least over the range from 2 to 144 hours. These data may be expressed in terms of total dose by plotting induced mutation frequency against the product of concentration (ppm) and duration of exposure (hours) (Figure 5). For purposes of comparison, a standard curve for X-ray effect is shown in rads. Slope and shape of the curve for DBE induction of color change resemble those for radiation injury.

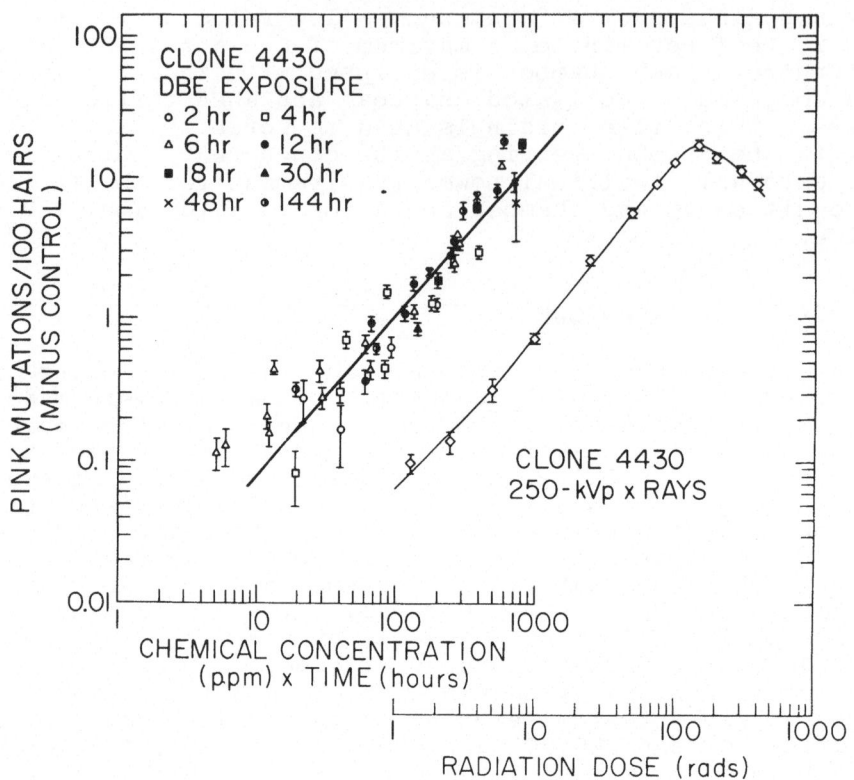

Figure 5. Stamen hair mutation frequencies from several
experiments are plotted against total dose of 1,2-dibromo-
ethane (DBE) (ppm x hours of exposure). A linear response
curve fits all data points from 2- to 144-hour exposures.
The standard acute X-ray curve is shown for comparison.

        Although a large percentage of the effort of this group
has been spent on the development of the mobile monitoring
vehicle, a number of chemicals have been tested in the laboratory to validate the system as a monitor for gaseous mutagens. Typical dose-reponse curves for several chemicals

are shown in Figure 6. Chemicals such as the gasoline additives 1,2-dibromoethane (DBE) and trimethyl phosphate (TMP) were found to be potent mutagens while SO_2, NO_2, vinyl chloride, and freon-12 were weak mutagens according to this test system. Other chemicals or air pollutants tested are listed in Table 1. The concentration listed is the lowest value tested that showed a significant mutagenic response.

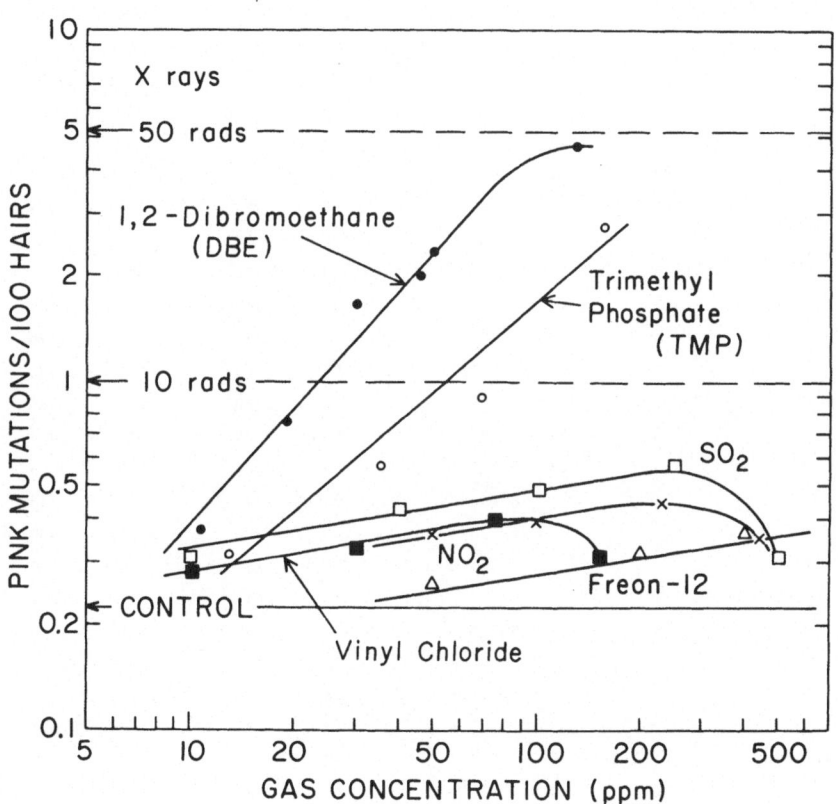

Figure 6. Typical dose-response curves for pink events in *Tradescantia* clone 4430 are shown following 6-hour exposures to various gaseous compounds.

Table 1

Summary of Mutation Response Data for Various Chemicals Used on Clone 4430 in Terms of Lowest Concentration Giving Significant Effect

Chemical	Exp. Time (hr)	Min.* Conc. (ppm)	Hairs Scored $(x10^3)$	Total Pink Events	Pink Events per 100 Hairs (-Control)	± SE	Stat. Sig. (%)
Air Pollutants							
Ozone (O_3)	6	5.0	48	153	.098	.040	2%
Sulfur Dioxide (SO_2)	6	40	41	170	.222	.041	1%
Nitrogen Dioxide (NO_2)	6	50	24	87	.112	.056	5%
Nitrous Oxide (N_2O)	6	250	29	115	.117	.055	1%
Industrial Chemicals							
Ethyl Methanesulfonate (EMS)	6	5	20	246	1.012	.133	1%
1,2-dibromoethane (DBE)	6	1	258	1088	.118	.027	1%
	144	0.14	148	1119	.315	.035	1%
Trimethylphosphate (TMP)	6	13	32	115	.125	.051	2%
Trichloroethylene (TCE)	6	0.5	44	148	.112	.036	1%
Vinyl Chloride (VC)	6	75	34	133	.112	.046	2%
	24	25	56	281	.151	.041	1%
Vinylidene Chloride (VDC)	6	86	30	130	.064	.056	Insig
	24	22	100	338	.057	.028	5%
Vinyl Bromide (VB)	24	50	49	201	.159	.048	1%
2-Bromoethanol (2BE)	6	24	33	131	.107	.046	2%
Freon-12 (Fr-12)	6	392	32	103	.095	.059	Insig
Freon-22 (Fr-22)	6	194	66	249	.100	.039	2%
Hexamethylphosphoramide (HMPA)	6	?	48	314	.277	.051	1%
Benzene	6	4000	43	292	.287	.063	1%
Caffeine	Chronic	$10^{-4}M$	39	142	.047	.040	Insig
Atrazine	Chronic	0.045g/pot	93	260	.0	.0	Insig
Sodium Azide	3	$10^{-4}M$	19	96	.269	.055	1%
1,1-dibromoethane	6	58	56	219	.073	.039	Insig
Dimethylamine Hydrochloride	2	$10^{-2}M$	16	83	.151	.080	Insig
Vapona	6	Sat?	81	278	.0	.0	Insig

*Minimum concentration used which showed a significant increase over background mutation rate.

RESULTS OF EXPOSURE TO AMBIENT AIR POLLUTION

The first field trials for the mobile monitoring vehicle (MMV) were conducted in the summer of 1976. A location was sought which had high levels of a mixture of pollutants and was within about a two-hour drive from Brookhaven National Laboratory.

The first test site selected was Elizabeth, NJ beside a NJ air pollution monitoring station. The NJ Turnpike, toll plaza, petroleum refineries, Newark Airport, and other indus- trial pollution sources surrounding this test site are shown diagrammatically in Figure 7. When two-week exposures were made in July and October 1976 and January 1977, the data indicated increases in mutation frequencies, following expo- sure to ambient air, which were significant at the 1% level for all three periods (Table 2). In the third two-week exposure, January 1977, two chambers were exposed to ambient air to demonstrate that the induced effects observed in the previous two runs were real and not a unique chamber effect in the third control chamber. Data from the ambient air samples were not different from each other, but both were significantly higher than the concurrent control. Apparently no unique chamber effect exists between chambers, even under field conditions.

Wind direction is an important factor in the location of a mobile monitoring unit. The high induced mutation rate in July occurred with prevailing southwesterly winds, while the October run had prevailing northwesterly winds (Figure 7). Pollution sources were certainly different in these two exposures, but a much more sophisticated air monitoring facility and a detailed map of industrial and natural pollu- tion sources in the greater Elizabeth area would be required to identify the environmental mutagen(s) and its probable source.

These data were encouraging and supported the use of the Tradescantia test system as a field monitor for air pol- lution. To continue the study, a series of exposures was planned in collaboration with the U.S. EPA Epidemiology and Measurements Sections. Test sites were selected because of high cancer mortality or presumed exposure to high levels of carcinogens. The MMV experiments were to look for bio- logical activity, while an EPA mobile monitoring van made real-time measurements of the pollution levels. Organic vapors were collected on Tenax absorbers for subsequent identification by Dr. Edo Pellizzari. The sites selected

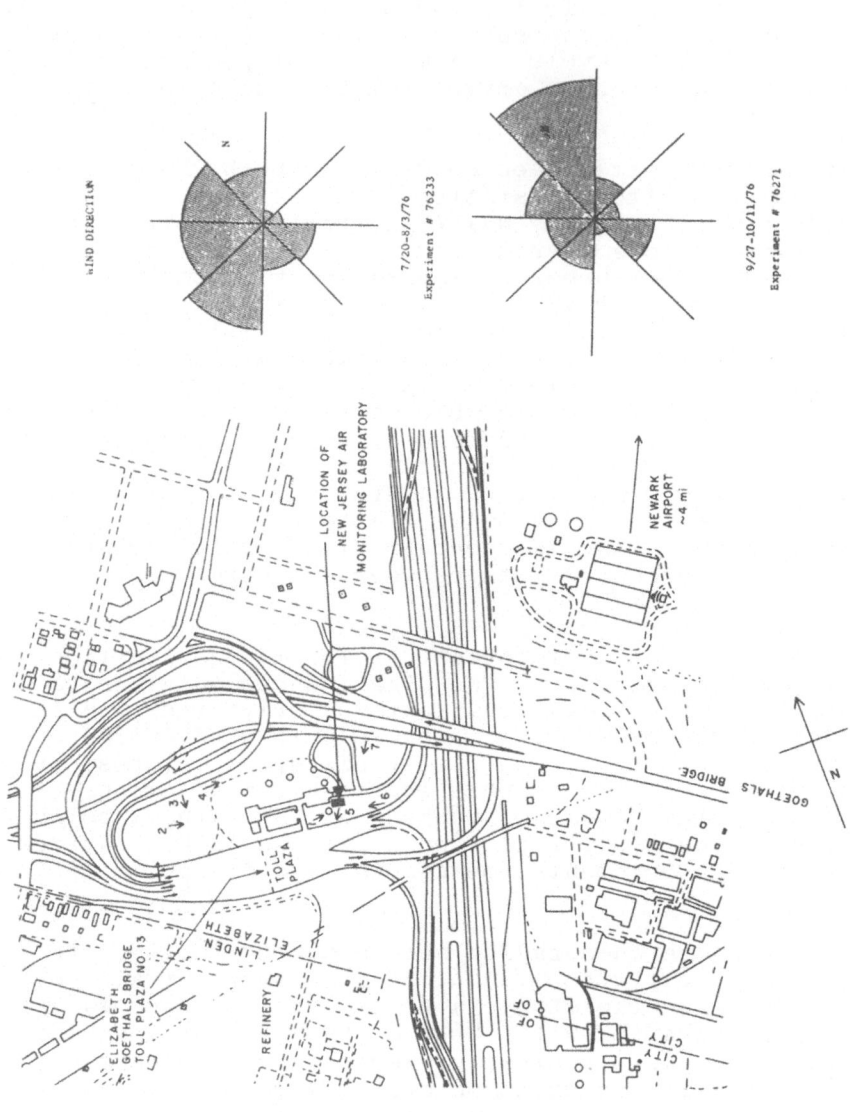

Figure 7. Diagrammatic view of New Jersey test site showing some of the sources of high ambient air pollution. MMV was located beside New Jersey Air Monitoring Laboratory (see arrow). Prevailing winds are shown for two experiments.

Table 2

Mutagenicity of Ambient Air at Elizabeth, NJ
as Measured by <u>Tradescantia</u> Stamen Hairs

Treatment	No. Flowers	No. Hairs	No. Pink Events	Events/Hair ± S.E.
Control	726	299,475	1182	.00395 ± .00013
Ambient Air	658	268,464	1386	.00516 ± .00016
7/20-8/3/76	Ambient Air Minus Control			.00122 ± .00021*
Control	892	350,824	1487	.00424 ± .00012
Ambient Air	890	358,047	1727	.00482 ± .00012
9/27-10/11/76	Ambient Air Minus Control			.00058 ± .00012*
Control (1)	689	266,023	872	.00328 ± .00012
Ambient Air (2)	742	291,161	1146	.00394 ± .00013
	Ambient Air (2) Minus Control			.00066 ± .00017*
Ambient Air (3)	617	231,557	873	.00377 ± .00014
	Ambient Air (3) Minus Control			.00049 ± .00018*
Ambient Air (2+3)	1359	522,718	2019	.00386 ± .00009
1/21-2/4/77				
	Ambient Air (2+3) Minus Control			.00058 ± .00015*

*Significant at the 1% level.

for this phase of the study were: Charleston, WV, Birming-
ham, AL, Baton Rouge, LA, Houston, TX, Upland, CA, Magna, UT,
and Grand Canyon, AZ. The latter site at Grand Canyon served
as a clean air control study.

The results of these field exposures are summarized
graphically in Figure 8. The pollution sources indicated
here are only general categories under the heading of the
major industries in the areas and do not imply a known correlation between mutation response and specific industrial
effluent. Statistically significant increases in mutant
event frequencies above control levels were observed at
Elizabeth, Charleston, Baton Rouge, and Houston. The

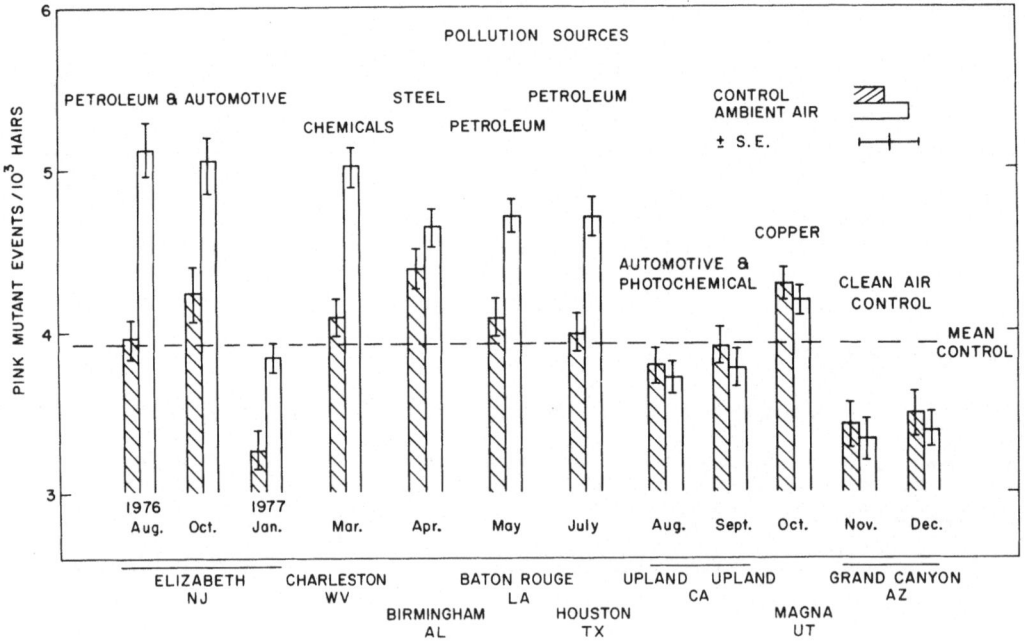

Figure 8. The mutagenicity of ambient air as measured by
Tradescantia in the mobile monitoring vehicle is summarized
for the eight test sites visited.

remaining locations, especially the clean air site at Grand
Canyon, showed no significant response to ambient air.
These data are also shown in Table 3 arranged by pollution
source and presented as a pollution-induced increase in
mutation rate as percent of control. Locations associated
with petroleum refining and mixed chemical processing gave
increases ranging from 31% down to 17%. The real-time measurements of both organic and inorganic compounds are being
analyzed at the present time, and when completed, these
results may provide more specific identification of compounds
common to those sites showing induced mutations. If suspect
compounds are identified they can be tested individually
under controlled laboratory conditions using existing techniques.

It should be emphasized that a negative response in a
single exposure of a test organism may provide inadequate
assurance of absence of a health hazard. As pointed out in
the Elizabeth experiment, the prevailing wind direction
changed from summer to fall and the induced mutation frequency
dropped from 31% to 18%. Wind direction, amount of precipitation, industrial complex work schedule, etc., all have a
direct bearing on the pollution mixture and level at a fixed
monitoring location.

CONCLUSION

The body of evidence is growing for a meaningful extrapolation from cytological and genetic effects in microorganisms, cell cultures, plants, insects, and mammals to health
hazards in man. The high correlation between mutagenicity
and carcinogenicity supports the use of visible genetic markers in test organisms as monitors for carcinogens. The
observation of similar chromosome aberrations in both gametic
and somatic tissues gives cytological evidence for the effectiveness of somatic mutation markers as an assay for chemical
mutagenicity and hence health hazard potential. The Tradescantia stamen hair system encompasses the cytogenetic and
somatic potential to make the system a useful tool for mutagenicity monitoring of ambient air pollution mixtures or isolated fractions. This plant is uniquely adapted to field exposures, hardy enough to tolerate a broad range of environmental conditions, and requires no elaborate sterile culture
conditions. The data presented above demonstrate the high
sensitivity of the system to gaseous compounds and the relatively short time from start of exposure to definition of results (3 weeks). In the absence of hard genetic evidence for

Table 3

Mutagenicity of Ambient Air at Various Industrial Sites
as Measured by Tradescantia Stamen Hairs

Site Location	Date of Exposure	Pollution Source	Pollution-induced Events/10^3 Hairs ± S.E.	Increase (% of Control)
Elizabeth, NJ	July 1976	Petroleum & Automotive	1.22 ± 0.21	30.9
	Oct. 1976	Petroleum & Automotive	0.80 ± 0.23	18.8
	Jan. 1977	Petroleum & Automotive	0.58 ± 0.15	17.7
Baton Rouge, LA	May 1977	Petroleum	0.61 ± 0.16	15.0
Houston, TX	July 1977	Petroleum	0.72 ± 0.17	18.1
Charleston, WV	March 1977	Chemicals	0.71 ± 0.16	17.1
Birmingham, AL	April 1977	Steel	0.24 ± 0.17	5.5
Magna, Utah	Oct. 1977	Copper	-0.10 ± 0.14	-2.3
Upland, CA	Aug. 1977	Automotive & Photochemical	-0.08 ± 0.15	-2.1
	Sept. 1977	Automotive & Photochemical	-0.15 ± 0.17	-3.8
Grand Canyon, AZ	Nov. 1977	Clean Air	-0.10 ± 0.14	-2.9
	Dec. 1977	Clean Air	-0.10 ± 0.18	-2.9

extrapolation from plants to man, at least this system can become part of a battery of tests which can provide early warning of the potential health hazard of exposure to mixed air pollutants.

ACKNOWLEDGMENTS

This work was supported jointly by the U.S. Department of Energy, National Institute of Environmental Health Sciences, and U.S. Environmental Protection Agency. The authors acknowledge with thanks the special efforts of: Mr. N.R. Tempel for MMV assembly, deployment, and instrumentation and Mr. R.C. Sautkulis for supervision of stock plants and field exposures; Mr. W. Barnard, R. Baxter, and R. Ballard for aerometric instrumentation development and field operation; and Mr. J. Dame and D. Brashear of Xonics, Inc. for operation of the CHAMP-van and assistance in the field operation. The many hours of flower analysis by Mr. E.E. Klug, Ms. A. Nauman, Ms. M.M. Nawrocky, Ms. V. Pond, Mr. R.C. Sautkulis, and Ms. R.C. Sparrow also gratefully acknowledged.

REFERENCES

1. Mericle LW, Mericle RP: Genetic nature of somatic mutations for flower color in Tradescantia, clone 02, Radiation Botany 7:449-464, 1967

2. Nauman CH, Klotz PJ, Sparrow AH: Dosimetry of tritiated 1,2-dibromoethane in floral tissues of Tradescantia. Mutat Res 38:406, 1976

3. Schairer LA, Van't Hof J, Hayes CG, Burton RM, de Serres FJ: Exploratory monitoring of air pollutants for mutagenicity activity with the Tradescantia stamen hair system. Environmental Health Perspectives, in press

4. Sparrow AH, Baetcke KP, Shaver DL, Pond V: The relationship of mutation rate per roentgen to DNA content per chromosome and to interphase chromosome volume. Genetics 59:65-78, 1968

5. Sparrow AH, Schairer LA: Mutational response to Tradescantia after accidental exposure to a chemical mutagen. EMS Newsletter 5:16-19, 1971

6.   Sparrow AH, Schairer LA: Response of somatic mutation
     frequency in Tradescantia to exposure time and concen-
     tration of gaseous mutagens. Mutat Res 38:405-406, 1976

7.   Sparrow AH, Schairer LA, Marimuthu KM: Radiobiologic
     studies of Tradescantia plants orbited in Biosatellite
     II. In: The experiments of Biosatellite II, (Saunders
     JF, ed.). NASA Special Publication 204, 99-122.
     Scientific and Technical Information Office, NASA,
     Washington, DC, 1971

8.   Sparrow AH, Schairer LA, Nawrocky MM, Sautkulis RC:
     Effects of low temperature and low level chronic gamma
     radiation on somatic mutation rates in Tradescantia.
     Radiation Res 47:273-274, 1971

9.   Sparrow AH, Schairer LA, Villalobos-Petrini R: Compari-
     son of somatic mutation rates induced in Tradescantia
     by chemical and physical mutagens. Mutat Res 26:265-276,
     1974

10. Sparrow AH, Underbrink AG, Rossi HH: Mutations induced
     in Tradescantia by small doses of X-rays and neutrons:
     analysis of dose-response curves. Science 176:916-918,
     1972

11. Underbrink AG, Schairer LA, Sparrow AH: Tradescantia
     stamen hairs: a radiobiological test system applicable
     to chemical mutagenesis. In: Chemical Mutagens: Prin-
     ciples and Methods for Their Detection, Vol. 3
     (Hollaender A, ed.). New York, Plenum Press, 1973,
     171-207

12. Underbrink AG, Schairer LA, Sparrow AH: The biophysical
     properties of 3.9-GeV nitrogen ions. V. Determinations
     of the relative biological effectiveness for somatic
     mutations in Tradescantia. Radiation Res 55:437-446,
     1973

13. Underbrink AG, Sparrow RC, Sparrow AH, Rossi HH: Rela-
     tive biological effectiveness of X-rays and 0.43-MeV
     monoenergetic neutrons on somatic mutation and loss of
     reproductive integrity in Tradescantia stamen hairs.
     Radiation Res 44:187-203, 1970

PHYSICAL AND BIOLOGICAL STUDIES OF COAL FLY ASH

Gerald L. Fisher and Clarence E. Chrisp
Radiobiology Laboratory
University of California
Davis, California

In our initial studies of the potential health impact of energy technologies, we have performed physical, chemical, and mutagenic studies with coal fly ash. Although the vast majority (95-99%) of the fly ash produced in coal combustion for electric power generation is retained in the power plant, we (5) have estimated that 2.4 million metric tons of fly ash were emitted in the atmosphere from U. S. coal-fired electric plants in 1974. Because the principal particulate emission control technologies, electrostatic precipitators (ESP) or wet scrubbers, have low collection efficiency for smaller particles (34), much of the released fly ash is in the "respirable" size range (aerodynamic diameters <10 m) (11). This fine particle fraction presents the greatest potential health hazard because fine particles have the longest atmospheric residence times, and thus the greatest potential for ultimate human inhalation (21), and are generally most efficiently deposited in deep lung and least efficiently removed by mucociliary transport (35).

FLY ASH COLLECTION

To obtain sufficient quantity of size-classified fly ash for detailed physical and biological testing, a specially designed in-stack fractionator was constructed (20). The apparatus was mounted in the stack breeching downstream from the electrostatic precipitator (ESP) of a modern western U. S. power plant burning high ash, low sulfur pulverized coal. At the time of stack sampling, ESP hopper fly ash was also collected. The apparatus consisted of a heated enclosure containing two cyclone separators in series followed by a 25-jet

centripeter (virtual dichotomous impactor). The stack gasses
were drawn through the inlet probe into the heated enclosure,
which was maintained at 95°C to prevent moisture condensation
associated with the high dew point of the stack effluent.
The serial arrangement of the two cyclones and the centripeter
provided in situ size-classification of four size fractions.
The two cyclone fractions had volume median diameters (VMDs)
of 20 (cut 1, coarsest) and 6.3 μm (cut 2) and the centripeter
fractions had VMDs of 3.2 (cut 3) and 2.2 μm (cut 4, finest)
(Table 1). All fractions had geometric standard deviations
(σ_g) of approximately 1.8. The fractionator was operated for
30 days at a flow rate of 30 cfm. Approximately 16 kg of material was classified with approximately 67%, 16%, 7%, and 10%
of the mass in cuts 1, 2, 3, and 4, respectively. The size
distributions of the four sized fractions were compared (after
conversion to aerodynamic equivalent size) to samples collected isokinetically from the stack (5). This approach allowed for direct comparison of the size-fractionated material
to fly ash representative of normal stack emissions. The
comparison indicated the enhancement of fine particles and
the depletion of coarse particles in cuts 3 and 4 relative
to the isokinetically collected sample. Cut 1 was enhanced
in coarse particles, while cut 2 approximated the isokinetic
sample fairly well from 1.4 to 20 μm. Specifically, cuts 3
and 4 displayed six- to ten-fold and ten- to twenty-fold
increases, respectively, in the relative mass contributions
from 1 to 2 μm, while cut 1 contained less than one-tenth the
relative mass in this size interval when compared to the isokinetic data. Therefore, with regard to subsequent chemical
and biological studies, it is important to note the size-
classification procedure resulted in extensive enhancement of
the fine particles (1-2 μm) in cuts 3 and 4, relative to the
total particulate emission.

Physical and Morphological Studies

     The average particulate density in the four size frac-
tions was found (5) to correlate negatively ($p < 0.05$) with
the VMDs (Table 1). A detailed morphological analysis of
particle types indicated that the variation in density could
be explained by the size dependence of the relative abundance
of the morphological classes.

     We used light microscopy to define eleven major classes
of particulate morphology (5). On the basis of opacity and
particle shape, a fly ash morphogenesis scheme was developed
(Figure 1). The morphological classes included particles

Table 1

Physical Properties of Size-Classified
Stack-Collected Coal Fly Ash

Fraction	Cut #	Volume Median Diameter	Geometric Standard Deviation	Percent of Total Mass Collected	Mean Particle Density (g/cm³)
First cyclone	1	20	1.8	67	1.85
Second cyclone	2	6.3	1.8	16	2.19
Centripeter-large fraction	3	3.2	1.8	7	2.36
Centripeter-small fraction	4	2.2	1.9	10	2.45

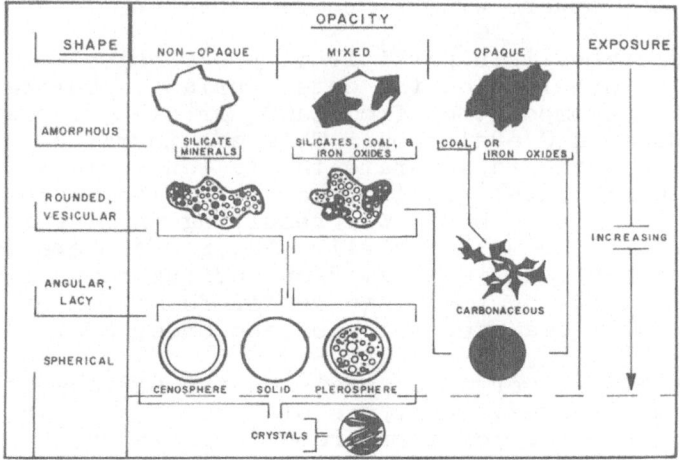

Figure 1. Morphogenesis scheme indicating probable relation-ship between particle morphology and chemical composition. Opacity and shape are used as primary characteristics for morphological classification.

that appeared amorphous and either opaque or non-opaque with
relatively limited exposure to combustion conditions within
the boiler. With further exposure to combustion conditions,
these particles developed somewhat rounded surfaces and contained vesicles. Continued exposure to combustion conditions
resulted in formation of spherical particles derived from molten inorganic minerals or soot particles from incomplete coal
combustion.

We have defined five classes of spherical particles, the
most abundant morphological type. Solid, non-opaque spheres
and hollow, non-opaque spheres (cenospheres) are predominantly
aluminosilicates derived from clay minerals within the coal
(3). Spheres may range in color from water-white through yellow to dark red to opaque. Opaque spheres are mostly magne-
tite and are easily identified in microscopic studies by
taking advantage of their magnetic properties (5). Some
spheres contain large numbers of smaller spheres (Figure 2).
These plerospheres are most abundant in the coarser fly ash
fractions. Careful examination of the plerospneres indicates
that often the encapsulated spheres within the plerosphere
are themselves plerospheres. We (3) have demonstrated that
the gases within the plerospheres are H_2O and CO_2. On the
basis of the morphological appearance, bulk chemical composition and gaseous content we have postulated a mechanism to
account for the sphere-within-sphere structure.

As a noncombustible particle is progressively heated,
a molten layer develops on the outer surface. During that
time, mineral decomposition from $CaCO_3$ or clay minerals may
result in CO_2 or H_2O evolution. This gas formation serves
as the driving force to separate the molten surface from the
solid particulate core. Further gas formation causes the
surface of the core to boil away resulting in microsphere
formation within the molten shell. The plerosphere is frozen after the particle is carried out of the combustion zone.
We have calculated the time require for formation of a plerosphere of 50 μm diameter to be on the order of 1000 μsec.

We have also observed crystals on the surface of and
within fly ash spheres. Analysis of some of the large surface crystals by electron microprobe indicated high concentrations of calcium and sulfur with no other elements detected. On the basis of the SEM appearance of these crystals, we (3) concluded that they were anhydrite ($CaSO_4$) or
gypsum ($CaSO_4 \cdot 2H_2O$) resulting from interaction of surface
formed or deposited H_2SO_4 with particulate calcium oxide.
Interiorized crystals generally appeared to radiate from one

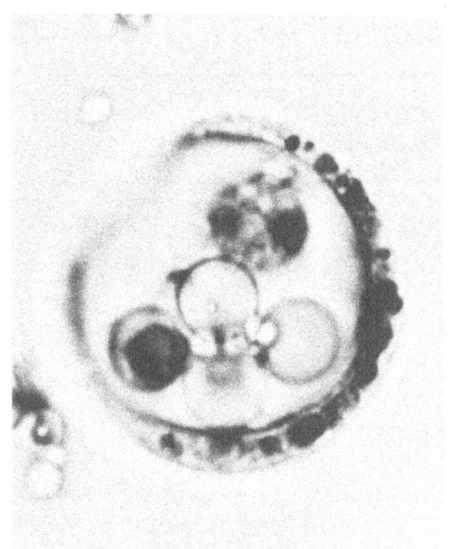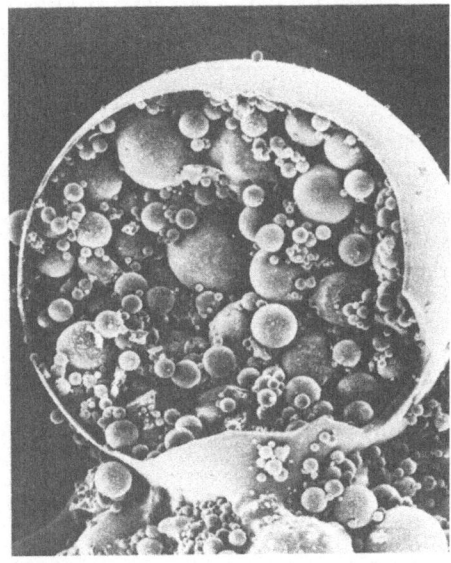

Figure 2. Micrographs of plerospheres indicating the sphere within sphere structure of these fly ash particles. The plerosphere in the light photomicrograph (left) is 20 μm in diameter; the scanning electron micrograph (right) depicts an 80 μm diameter plerosphere.

or two points on the sphere surface through the sphere. These "quench" crystals have been reported to form from heterogeneous nucleation at the surface of molten silicate droplets during rapid quenching (14).

We have quantified the relative abundance of the morphological types of particles in the four fly ash fractions (Table 2). The relative abundance of most particle types appears to be positively correlated with particle size. In contrast to this observation, non-opaque spheres were correlated negatively with particle size. The most striking differences in frequency distributions were observed between cut 1 and cut 4. Cut 1 was composed of 41% cenospheres and

Table 2

Frequency (%) Distribution of Particle Classes in
Size-Classified Coal Fly Ash

Particle Class	Cut 1 (20 μm)	Cut 2 (6.3 μm)	Cut 3 (3.2 μm)	Cut 4 (2.2 μm)
Combined amorphous, opaque and non-opaque	7.4	2.4	0.8	0.3
Combined vesicular, opaque and non-opaque	14.7	6.9	2.9	3.0
Sooty	1.3	0.6	0.3	0.3
Cenosphere	41.4	26.2	13.2	7.9
Plerosphere	0.5	0.2	--	--
Opaque sphere	1.6	0.9	0.3	0.2
Non-opaque sphere	25.6	56.0	79.2	87.2
Sphere with crystals	6.8	6.8	3.2	0.9

26% non-opaque spheres while cut 4 was composed of 8% cenospheres and 87% non-opaque spheres. The greater amount of
solid spheres and lesser amount of vesicular particles appears to explain the observed trend of increased average
particle density with decreased particle size.

Elemental and Chemical Analysis

     Because of the observed morphological heterogeneity,
we initiated elemental analysis of individual particles. In
our initial study (29), we used three-color X-ray mapping
techniques with a scanning electron microscope (SEM). We
analyzed fly ash provided by the NBS as a standard reference
material (NBS-SRM 1633) for 12 trace elements. Analysis of
fly ash particles with similar SEM morphologies indicated
extreme elemental heterogeneity, i.e., morphologically similar

particles were found to contain high concentrations of Ti, S, Al, K, Ca, or Fe. Further studies are now underway to evaluate elemental composition of the eleven light-microscopically defined morphological classes. Preliminary results indicate that the pigmentation in non-opaque spheres from water-white to yellow to red is associated with iron concentrations (4). Analysis of opaque, amorphous particles indicates these particles are composed primarily of low atomic number elements, reflecting the organic components of coal. Particles rich in Ni, Cr, Zn, or Mn have been observed.

Detailed elemental analyses of the four fly ash fractions were performed by instrumental neutron activation analysis (INAA) and atomic absorption spectrophotometry (AAS). Prior to analysis of the fly ash fraction, the accuracy and precision of the two techniques were evaluated using NBS fly ash (SRM-1633) (26). The AAS analysis involved a room temperature digestion in hydrofluoric acid followed by addition of a saturated boric acid solution (32). This digestion technique resulted in quantitative dissolution of all elements except selenium and barium. Comparison of INAA and AAS determination of Al, Ba, Co, Cr, Fe, K, Mn, Na, Ni, Ti, and Zn indicated excellent agreement between the two techniques as well as with previously published literature values (27). Be, Cu, Cd, Mg, Ca, and Pb analyses by AAS and As, Ce, Cs, Eu, Hf, La, Rb, Sb, Sc, Se, Sm, Sr, Ta, Tb, Th, U, V, W, and Yb analyses by INAA also agreed well with previously published literature values.

Summary tables of the analytical results are presented for those elements displaying concentrations independent of particle size (Table 3) and dependent on particle size (Table 4). For elements analyzed by both INAA and AAS, the data reported are the results of the analytical technique with the smaller coefficient of variation. Data from atomic absorption analyses are the average of two independent determinations; the INAA data are the weighted averages of three independent determinations. Concentration dependence on particle size was determined qualitatively with the criteria that consistent concentration trends beyond experimental uncertainty were observed for each fraction, although significantly higher concentrations of the element may have been observed in the finest fraction relative to the coarsest fraction. The enhancement factor is defined as the ratio of the element concentration in cut 4 to its concentration in cut 1.

Table 3

Elemental Concentrations Independent[1] of Particle Size

Element	Technique	Cut 1 (VMD = 20 μm)	Cut 2 (VMD = 6.3 μm)	Cut 3 (VMD = 3.2 μm)	Cut 4 (VMD = 2.2 μm)
		Concentration in %			
Al	AAS[2]	13.8(0.1)	14.4(0.1)	14.2(0.8)	14.1(0.3)
Fe	INAA[3]	2.5(0.1)	2.9(0.2)	3.0(0.1)	3.2(0.1)
Ca	AAS	2.12(0.14)	2.23(0.08)	2.30(0.14)	2.38(0.09)
Na	AAS	1.19(0.13)	1.75(0.05)	1.83(0.06)	1.85(0.03)
K	AAS	0.74(0.01)	0.80(0.07)	0.82(0.08)	0.81(0.03)
Ti	AAS	0.62(0.05)	0.76(0.05)	0.77(0.11)	0.78(0.06)
Mg	AAS	0.47(0.01)	0.56(0.01)	0.60(0.02)	0.63(0.01)
		Concentration in μg/g			
Sr	INAA	410(60)	540(140)	590(140)	700(210)
Ce	INAA	113(4)	122(5)	123(6)	120(5)
La	INAA	62(3)	68(4)	67(11)	69(3)
Rb	INAA	51(3)	56(4)	57(3)	57(8)
Nd	INAA	45(4)	47(4)	49(7)	52(6)
Th	INAA	25.8(0.6)	28.3(0.6)	29(1)	30(2)
Ni	AAS	25(3)	37(1)	43(4)	40(2)
Sc	INAA	12.6(0.5)	15.3(0.6)	15.8(0.6)	16.0(0.2)
Hf	INAA	9.7(0.4)	10.3(0.3)	10.5(0.3)	10.3(0.5)
Co	INAA	8.9(0.2)	16.3(0.8)	19(1)	21(1)
Sm	INAA	8.2(0.3)	9.1(0.4)	9.2(0.4)	9.7(0.4)
Dy	INAA	6.9(0.3)	8.5(0.9)	8.1(0.3)	8.5(0.8)
Yb	INAA	3.4(0.4)	4.1(0.4)	4.0(0.2)	4.2(0.3)
Cs	INAA	3.2(0.1)	3.7(0.2)	3.7(0.2)	3.7(0.2)
Ta	INAA	2.1(0.1)	2.3(0.2)	2.5(0.3)	2.7(0.1)
Eu	INAA	1.0(0.1)	1.2(0.2)	1.2(0.2)	1.3(0.4)
Tb	INAA	0.90(0.05)	1.06(0.06)	1.10(0.07)	1.13(0.06)

[1]Concentration dependence with particle size was determined qualitatively with the criteria that consistent concentration trends beyond experimental uncertainty were observed for each fraction.

[2]AAS values are the averages of two independent determinations; the ranges are given in parentheses.

[3]INAA values are the weighted averages of three independent determinations; uncertainties (in parentheses) are the largest of twice the weighted standard deviation, the range, or an estimate of the accuracy.

Table 4

Elemental Concentrations Dependent[1] on Particle Size

Element	Technique	Cut 1 (VMD = 20 μm)	Cut 2 (VMD = 6.3 μm)	Cut 3 (VMD = 3.2 μm)	Cut 4 (VMD = 2.2 μm)	Enhancement Factor
			Concentration in μg/g (unless indicated)			
Cd	AAS	0.4(0.2)	1.6(0.3)	2.8(0.4)	4.6(0.2)	11.5
Zn	AAS	68(1)	189(4)	301(10)	746(218)	11.0
Se	INAA	19(2)	59(2)	78(2)	198(20)	10.4
As	INAA	13.7(1.3)	56(14)	87(9)	132(22)	9.6
Sb	INAA	2.6(0.1)	8.3(0.4)	13.0(0.7)	20.6(0.7)	7.9
W	INAA	3.4(0.2)	8.6(1.6)	16(2)	24(2)	7.1
Mo	INAA	9.1(2.5)	28(1.4)	40(5)	50(9)	5.5
Ga	INAA	43(12)	116(52)	140(23)	178(90)	4.1
Pb	AAS	73(3)	169(2)	226(4)	278(3)	3.8
V	INAA	86(44)	178(17)	244(18)	327(40)	3.8
U	INAA	8.8(1.9)	16(3)	22(4)	29(4)	3.3
Cr	AAS	28(3)	54(3)	66(3)	71(4)	2.5
Ba(%)	AAS	0.168(0.001)	0.245(0.002)	0.320(0.013)	0.409(0.018)	2.4
Cu	AAS	56(1)	89(1)	107(4)	137(1)	2.4
Be	AAS	6.3(0.2)	8.5(0.2)	9.5(0.3)	10.3(0.5)	1.6
Mn	AAS	209(7)	231(5)	273(7)	309(3)	1.5
Si(%)	AAS	29.6(0.7)	28.0(0.1)	27.5(0.3)	26.8(0.1)	0.90
F$^-$	IC	60	420(40)	840(60)	2400(300)	40
SO$_4$ $^{2-}$ (%)	IC	0.15	0.67(0.02)	0.99(0.15)	1.62(0.07)	10.8

[1] Concentration dependence with particle size was determined qualitatively with the criteria that consistent concentration trends beyond experimental uncertainly were observed for each fraction.

[2] Ratio of concentration in cut 4 to that in cut 1.

[3] AAS values are the averages of two independent determinations; the ranges are given in parentheses.

[4] INAA values are the weighted averages of three independent determinations; uncertainties (in parentheses) are the largest of twice the weighted standard deviation, the range, or an estimate of the accuracy.

[5] Ion chromatography values are the mean and standard deviations of three water extracts with the exception of cut 1, for which only one determination was made.

The major element composition of the fractionated fly ash is relatively independent of particle size with the exception of silicon, which appears to decrease with decreasing particle size. Greater than 92% of the mass of the fractionated fly ash can be accounted for by oxides of Si, Al, Fe, and Ca. The more volatile elements (or their oxides), Cd, Zn, Se, As, Sb, Mo, Ga, Pb, and V display clear-cut increases in concentration with decreasing particle size, in agreement with the vapor-condensation mechanism of Natusch and Wallace (24). It is important to note, however, that refractory elements also display concentration trends inversely dependent on particle size. Therefore, processes other than vapor condensation are involved in the concentration-size relationship. The elements U and Cr are associated with the organic fraction of coal (22) and may be released in the combustion process as fine particles that may agglomerate with other particles. The elements Fe, Mn, Ba, and Sr (22) may in part be present as carbonate minerals which decompose to form fine particles during coal combustion and again agglomerate with other particles. Copper is probably present in part as the sulfide and Be as the aluminosilicate in the coal (22). Thus, mineral decomposition and elemental distribution may in part explain the elemental trends of the high boiling chemical species.

Analyses of H_2O extracts of the fly ash fractions by ion chromatography (10) indicated an inverse concentration dependence on particle size for sulfate and fluoride (Table 4). Sulfite was not detected in the samples by either ion chromatography or thermometric titration calorimetry.

Filtration studies with neutron activated fly ash indicated that the elements Mo, Ca, Se, Ba, Co, As, and Sb display significant solubility at physiological pH (6). The elements displaying the greatest solubilities relative to the initial fly ash concentrations were Mo, Ca, and Se with relative solubilities of 55%, 30%, and 20%, respectively.

Analysis of the organic compounds in the fly ash has been initiated using gas chromatography with high resolution glass capillary columns and mass spectrometry (17). Chromatograms clearly demonstrate the presence of over 120 well-resolved peaks. To date, the following polynuclear aromatic hydrocarbons have been tentatively identified based on retention data and Kovat's Indices of Standard Compounds, and/or mass spectral data: dibenzofuran, pyrene, 1,2-benzoanthracene, 20-methylcholanthrene, benzo g,h,i(gi) perylene, naphthalene,

1-methylnaphthalene, fluorene, phenanthrene, anthracene, fluoranthene, and benzoanthracene 7,12-dione. Work is presently underway to substantiate these observations and to identify further the organic compounds in fly ash.

MUTAGENICITY TESTING OF COAL FLY ASH

This report will review recently published data (2) and illustrate our approach to biological testing of a complex mixture. Certain metals which are carcinogenic in man or animals (8,9,12,15,16,19,28,33) were shown to be concentrated in stack fly ash as described earlier in this report. The presence of carcinogenic substances as a result of fossil fuel combustion has been suspected since scrotal cancers were first observed in chimney sweeps in 1775 by Percival Pott (30). Subsequently, organic compounds from coal tar products proved to be carcinogenic (18).

Because a high positive correlation between carcinogenicity of substances for animals or man and mutagenicity in a bacterial test system has been shown by Ames (25), we decided to use this simple and economical test for the detection of putative carcinogens on the surface of cut 4 fly ash. Briefly, all five strains of histidine requiring auxotrophs or Salmonella typhimurium, TA1535, TA100, TA1537, TA1538, and TA98, kindly supplied to us by B.N. Ames, were used in testing cut 4 of fly ash collected from the stack of a coal burning power plant. The genetic background and testing methods for these strains have been previously described (1).

Care was taken in the selection of the proper solvent for the extraction of possible mutagens from the surface of fly ash. Several laboratory solvents were tested for toxicity and for mutagenicity. One must be careful to distinguish between toxicity and mutagenicity in this test system. It is necessary to incorporate a small amount of histidine into the medium so the bacteria may undergo several replications. Resultant tiny colonies are seen as a background lawn. However, if a solvent or mutagen is toxic, some of the bacteria may be lysed, leaving others with a greater amount of histidine per bacterium. This may be enough so that visible colonies are formed that may be mistaken for his[+] revertants. If small colonies are seen, it is necessary to examine the plate under a microscope to see if the background lawn is sufficient. If not, either the solvent or test mutagen is toxic. In addition, laboratory solvents can also be a source of mutagens, either because of solvent impurities in manufacture, or contamination with mutagens in the laboratory environment.

In initial studies cyclohexane, a nonmutagenic, nonpolar, organic solvent was used. Cut 4 fly ash was. extracted with four 10 ml volumes of cyclohexane at room temperature and the supernatant was passed through a 0.45 μm filter to remove fly ash particles. Results of pour plate tests are shown in Table 5. His revertants were seen with strains TA98 and TA1538, but not with TA1537, TA1535, and TA100. These results indicated the probable presence of nonpolar, organic, frameshift mutagens.

Two media were selected for further studies with cut 4 fly ash. Dulbecco's phosphate buffered saline was used because it has the pH and tonicity of physiological fluids. Horse serum was selected because serum has a chemical constituency similar to lung alveolar fluid and forms soluble complexes with some carcinogenic heavy metals (13). Fly ash samples were incubated with each of these media for a minimum

Table 5

Number of TA1538 His$^+$ Revertants/Plate

| Test Mixture | S-9 Not Added | | S-9 Added | |
	Fly Ash	Control	Fly Ash	Control
Cyclohexane extract	62 + 2	5 + 2	152 + 8	27 + 5
Serum filtrate	154 + 32	10 + 2	202 + 18	12 + 5
Saline filtrate	17 + 3	4 + 1	40 + 9	16 + 2

S-9 is the supernatant fraction of Aroclor-induced rat liver homogenate, centrifuged at 9000 g. Positive controls were spot tests with 4-nitro-quinoline-N-oxide without S-9 and with 2-aminofluorene and S-9 added. The mean number of spontaneous revertants per plate was 7 + 1 without S-9 and 20 + 1 with S-9. The numbers given present the mean number of colonies + the standard deviation on 3 replicate plates. Concentrations of fly ash are equivalent in all 3 test mixtures (78 mg/ml). Filtrate (100 μl) was added to 2 ml of soft top agar before plating.

of one week at 37°C. After incubation, the fly ash mixtures were centrifuged at 35,000 g and the supernatants were passed through a 0.45 μm membrane filter to remove particulate matter. Media controls of serum or saline were treated in the same fashion as the fly ash mixtures. No mutagenic activity was found with spot tests, but his[+] revertants were found with the pour plate technique (Table 5). This was evidence that the mutagen or mutagens did not readily diffuse into the media from the paper discs. Again, only the frame shift mutants TA-98 and TA1538 showed his[+] revertants. More revertants were seen with strain TA1538 than TA98, so the former was used in subsequent studies. A small increase in his revertants was seen when optimal concentrations of rat liver homogenates from rats treated with polychlorinated biphenyl (Aroclor 1254) was added to pour plates (Table 5). Repetition of these tests has shown that there is a small but highly significant (p < 0.001) increase in his[+] revertants with metabolic activation. At first the fly ash was autoclaved before incubation with the various solvents in order to avoid bacterial contamination. Later it was found that the fly ash was sterile and autoclaving prior to incubation did not change the number of revertants.

A dose response curve for mutagenicity of cut 4 fly ash filtrates in strain TA1538 is shown in Figure 3. Serum filtrates had approximately a ten-fold greater activity than saline filtrates. All mutagenic activity was found in the aqueous fraction after extraction of saline filtrates with cyclohexane. Solubility of substances responsible for mutagenic activity in saline, a polar solvent, suggested the presence of a polar organic or an inorganic mutagen. In addition these data imply that horse serum might be a useful extract for complex mixtures of mutagens.

Reproducibility of the Ames test with fly ash serum filtrates was examined. The ratio of his[+] revertants to spontaneous revertants ranged from 20 to 60 when fly ash serum filtrates were incubated at different times and the same filtrates stored and tested on different days. This variability was greater than that observed when samples were incubated at the same time and tested on the same day.

It is well known that serum protein can bind to both organic (31) and inorganic (13) compounds. Fly ash serum filtrates were fractionated on a Sephadex G-25 column with a cut-off of 25,000 daltons. Figure 4 shows the protein pattern for three fractions collected from the column. Approximately 80% of the mutagenic activity was associated with the

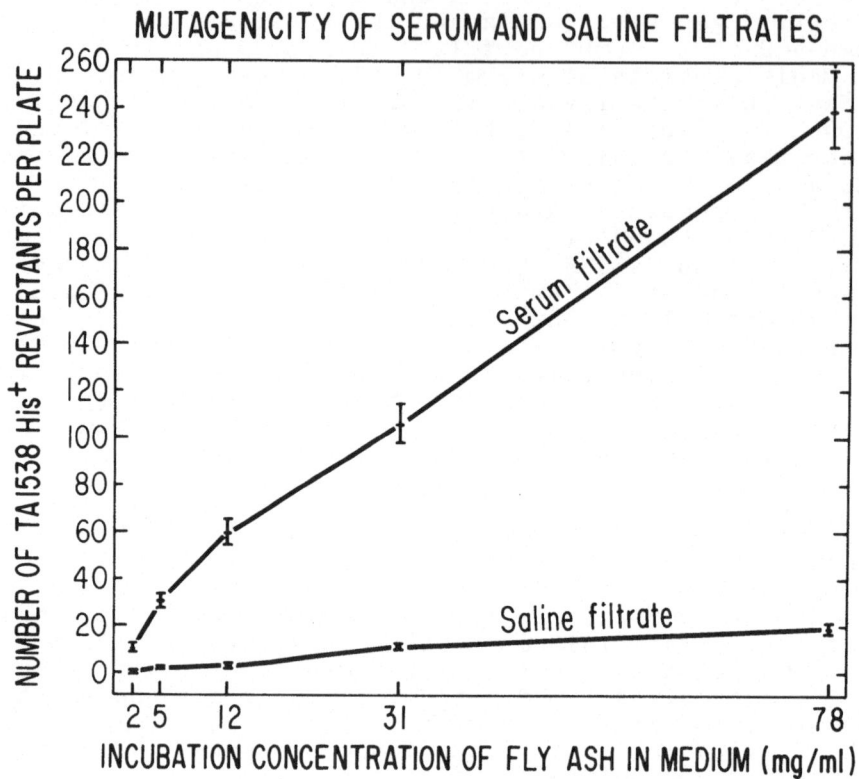

Figure 3.   Mutagenicity of fly ash serum and saline. Filtrates with strain TA1538. The number of his revertants
per plate is the mean of 5 to 20 determinations minus the
mean of the appropriate background revertants (serum or
saline). The background reversion was defined as the group
mean of the spontaneous revertants and the appropriate media
control after it was determined that the number of his[+] revertants in all negative controls was not significantly different from that of spontaneous revertants. The means (±
SEM) of the background revertants 5.8(± 0.4), 6.9(± 0.9),
4.0(± 0.6) for the spontaneous revertants, serum controls,
and saline controls respectively. Filtrate (100 µl) was
added to 2 ml of soft top agar before plating. Plates were
incubated for 2 days at 37°C. The vertical bars are 1 SEM.

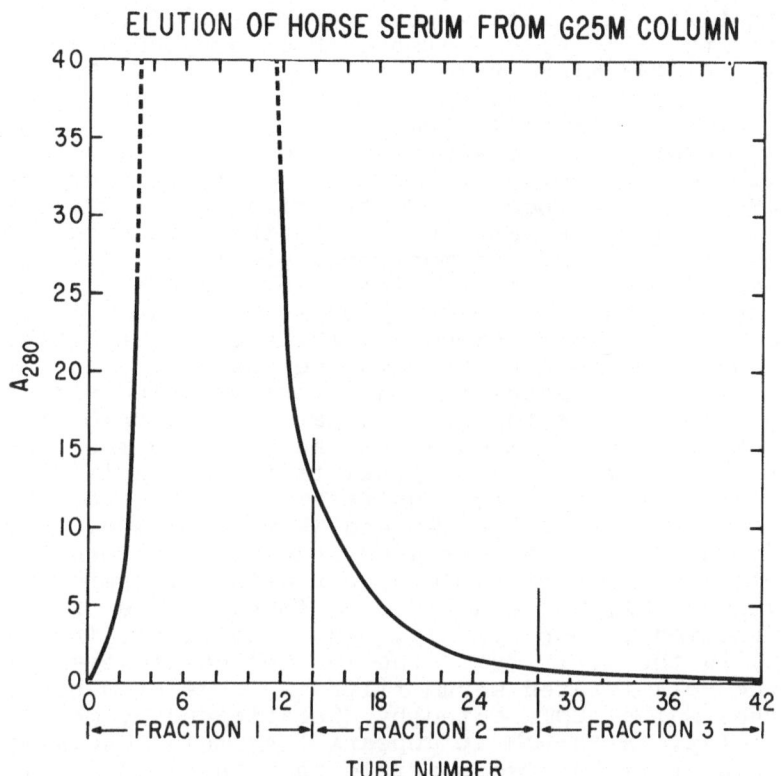

Figure 4. Pattern of elution of horse serum from a molecular weight exclusion column with a cutoff of 25,000 daltons.
The first fraction contains 95% of the serum protein, the
secod less than 5%, and the third only low molecular weight
compounds.

first fraction which contained 95% of the total serum protein. This indicated that most of the substances accounting
for mutagenic activity were probably bound to serum proteins.

Mutagenicity of EDTA-Treated Fly Ash Filtrates

The Ames test has not been very useful in testing known
carcinogenic heavy metals for mutagenicity; however, a few
have been shown to be mutagenic in this system (7). It was

decided that if heavy metals were responsible for any mutagenic activity, a metal chelator such as ethylenediamine-
tetraacetic (EDTA) might remove this activity. EDTA-treated
and untreated serum filtrates were fractionated on a column
as illustrated in Figure 4. EDTA (2 mM) was added to one
portion of serum filtrate and stirred overnight at 4°C before elution on the column. A second portion was prepared
in the same manner without prior treatment with EDTA. Each
of these two filtrates was eluted with three void volumes of
double distilled water. As mentioned previously, the first
fraction contained most of the total serum protein (Figure
4). The second had the remaining protein and a small amount
of low molecular weight compounds, while the third fraction
contained only low molecular weight components. Each of the
three fractions was lyophilized and reconstituted with double
distilled water before testing. Regardless of prior treatment with EDTA, the total mutagenic activity in the fractions
was lower than that in the original filtrate (Table 6). Of
the total net activity after subtraction of background revertants (5.0 $\pm$ 1.0), 79%, 18%, and 3% were present in the
first, second, and third untreated fractions, respectively.
Of the total net activity after subtraction of appropriate
control values 83%, 0%, and 17% were found in the three EDTA-
treated fractions, respectively. The significant increase
(p < 0.01) in the activity of the low-molecular-weight fraction of the EDTA-treated serum filtrate lends credence to
the hypothesis that EDTA acted by chelating heavy metals from
serum proteins. Although it appears that metal chelation
is responsible, it is also possible that the EDTA may act to
increase bacterial cell permeability to mutagens. In addition, there may be synergism between metals and organic compounds. The fact that the mutagenic activity of the fractions is less than the total, regardless of EDTA treatment,
is partially explained by the necessity to subtract the control revertants from each fraction.

Studies are underway to evaluate the carcinogenic potential of coal fly ash as well as the possible role of fly
ash inhalation in respiratory disorders.

Table 6

Column Chromatograpy
Number of TA1538 His$^+$ Revertants/Plate

	Fly Ash	Fly Ash + EDTA	Control
Unfractionated serum filtrate	162 ± 18	261 ± 25	8 ± 2
Serum filtrate fraction 1	78 ± 11	94 ± 10	7 ± 1
Serum filtrate fraction 2	21 ± 4	11 ± 4	11 ± 2
Serum filtrate fraction 3	7 ± 2	22 ± 3	4 ± 1

Concentrations of fly ash were 78 mg/ml. The number given represents the number of revertants ± the standard deviation on 5 replicate plates. Filtrate (100 µl) was added to 2 ml of soft agar before plating.

REFERENCES

1. Ames BN, McCann J, Yamasaki E: Methods for detecting carcinogens and mutagens with the Salmonella mammalian microsome mutagenicity test. Mutat Res 31:347-363, 1975

2. Chrisp CE, Fisher GL, Lammert JE: Mutagenicity of filtrates from respirable coal fly ash. Science 199:73-75, 1978

3. Fisher GL, Chang DPY, Brummer M: Fly ash collected from electrostatic precipitators: Microcrystalline structures and the mystery of the spheres. Science 192:553-555, 1976

4. Fisher GL, Hayes T: unpublished data

5. Fisher GL, Prentice BA, Silberman D, Ondov JM, Biermann AH, Ragaini RC, McFarland AR: Physical and morphological studies of size-classified coal fly ash. Environ Sci Tech, in press, 1978

6. Fisher GL, Silberman D, Heft RE, Ondov JM: Fly ash fil-
    terability, differential solubility and elemental dis-
    tribution studies. In: Radiobiology Laboratory Annual
    Report, University of California, Davis California 34-
    40, 1977

7. Flessel CP: Metals as mutagens. Adv Exp Biol Med 91:
    117-128, 1978

8. Furst A, Schlauder M, Sasmore DP: Tumorigenic activity
    of lead chromate. Cancer Res 36:1779-1783, 1976

9. Furst A: An overview of metal carcinogenesis. Adv Exp
    Biol Med 91:1-12, 1978

10. Hansen LD, Fisher GL: unpublished data

11. Hatch TF, Gross P: Pulmonary Deposition and Retention
    of Inhaled Aerosols. New York, Academic Press, 1964

12. Heath JC: Carcinogenic action of metals. Brit Emp
    Cancer Campaign Rep, Part II:389, 1963

13. Heath JC, Webb M, Caffrey M: The interaction of carci-
    nogenic metals with tissues and body fluids: Cobalt and
    horse serum. Br J Cancer 23:153-166, 1969

14. Hurt J, Biechnicki DJ: Ultrafine-grain ceramics from
    melt phase. In: Ultrafine-Grain Ceramics (Burke JJ,
    Reed NL, Weiss V, eds.). Syracuse, Syracuse University
    Press, 1970, pp 286-287

15. International Agency for Research on Cancer. Evaluation
    of Carcinogenic Risk of Chemicals to Man, Vol I. Lyon,
    184, 1972

16. International Agency for Research on Cancer. Some In-
    organic and Organic Metallic Compounds, Vol II. Lyon,
    181, 1973

17. Jennings WG, Sucre L, Fisher GL, Raabe OG: Analysis of
    the organic constituents of coal, fly ash, coke and
    coal tar. In: Radiobiology Laboratory Annual Report,
    University of California, Davis, California, 1977, pp
    24-33

18. Kubota H, Griest WH, Guerin MR: Determination of carcinogens in tobacco smoke and coal-derived samples - trace polynuclear aromatic hydrocarbons. In: Trace Substances in Environmental Health IX (Hemphill DD, ed.), Columbia, University of Missouri, 1975, pp 281-289

19. Lau TJ, Hackett RL, Sunderman FW: The carcinogenicity of intravenous nickel carbonyl in rats. Cancer Res 32: 2253-2258, 1972

20. McFarland AR, Bertch RW, Fisher GL, Prentice BA: A fractionator for size-classification of aerosolized solid particulate matter. Environ Sci Tech 11:781-784, 1977

21. Mercer TT: Aerosol Technology in Hazard Evaluation. New York, Academic Press, 1973, pp 21-62

22. Murchison D, Westoll, TS: Coal and Coal-Bearing Strata. New York, American Elsevier, 1968, p 418

23. Natusch DFS: Potentially carcinogenic species emitted from fossil fuel power plants. Environ Health Perspectives, in press, 1978

24. Natusch DFS, Wallace JR: Urban aerosol toxicity: The influence of particle size. Science 186:695-699, 1974

25. McCann J, Choi E, Yamasaki E, Ames BN: Detection of carcinogens as mutagens in the Salmonella microsome test. Assay of 300 chemicals. Proc Nat Acad Sci 72: 5135-5139, 1975

26. Ondov JM, Ragaini RC, Heft RE, Fisher GL, Silberman D, Prentice BA: Interlaboratory comparison of neutron activation and atomic absorption analyses of size-classified stack fly ash. Proc NBS 8th Materials Research Symposium, Gaithersburg, MD, 1977, pp 565-572

27. Ondov JM, Zoller WH, Omez I, Aras NK, Gordon GE, Rancitelli LA, Abel KH, Filby RH, Shah KR, Ragaini RC: Elemental concentrations in the National Bureau of Standards' environmental coal and fly ash standard reference materials. Anal Chem 47:1102, 1975

28. Ottolenghi AD, Haseman JK, Payne WW, Falk HL, MacFarland HN: Inhalation studies of nickel sulfide in pulmonary carcinogenesis. J Nat Can Inst 54:1165-1172, 197

29. Pawley JB, FIsher GL: Using simultaneous three color
     X-ray mapping and digital-scan-stop for rapid elemental
     characterization of coal combustion by-products. J
     Micros 110:87-101, 1977

30. Pott P: The Chirurgical Works of Percival Pott, Vol II.
     Philadelphia, James Webster, 1819, p 291

31. Rosenor VM, Oratz M, Rothschild MA: Albumin Structure,
     Function and Uses. New York, Pergamon Press, 1977, pp
     143-158

32. Silberman D, Fisher GL: Analysis of coal fly ash by
     atomic absorption spectroscopy. Pacific Conference on
     Chemistry and Spectroscopy, October 1977, Anaheim,
     California

33. Stone GD, Shimkin MB, Troxell MC, Thompson TL, Terry LS:
     Test for carcinogenicity of metallic compounds by the
     pulmonary tumor response in strain A mice. Cancer Res
     36:1744-1747, 1976

34. Vandergrift AE, Shannon LF, Gorman PG: Controlling fine
     particles. Chem Eng 80:107-114, 1973

35. Yeh, HC, Phalen RF, Raabe OG: Factors influencing the
     deposition of inhaled particles. Environ Health Per-
     spect 15:147-156, 1976

MUTAGENICITY OF SHALE OIL COMPONENTS

R.A. Pelroy and M.R. Petersen
Biology Department
Battelle-Northwest
Richland, Washington

Raw shale oil is a complex chemical mixture differing from most crude petroleums in having comparatively high concentrations of basic (nitrogen-containing) and phenolic compounds, in addition to having neutral compounds and polynuclear aromatic hydrocarbon (PNA) constituents more commonly found in crude oils (7). In the work described below, we have investigated the mutagenicity of a pilot plant sample of a crude shale oil (designated LO1) and two subfractions derived from this material.

Although the Ames assay has been widely used for mutagenic screening of pure chemicals (1,3,4), its use for bioassay of complex chemical mixtures has been more limited. Cigarette smoke condensate (2,5), complex mixtures of polycyclic compounds associated with airborne pollutants (9,10), and to a lesser extent, some synthetic fuels, have been assayed in this way (8).

In the work to be reported at this symposium, we have directed our attention to two problems that can arise during the Ames testing of complex chemical mixtures. First, we have estimated the ability of known chemical mutagens (premutagens requiring metabolic activation) to express themselves in the chemical environment to be represented by a raw shale oil or its subfractions. Second, we have estimated the degree of cell killing that occurs or is the result of exposing the Salmonella typhimurium test strains to these complex fractions under the conditions employed for the Ames assay, and the possible importance of such killing on the sensitivity of this assay.

MUTAGENIC PROPERTIES OF A SHALE OIL SAMPLE

The raw shale oil, L01, was fractionated into five subfractions: acidic (phenolic), basic, neutral, PNA, and a
complex residual mixture defined as a tar fraction. The
raw shale oil was mutagenic in the standard Ames assay.
The basic and PNA subfractions contained most of the mutagenic activity recoverable after separation of the shale
oil into its various chemical classes. In all cases,
mutagenicity was dependent on metabolic activation cata-
lysed by postmitochrondrial, microsomal enzymes. As
shown in Figure 1, the mutational response of S. typhimurium
TA100 was comparatively low for the crude product, the
basic, and PNA fractions. Comparable results were obtained
for the other test strain that we used for most of this
work, S. typhimurium TA98. In general, the mutational
responses for the basic and PNA range from 0.1 to 1 revertant colony per μg per 10^9 test cells added to the assay
system. In some instances, the response curves for the
two subfractions were linear for a greater concentration
range than shown in Figure 1. However, nonlinear mutagenic responses shown here are typical for both the crude
product and its subfractions.

Mutagenicity of Pure Chemical Plus Complex Fraction
Mixtures

A potential problem in interpreting the results of
the standard Ames test of complex chemical mixtures is the
possibility that the mutagenicity of the whole will be
significantly different than the sum of the individual
components.

One method of estimating the importance of chemical
composition on the Ames assay is to add a known mutagen
or premutagen to a complex fraction, and then compare
the mutagenicity of the mixed-system (chemical + fraction)
with the mutagenicity of the chemical alone. This experimental approach was followed with the raw shale oil, the
basic and PNA subfractions as complex materials, and
2-aminoanthracene, benzo(a)pyrene (benzopyrene) and 7,9-
dimethylbenz(c)acridine (dimethylbenzacridine) as known
premutagens.

In these experiments, the concentration of the pure
chemical was held constant at a value sufficient to yield
a strong mutational response when assayed alone, i.e.,

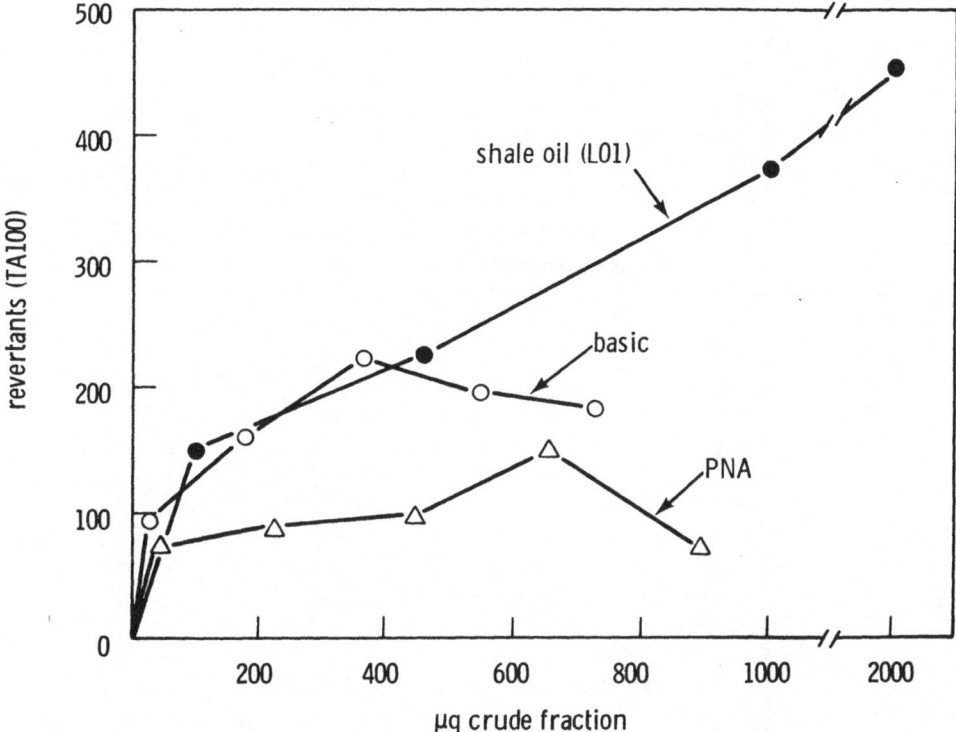

Figure 1. Mutagenicity of shale oil (L01) and the basic
and PNA fractions derived from L01. Salmonella typhimurium
TA100 was the test strain and each sample plate contained
50 μl of the S-9 enzymes.

1 μg 2-aminoanthracene, or 20 μg for benzopyrene and dibenzanthracene per assay plate (Figure 2). The concentration of the S-9 enzymes for the mixing experiments was
determined on the basis of that required for activating
the raw shale or its subfractions to form mutagens against
TA100. The data comparing S-9 requirements for the three
pure chemicals, the shale oil, and the four subfractions
is shown in Figure 3. For all of the mixing experiments
reported here, a constant value of 50 μl of S-9 per plate was
used. It should be noted (Figure 3) that the optimum S-9 concentrations for metabolic activation of dimethylbenzacridine
and benzopyrene, and for the crude fractions were approximately the same, while the optimum concentrations of S-9

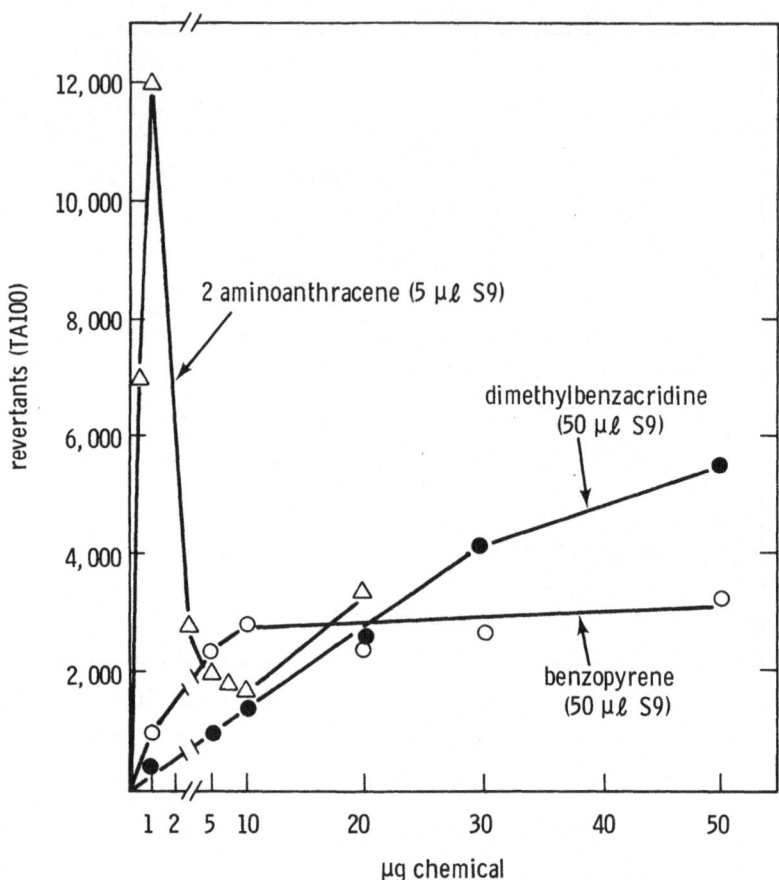

Figure 2. Mutagenicity of three chemicals as a function of concentration of pure or complex chemicals.

for activation of 2-aminoanthracene (alone) was considerably less.

    Three patterns of response were observed in the mixing experiments, depending on the chemical in question. For 2-aminoanthracene, addition of the raw shale oil or either the basic or the PNA fractions derived from the crude product, led results in a sharp increase in the number of revertants formed from TA100 (Figure 4) over that expected for the sum of the fraction plus chemical.

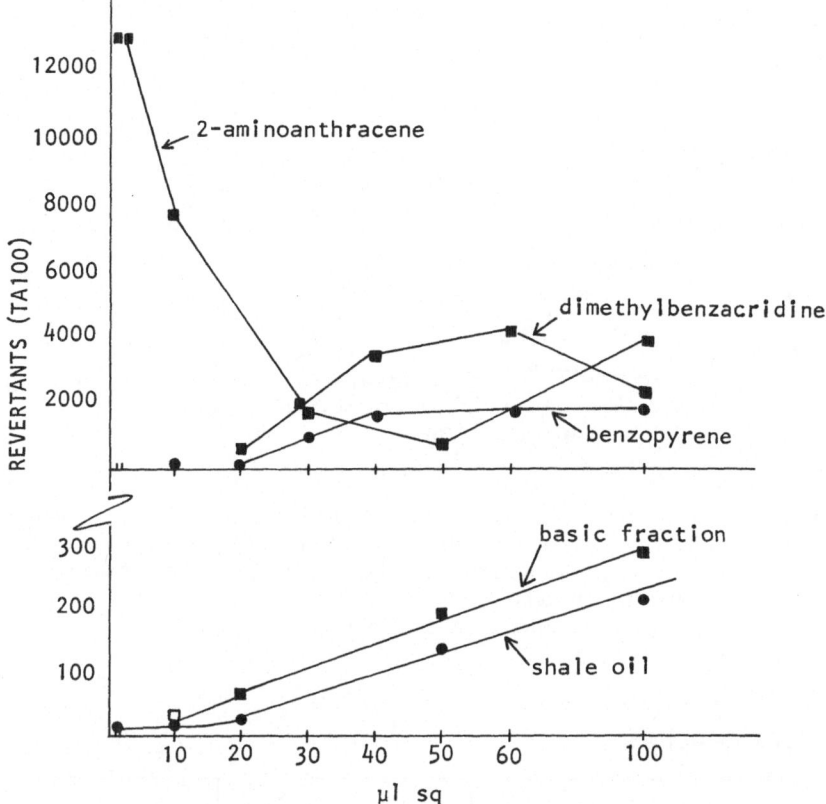

Figure 3. Mutagenicity of three chemicals and two complex fractions as function of S-9 concentration.

The increase in the mutagenicity of the mixture was greater than four times the maximum response for any one of the crude fractions assayed separately, and was equal to about 17% of the maximum mutagenic response observed for 2-aminoanthracene assayed alone at its optimum S-9 concentration (Figure 2).

In contrast to the results for 2-aminoanthracene the mutagenicity of benzopyrene steadily diminished with increasing concentrations of the three crude mixtures (Figure 5). In each case, the mixture yielded approximately the same number of revertant colonies per plate as the crude fraction alone and the mutagenicity of benzopyrene was marked.

2 aminoanthracene (1 µg) vs:

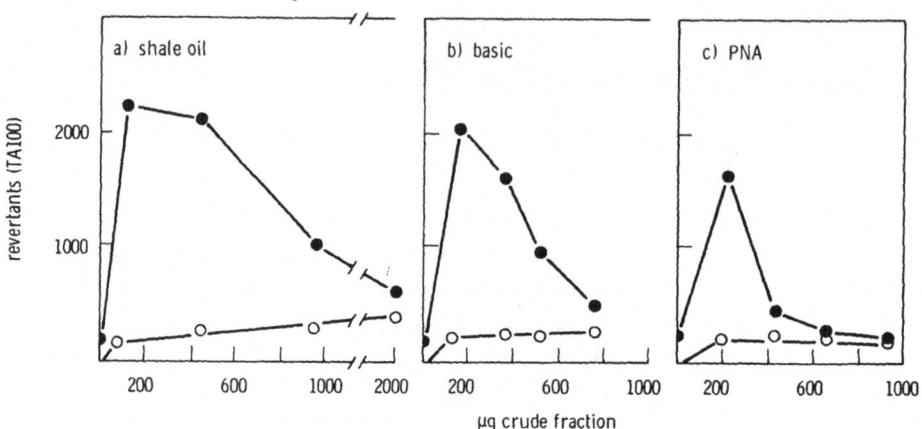

Figure 4.   Mutagenicity of 2-aminoanthracene (1 µg) and increasing concentrations of shale oil (LO1), basic, or PNA
fraction.   The concentration of the S-9 enzymes was constant
at 50   µl per assay plate.
    ● - combination pure chemical and crude fraction
    o - crude fraction

benzopyrene (9 µg) vs:

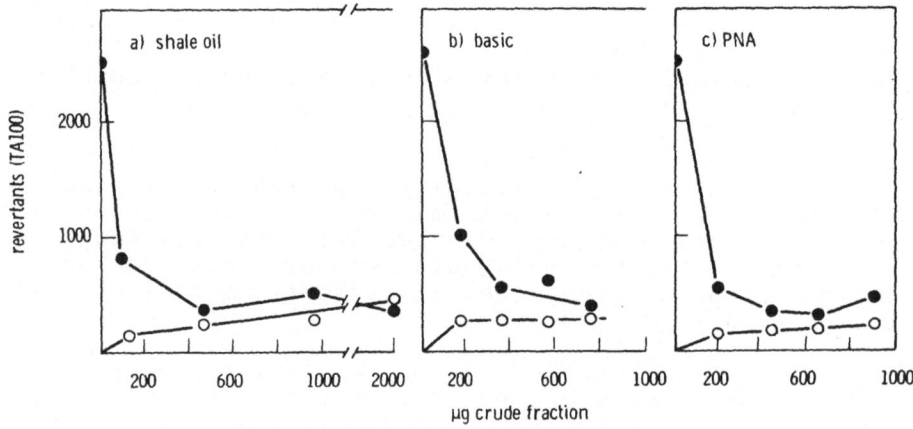

Figure 5.   Mutagenicity of benzopyrene (9 µg) and increasing
concentrations of shale oil (LO1), basic, or PNA fraction.
Conditions same as Figure 4.

The mixing experiments for dimethylbenzacridine showed
a third pattern. Here the mutagenicity of the mixture was
only slightly less than the sum of responses for the chemical
alone and crude fractions assayed at various concentrations (Figure 6). Addition of shale oil had the least
effect on the combined system, while the basic and PNA
fractions showed little inhibitory effect up to approximately 200 µg per assay plate.

Toxicity to Test Cells

In the standard Ames assay, the level of cell killing
due to formation of toxic metabolites or due to chemical
composition is not directly measurable. Since complex
hydrocarbon mixtures are generally toxic to bacteria, the
Ames assay of shale oil should take this into account.
In the work described here we have used an indirect method
to estimate the toxicity that occurs during mutagenesis
caused by the pure chemicals and complex fractions studied
above in the mixing experiments.

A revertant of TA100 was isolated from an assay plate.
This organism, designated TA100 rev, was added to the
standard Ames assay system at a range of dilutions from
10^{-4} to 10^{-7} from nutrient broth cultures containing
approximately 2×10^9 viable cells per ml. Because TA100
was wild type with respect to the biosynthesis of histidine,
it was able to grow on the assay plates used in the Ames
assay (i.e., on a glucose mineral base containing biotin
for which TA100 rev was still auxotrophic).

Addition of TA100 rev to the standard Ames assay
system showed that survival of this strain differed greatly
depending on the complex material or pure chemical being
assayed. For the three pure chemicals studied above, only
2-aminoanthracene showed a strong killing effect on TA100
rev (Figure 7). The concentration dependence for 2-amino-
anthracene induced toxicity closely followed the concentration dependence observed for mutagenesis (Figure 2), so, at
least in qualitative terms, loss of viability for TA100
rev was correlated with decreased mutational response by
the histidine auxotroph, TA100. On the other hand, neither
benzopyrene nor dimethylbenzacridine gave rise to killing
of TA100 rev over the concentration range used in the
standard Ames mutagenesis assays.

dimethybenzacridine (5 µg) vs:

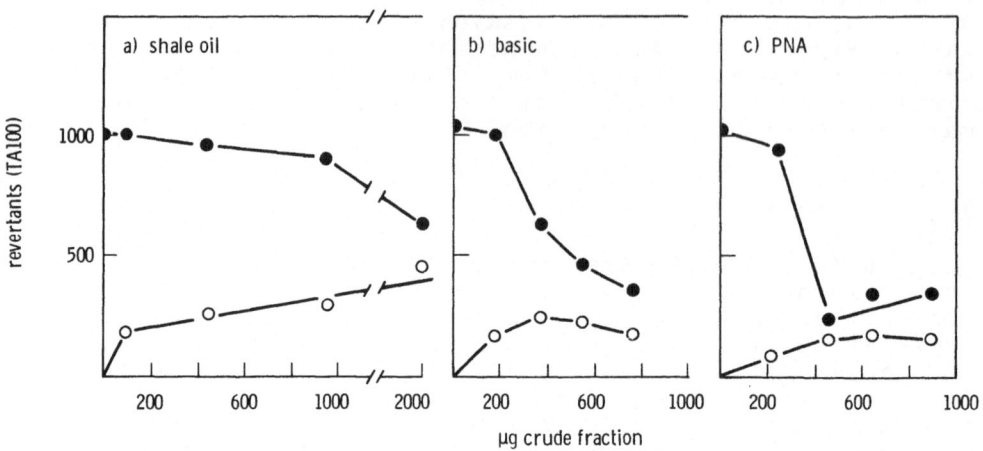

Figure 6. Mutagenicity of dimethylbenzacridine and increasing concentrations of shale oil (L01), basic, or PNA fraction. Conditions same as Figure 4.

Contrasting results were also obtained for the crude fractions. Shale oil (L01) and the PNA fraction showed little or no killing of TA100 rev in the Ames assay system at concentrations (per plate) approaching 1,000 µg for PNA fraction and nearly 2,000 µg for this raw shale oil (Figure 8).

The basic fraction, however, was highly cytotoxic for TA100 rev (Figure 8). The highest level of toxicity was observed for the complete assay system which contains all the necessary components for metabolic activation. Omission of reduced pyridine dinucleotide phosphate (NADPH$_2$), required for metabolic activation, reduced the killing of TA100 rev. For example, at approximately 400 µg basic fraction, killing of TA100 rev was nearly five times greater for assay plates containing NADPH$_2$ relative to those assay plates without this cofactor. At approximately 600 µg per plate, this relative increase was 18-fold and at slightly less than 800 µg per plate NADPH$_2$ dependent killing was more than 25 times greater than toxicity observed for the assay system minus the cofactor.

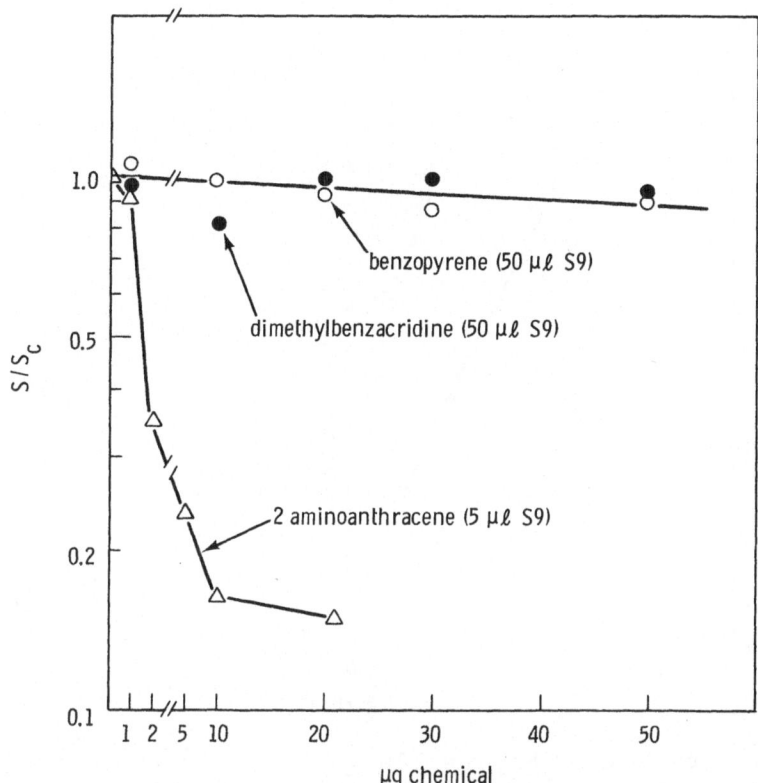

Figure 7. Survival of Salmonella typhimurium TA100 rev vs. premutagen. Concentration per plate of untreated (control) cells, Sc; exposed cells, S. The titer of TA100 rev on control plates was 1.7 x 10⁹ cells per ml of nutrient broth culture. The concentration of S-9 per assay plate is indicated in the figure.

In previous work we showed that formation of metabolite mutagens from 2-aminoanthracene and benzopyrene in the presence or absence of crude fractions is limited to the initial stages of the Ames assay, i.e, within the first 90- 120 min (6). Thus, the extensive killing demonstrated here for TA100 rev exposed to basic fraction might seriously reduce the mutagenic response for the system for this material.

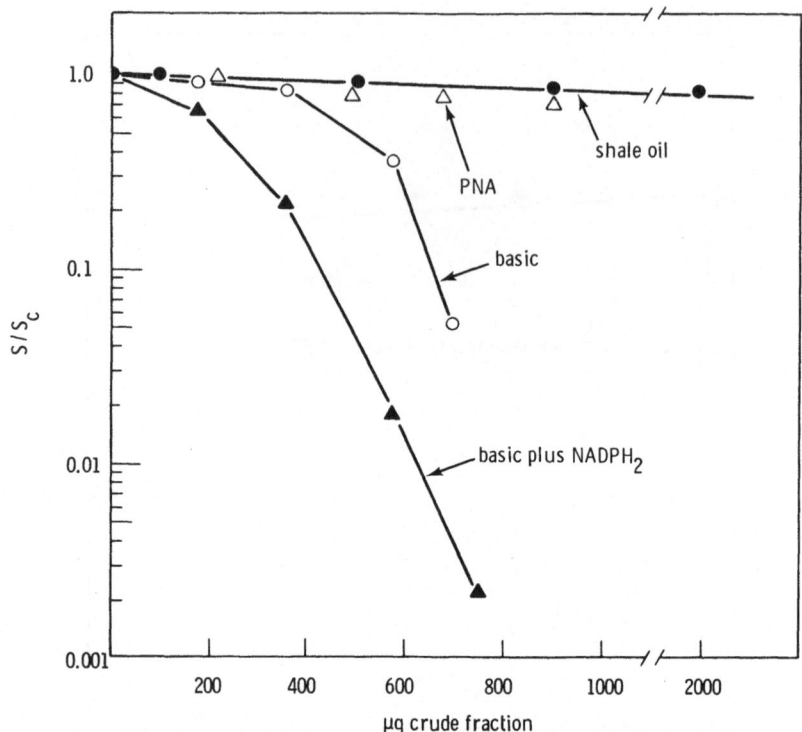

Figure 8. Survival of <u>Salmonella</u> <u>typhimurium</u> TA100 rev vs. complex fractions. The assay system for the basic fraction was complete for one set of plates and lacked a NADPH$_2$ generating system in a second set. The other samples contained the NADPH$_2$ generating system. S-9 concentration was fired at 50 µl per plate.

In summary, of the three chemical premutagens tested, 2-aminoanthracene and dimethylbenzacridine expressed more of their mutagenicity in the presence of shale oil than did benzopyrene. The mutagenicity of the latter compound was strongly suppressed by each of the complex fractions tested. The basic fraction in addition to being mutagenic was highly toxic to a revertant strain of <u>S. typhimurium</u> TA100 over the same concentration of crude fraction required for mutagenesis of the auxotrophic parental strain. Toxicity by the basic fraction was enhanced in the presence of a complete system for metabolic activation.

REFERENCES

1. Ames BN, McCann J, Yamasaki E: Methods for detecting carcinogens and mutagens with the Salmonella/mammalian-microsome mutagenicity test. Mutat Res 31:347, 1975

2. Kier LD, Yamasaki E, Ames B: Detection of mutagenic activity in cigarette smoke condenstates. Proc Natl Acad Sci 71:4159, 1974

3. McCann T, et al.: Detection of carcinogens as mutagens: Bacterial tester strains with R factor plasmids. Proc Natl Acad Sci 72:979, 1975

4. McCann J, et al.: Detection of carcinogens as mutagens in the Salmonella/microsome test: Assay of 300 chemicals. Proc Natl Acad Sci 72:5135, 1975

5. Mizusaki S, Takashima T, Tomura K: Factors affecting mutagenic activity of cigarette smoke condensate in Salmonella typhimurium TA1538. Mutat Res 48:29, 1977

6. Pelroy RA and Petersen MR: Use of Ames test in evaluation of shale oil fractions. Environ Health Perspectives, in press

7. Petersen MR, Fruchter J, Laul JC: Characterization of substances in products, effluents and wastes from synthetic fuel production tests. Quarterly report for the US Energy Research and Development Administration. Battelle, Pacific Northwest Laboratories, Richland, WA 99352. BNWL-2131, 1976

8. Rubin I, et al.: Fractionation of synthetic crude oils from coal for biological testing. Environ Res 12:358, 1976

9. Talcott R, Wei E: Airborne mutagens bioassayed in Salmonella typhimurium. J Natl Cancer Inst 58:449, 1977

10. Tokiwa H, et al.: Detection of mutagenic activity in particulate air pollutants. Mutat Res 48:237, 1977

MUTAGENIC ANALYSIS OF DRINKING WATER

Colin D. Chriswell, Bonita A. Glatz,
James S. Fritz, and Harry J. Svec
Iowa State University
Ames, Iowa

As recently as ten years ago relatively little was known about organic contaminants in drinking water. The carbon absorption methods (3,12) and other techniques were used to provide an indication of the amount of organic matter in water. However, only a handful of the individual compounds had ever been identified. During the past ten years it has become possible to separate and identify many organic substances in drinking water using techniques such as gas chromatography-mass spectrometry (GC-MS). Nearly 500 compounds have now been positively identified (9,10) and the list of identifications is continuing to grow.

Despite the progress that has been made, much remains to be learned about organic contaminants in water. In particular, we must elucidate the potential health effects of these organic compounds.

Some compounds have been identified in drinking water that may pose a threat to human health. Chloroform is present in water from every utility using chlorine as a disinfectant (2,13) (Figure 1), and chloroform and other trihalomethanes are suspected carcinogens (7,11). Other suspected carcinogens have also been identified in drinking water, but these compounds are generally less widespread and are rarely found at as high concentrations as the trihalomethanes (4,14). Continued identification and characterization efforts will undoubtedly reveal the presence of additional potentially harmful organic contaminants. It has become possible to use bioassay procedures such as the Salmonella/mutagenicity assay to guide the identification efforts towards compounds of the greatest potential interest.

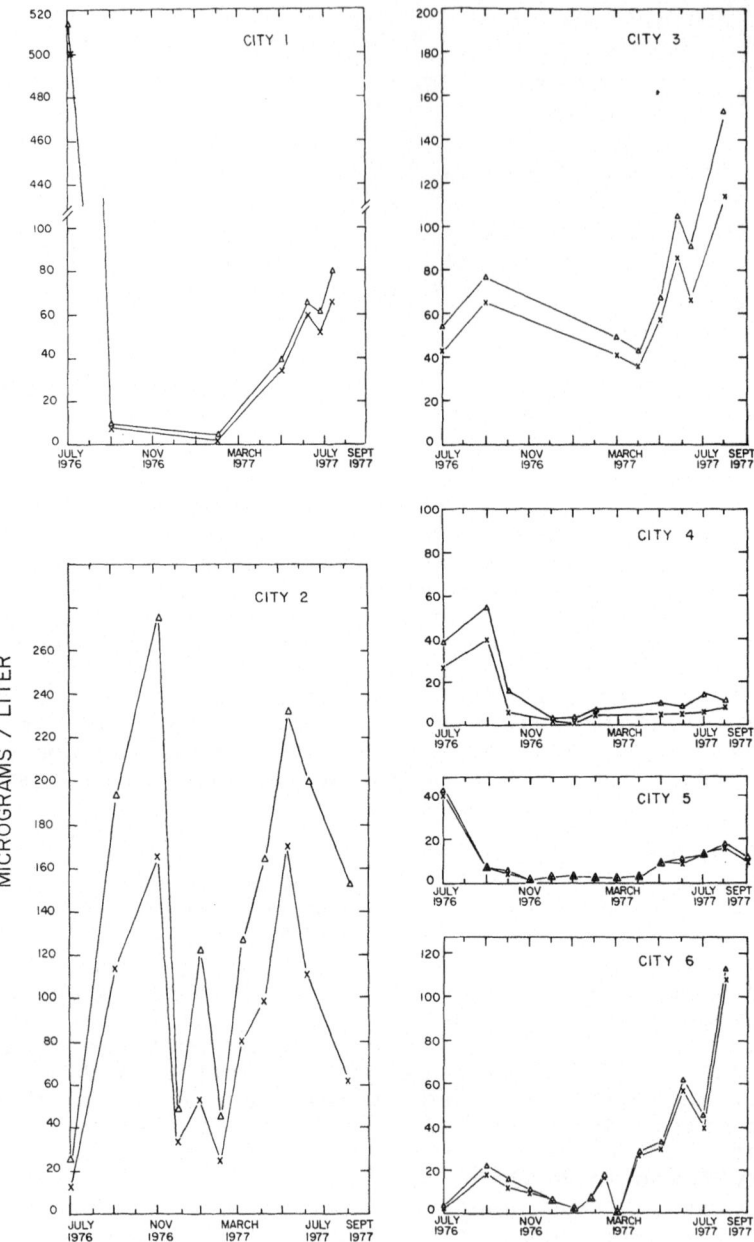

Figure 1. Levels of trihalomethanes found in drinking water
from fourteen cities. Upper line is total concentration of

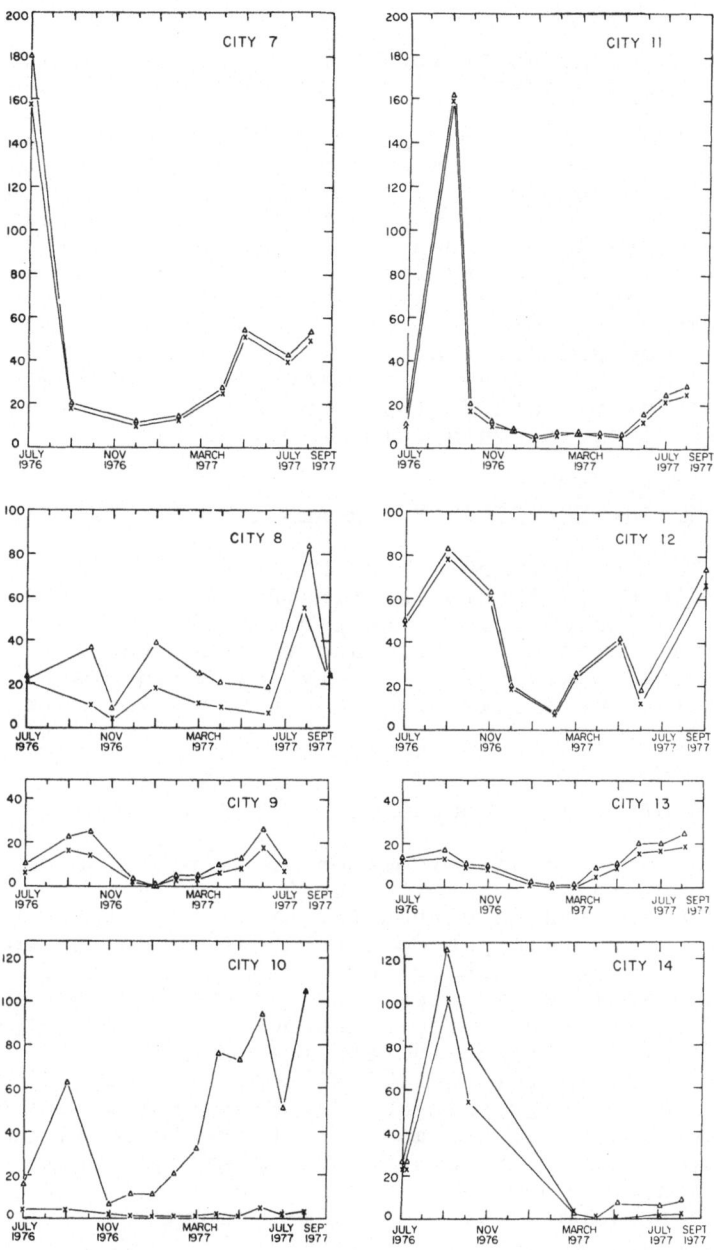

trihalomethanes expressed as chloroform equivalents. Lower
line is chloroform.

The research group at Iowa State University that I am representing has been involved in the development of analytical
methods for isolating, concentrating, and identifying organic
compounds. In the past, our group has consisted of analytical and physical chemists specializing in the areas of separations and mass spectrometry. During the past year bacte-
riologists, immunologists, sanitary engineers, and water
utility operators have joined our project. A multidisciplinary effort is being undertaken to determine more about
organic contaminants in drinking water and their potential
health effects. An immediate goal is to answer three questions: (1) How prevalent are mutagenic materials in drinking water? (2) What levels of mutagenic activity are
present in drinking water? (3) What are the chemical
characteristics of the mutagenic materials?

HOW PREVALENT ARE MUTAGENIC MATERIALS IN DRINKING WATER?

Since July of 1976 our group has been conducting a survey of organic contaminants in drinking water for the American Water Works Association. As part of that survey, organic
compounds are isolated from raw and finished water from each
of fourteen cities at monthly intervals. Aliquots of the
isolated organic materials have been assayed for mutagenic
activity.

Accumulation of Organic Compounds

Organic compounds are isolated by sorption on column
assemblies containing Amberlite XAD-2 resin in series with
Filtrasorb 200 activated carbon (Figure 2). With each sampling 200 l of water is passed through the sampling columns.
Both the primary and secondary columns are 6" x 1/2" i.d.
Accumulated organic substances are desorbed by elution with
100 ml of diethyl ether. The compounds are then further
concentrated by distilling the ether eluates to a final
volume of 1.00 ml. Of this 1.00 ml concentrate, 0.25 ml is
used for gas chromatographic and GC-MS determinations and
the remainder for mutagenic assays. Extracts obtained during the winter months of 1976 were composited, 300 μl of
dimethylsulfoxide (DMSO) added to each composite and the
residual ether evaporated. These DMSO concentrates contained
organic materials originally present in 15 l of water in each
10 μl of DMSO.

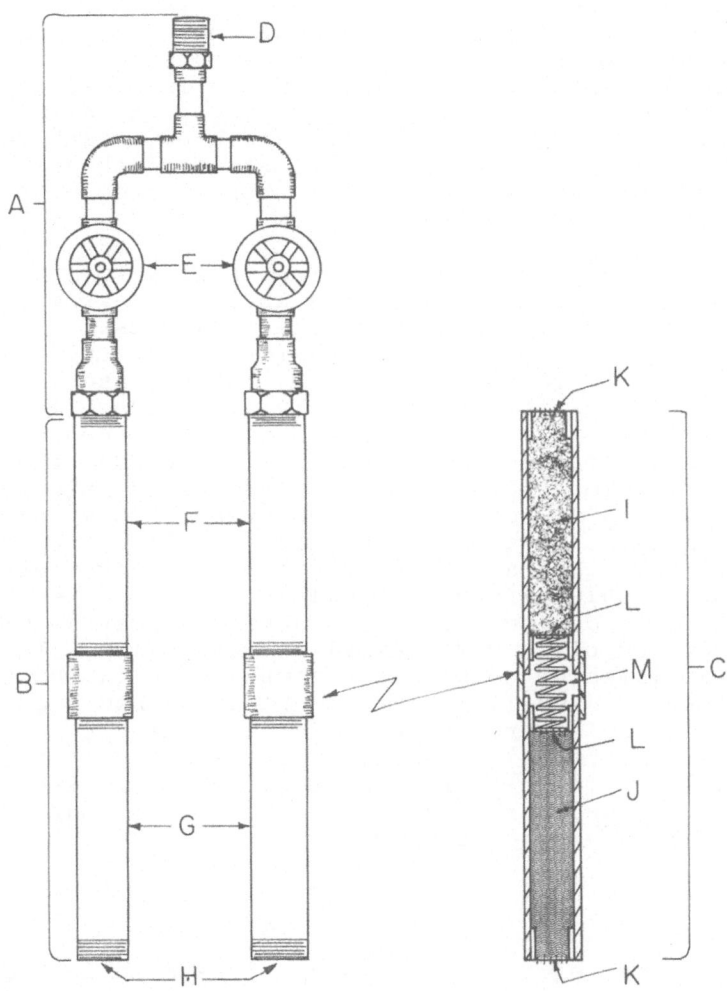

Figure 2.   A sampler used to accumulate organic materials from water.

Mutagenicity Assays

    Mutagenicity assays were performed using the spot test
procedure described by Ames, McCann, and Yamasaki (1).
Whatman No. 1 filter paper discs were soaked with 10 µl of
DMSO concentrate and placed in petri dishes on the surface
of agar seeded with approximately 10^8 cells of special
mutant strains of Salmonella tryphimurium. Strains TA98,
TA100, TA1535, TA1537, and TA1538 were used. Each sample
was tested at least twice with each strain with and without
the addition of the microsomal fraction of Aroclor 1254-
activated rat liver.

    The Salmonella strains lack the ability to grow without
added histidine but may regain the ability to grow in the absence of histidine by various mutagenic agents. Strains TA-
100 and TA1535 are reverted by substances causing base-pair
substitutions. Strains TA98, TA100, TA1537, and TA1538 are
reverted by frameshift mutagens of varying specificities.
Positive tests were defined in this work as a concentration
of revertant colonies in a circular array around the site of
sample application (Figure 3). The number of colonies in a
positive test is at least twice the number appearing in response to solvent controls. Marginal results were recorded
if only a small increase in colony count or a slight concentration of colonies around the sample were observed. The
liver fraction, designated S9, is added to provide many of
the key enzymes of in vivo mammalian metabolism. Thus, mutagenic metabolites of compounds not mutagenic in themselves
may be detected. Positive (known mutagens) and negative
(solvent) controls were included for each strain in each
experiment. No positive results were reported if replicate
determinations did not agree. A positive test with any one
strain of Salmonella indicates the presence of mutagenic
substances in the water sampled.

Results

    The results of these assays are presented in Figure 4.
Eleven of the fourteen finished and six of the raw water
sources exhibited some degree of mutagenic activity. The
greatest number of positive tests were obtained against
strain TA100. In contrast, the related strain, TA1535, was
not reverted by a single sample. This may in part be due to
the greater sensitivity bestowed on TA100 by the plasmid R
factor pKM101. In addition, TA100 is reverted by either
mutagens causing frame shift mutations or base-pair substi-

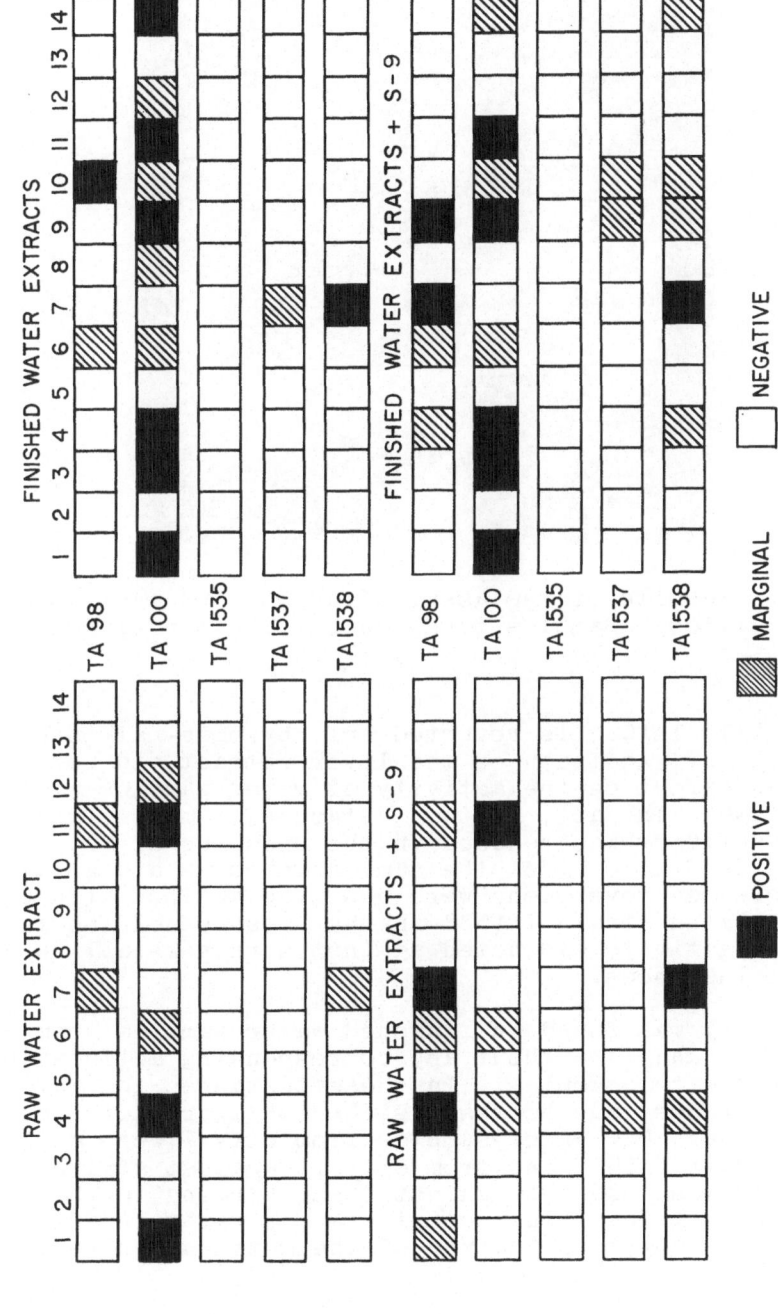

Figure 3.   A plate obtained from the testing of a water extract for mutagenic activity.

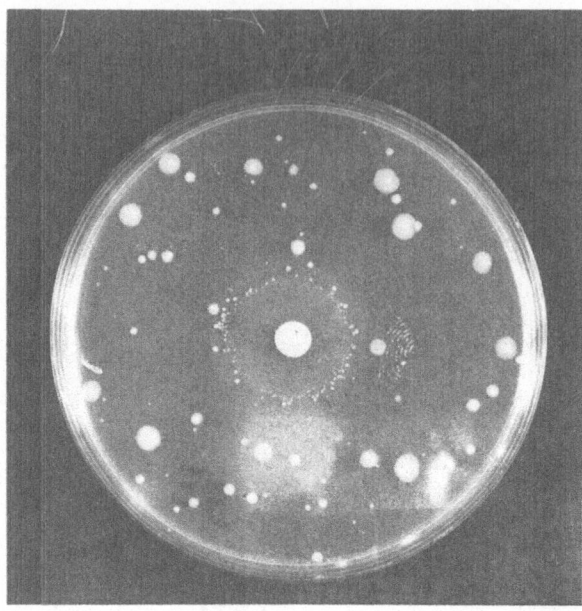

Figure 4. Results of the assay of composited samples from
fourteen cities. Samples taken during the winter months of
1976.

tutions while TA1535 is reverted only by base-pair substitution mutagens. Addition of the liver extract did not have a
pronounced effect on the activity of water samples against
strain TA100. In fact, in two instances, samples from cities
8 to 12, activity was reduced in the presence of S9. Strain
TA98 responded to many of the same samples as did strain TA-
100. Increased reversions were observed against strain TA98
and the related strain TA1538 in the presence of the liver
extracts. Activity was noted against strain TA1537 in a few
scattered instances.

      Activity was found in finished water samples from cities
3, 8, 9, 10, and 14 without any corresponding activity exhibited by raw water samples. In these instances the water
treatment process may be responsible for introducing mutagenic factors. Raw water samples from city 4 were active
against strain TA1537 and from city 11 against strain TA98.
No activity was observed against these strains in the finished water samples. In these instances water treatment may
have either altered the nature of the mutagenic materials or
removed them.

In this initial screening the prime goal was to determine if mutagenic materials were widespread ·in drinking water. We found such agents are very prevalent in finished water. We realize that the Ames test is not a perfect assay nor is the accumulation technique used perfect. Thus, these results may be only a conservative indication of the true prevalence of mutagenic materials in water.

WHAT LEVELS OF MUTAGENIC ACTIVITY ARE PRESENT IN DRINKING WATER?

The presence of any mutagenic materials in water is a cause for some concern. However, in order to evaluate the threat posed, the levels of mutagenic materials must be determined. In part, this is a matter of performing quantitative mutagenicity assays rather than the spot test procedures. It is also necessary to have confidence that all mutagenic materials are accumulated from water.

In the initial studies Amberlite XAD-2 resin was used as a primary accumulating agent. This sorbent is effective for recovering gas chromatographic organic compounds from water (5,6,8). However, it does not lead to the recovery of all organic materials from water. It is not known if mutagenic materials are of a nature such that they are recovered using XAD-2 resin.

To determine if other sorption techniques might remove more effectively mutagenic materials from water, sixteen columns, each containing a different test sorbent, were connected in parallel and used to sample finished water from four different Iowa utilities (Table 1). Some mutagenic materials were isolated from water using Amberlite XAD-2, XAD-4, XAD-7, XAD-8, Duolite S-761, and L-863 resins. Mutagenic materials were not isolated using activated carbons, weak base ion exchange resins, or a carbonaceous resin. The greatest amount of mutagenic activity was found in organic materials isolated using Amberlite XAD-4 resin. It is, of course, still not known that this sorbent is removing all mutagenic materials from water, but it is the most effective sorbent tested to date.

Studies are continuing in evaluating the effectiveness of other sorbents for removing organic mutagens from water, comparing mutagen recoveries at different water pH levels, and using reverse osmosis as the accumulating technique.

Table 1

Evaluation of Sorbents for Accumulating Mutagens

Sorbent	Sorbent Type	Activity of Isolated Organic Compounds
Amberlite XAD-2	PS-DVB Resin	+
Amberlite XAD-4	PS-DVB Resin	++
Amberlite XAD-7	Acrylic Ester Resin	+
Amberlite XAD-8	Acrylic Ester Resin	+
Duolite S-761	Phenol-Formaldehyde Adsorbent	+
Duolite L-863	PS-DVB Resin	+
Duolite S-37	Weak-Base Anion Exchange Resin	−
Duolite A-7	Weak-Base Anion Exchange Resin	−
Duolite ES-561	Weak-Base Anion Exchange Resin	−
Hydrodarco	Granular Activated Carbon	−
Filtrasorb 300	Granular Activated Carbon	−
Nuchar WVB	Granular Activated Carbon	−
Nuchar WVG	Granular Activated Carbon	−
NACAR G-216	Granular Activated Carbon	−
NACAR G-107	Granular Activated Carbon	−
Amberlite XE-340	Carbonaceous Resin	−

WHAT ARE THE CHEMICAL CHARACTERISTICS OF THE MUTAGENIC
MATERIALS?

A typical water source will contain on the order of one
part per million of total organic carbon. Most of this organic material, such as humic material, is believed to be of
natural origin. Other materials are introduced by man's
activities. Still other materials are produced during water
treatment. Identified organic compounds constitute only a
small fraction of the total amount of organic material in a
typical water supply.

It is of extreme importance that mutagenic materials
from drinking water be identified or at least characterized.
A very good correlation exists between mutagenicity and mammalian carcinogenicity, but the correlation is not perfect.
Thus, mutagenic materials from water should be characterized
so they can be tested for carcinogenicity. In addition,
effective measures for the control of mutagenic materials
can only be taken when the characteristics of the mutagens
are known.

The protocol adopted for the identification or characterization of mutagenic materials from water is based on successive, bioassay-guided fractionations until mutagenic activity
is isolated into a limited number of fractions with relatively
few components. Fractionation procedures must preserve the
integrity of the samples, be applicable to low amounts of
materials, and separate samples into fractions containing components of predictable characteristics.

Thus far we have evaluated fractionation procedures based
upon solvent extractions, thin-layer and column chromatography
on silica gel and alumina, column chromatography on Florisil,
high pressure liquid chromatography on Sephadex LH20, and
preparatory scale gas chromatography. Solvent extraction
procedures are not conveniently applicable to the ultra-trace
amounts of materials that can be isolated from drinking water.
We have found that organic compounds are lost or altered during fractionation procedures using silica or alumina. Polyaromatic hydrocarbons are irreversibly sorbed; alkenes and
some carbonyl-containing compounds undergo condensation or
polymerization reactions.

Fractions can be readily performed by preparatory scale
gas chromatography. This technique does, however, have the
serious limitation as it is applicable only to gas chromatographic compounds.

The initial step before performing a fractionation on Florisil is to transfer the sample from a diethyl ether to petroleum ether. Some materials precipitate during this solvent change. These materials are mutagenic, are not gas chromatographic, contain only very low levels of carbon, and give no characteristic IR or NMR spectra. In short, we have no idea what the material is but do know it is mutagenic. After materials are eluted from Florisil, the Florisil is dissolved in hydrofluoric acid to recover any very polar materials. The very polar materials have given no indication of mutagenic activity.

Fractionations performed using activated Florisil are based primarily on sample component polarities. A sample is introduced onto the top of a Florisil column and components are sequentially eluted with solvents of increasing polarities. In our current work we have found it desirable to elute organic materials isolated from water with 2% methylene chloride in petroleum ether, 60% methylene chloride in petroleum ether and 60% methylene chloride plus 2% acetonitrile in petroleum ether. Alkanes, alkenes, arenes, and halogenated hydrocarbons are eluted in the initial fraction. Despite the fact that this fraction would contain any polyaromatic hydrocarbons in samples, no mutagenic activity has been detected in it. The second elution fraction contains aldehydes, ketones, nitro-substituted compounds, nitriles, and some weaker phenols and amines. Mutagenic activity has been detected in this fraction. The third fraction contains alcohols, phthalates, amines, and phenols. Mutagenic activity has also been detected in this fraction (Table 2).

Fractionations of organic compounds on Sephadex LH20 are based both on sample component polarity and molecular size. A sample is introduced onto an LH20 column and eluted with 2-propanol. Alkanes are eluted first in order of decreasing molecular weights. Alkanes are followed by polar organic compounds. The last materials eluted are aromatic compounds. The aromatic compounds are eluted in order of increasing number of fused rings (Figure 5).

Identification or characterization efforts are based on the volatility of the isolated components. Gas chromatographic components are identified based on responses towards element specific detectors and GC-MS. We have not yet begun to characterize the nonvolatile components.

Table 2

Florisil Fractionation

Fraction	Types of Compounds	Mutagenic Activity
2% CH_2Cl_2 in petroleum ether	Aromatic and aliphatic hydrocarbons, halogenated aromatic and aliphatic hydrocarbons	–
60% CH_2Cl_2 in petroleum ether	Aldehydes, ketones, nitro-substituted compounds, nitriles, weak phenols, and amines	+
60% CH_2Cl_2 + 2% CH_3CN in petroleum ether	Alcohols, phthalates, amines, phenols	+
Hydrofluoric acid	Very polar	+

CONCLUSIONS

Mutagenic materials are widespread in drinking water and may be introduced into the water during treatment processes. This finding is cause for concern and we are continuing to monitor both finished and raw water in an attempt to determine what processes enhance the mutagenicity of water. We are also attempting to determine the characteristics of organic compounds in water that are responsible for the mutagenic activity. To date we have not identified a single compound that is responsible for a significant portion of the observed activity. Obviously a great deal remains to be done.

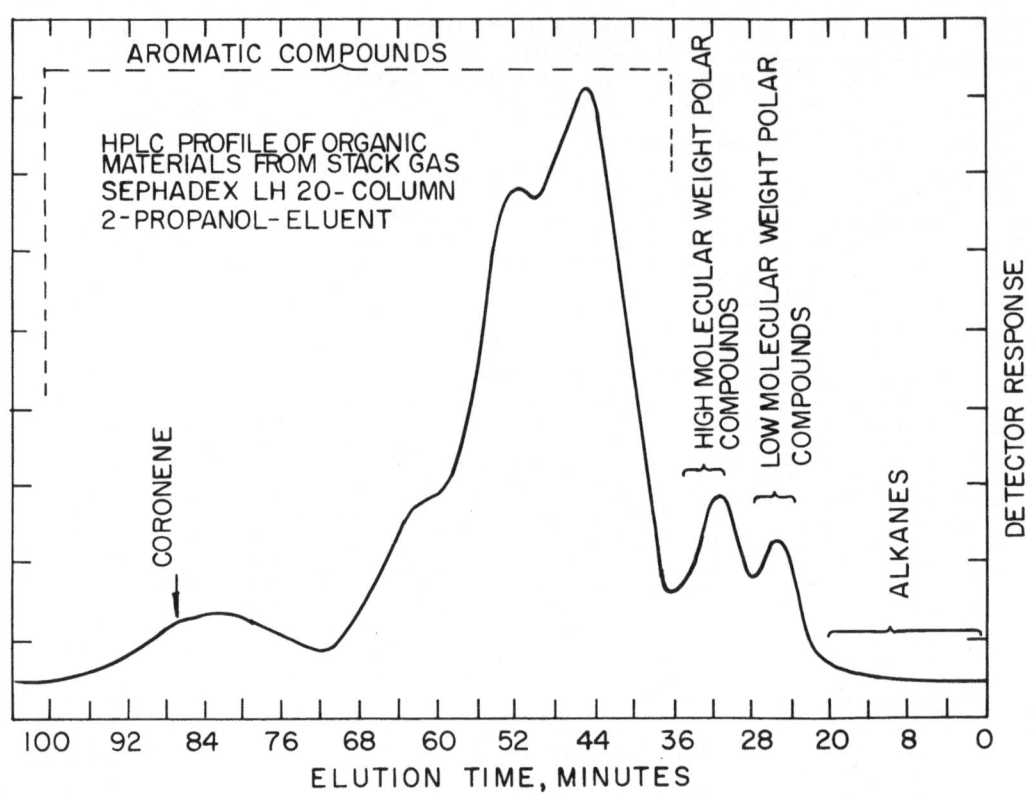

Figure 5.   Separation of organic materials on Sephadex.

Earlier I made a transition from talking about health effects of organic materials in water to talking about mutagens and potential carcinogens. A convenient procedure exists in the Ames test for determining if mutagenic materials are present in water. No convenient assay does exist for determining if organic materials from water possess other deleterious or beneficial properties.

REFERENCES

1. Ames BN, McCann J, Yamasaki E: Methods for detecting carcinogens and mutagens with the Salmonella/mammalian microsome mutagenicity test. Mutat Res 31:347, 1978

2. Arguello MD, Chriswell CD, Fritz JS, Kissinger LD, Lee KW, Richard JJ, Svec HJ: Trihalomethanes in water: A report on the occurrence, seasonal variations in concentrations, and precursors of trihalomethanes. Jour AWWA, submitted

3. Buelow RW, Carswell JK, Symons JM: An improved method for determining organics in water by activated carbon adsorption and solvent extraction. Jour AWWA 65:57 and 65:195, 1973

4. Chriswell CD, Arguello MD, Avery MJ, Ericson RL, Fritz JS, Junk GA, Kissinger LD, Lee KW, Richard JJ, Svec HJ, Vick R: Proceeding of the American Water Works Association Convention, May 1977

5. Chriswell CD, Ericson RL, Junk GA, Lee KW, Fritz JS, Svec HJ: Comparison of macroreticular resin and activated carbon as sorbents. Jour AWWA 69:56-69, 1977

6. Chriswell CD, Fritz JS, Svec HJ: Evaluation of sorbents as organic compound accumulators. AWWA Water Quality Technology Conference Proceedings, Dec. 1977

7. EPA Statement: Chlorinated and brominated compounds are not equal. Jour AWWA 69:5-12, 1977

8. Junk GA, Richard JJ, Grieser MD, Witiak D, Witiak JL, Arguello MD, Vick R, Svec HJ, Fritz JS, Calder GV: Use of macroreticular resins in the analysis of water for trace organic contaminants. Jour Chromatogr 99:745, 1974

9. Junk GA, Stanley SE: Organics in drinking water. Part
    1. Listing of identified compounds, Springfield, VA
    National Technical Information Service, 1975

10. McCabe LJ: Health effects of organics in water study.
    AWWA Water Quality Technology Conference, Dec. 1977

11. Report on the carcinogenesis bioassay of chloroform,
    Carcinogen Bioassay and Program Resources Branch,
    Carcinogenesis Program, Division of Cancer Cause and
    Prevention, National Cancer Institute

12. Standard methods for the examination of water and waste
    water, 13th ed., New York, NY, 1971

13. Symons JM, Bellar TA, Carswell JK, DeMarco J, Kropp KL,
    Roebeck GG, Seeger DR, Slocum CJ, Smith BL, Stevens AA:
    National organics reconnaissance for halogenated
    organics. Jour AWWA, 69:62, 1977

14. Von Rossum P, Webb RG: XAD resins and carbon for isola-
    tion of organic water pollutants, Anal Chem, 1978, in
    press

IN VITRO ACTIVATION OF CIGARETTE SMOKE CONDENSATE MATERIALS TO THEIR MUTAGENIC FORMS

R.E. Kouri, K.R. Brandt,
R.G. Sosnowski, L.M. Schechtman
Microbiological Associates
Department of Biochemical Oncology
Bethesda, Maryland

W.F. Benedict
Children's Hospital of Los Angeles
Los Angeles, California

INTRODUCTION

Cigarette smoke is a complex mixture composed of 5,000–10,000 different chemicals in the particulate phase, of which about 3,000 have been identified (1), and 1,000-2,000 chemicals in the gas phase. The particulate fraction contains many chemicals that are capable of inducing cancer in model test systems. Among these chemicals are certain polycyclic aromatic hydrocarbons (PAH) [e.g., benzo(a)pyrene (BP), dibenz(a,h)anthracene, and benz(a)anthracene]; certain nitrosamines (e.g., diethylnitrosamine and nitrosopiperidine); and certain aromatic amines [e.g., 2-napthylamine (2-NA) and 2-aminofluorenes (2-AF)] (see review 2). These chemicals are normally at levels approaching 0.5-20 ng/cigarette. The particulate phase also contains chemicals that are capable of promoting carcinogenesis (2-4). The level of these chemicals (e.g., catachol) are on the order of 10,000-100,000 ng/cigarette. Thus, there is a problem in determining not only whether cigarette smoke plays an active role in smoke-associated cancers in man, but also if this association occurs at the level of initiation and/or promotion of cancer.

One way to assess the potential initiating role that cigarette smoke may have is to test for the biological activity of certain cigarette smoke-derived fractions. This

[1]Supported in part through contracts from The Council for Tobacco Research USA, Inc., New York, NY 10002.

paper shows that measurement of the mutagenic potential of
cigarette smoke condensate (CSC) materials has some very
interesting ramifications. The fractions thought to contain
many biologically active chemicals (i.e., the PAH) have relatively weak, if any, mutagenic activity. The primary tissue
thought to be at risk to smoke effects is that of the lung.
However, of the lung systems employed, the mouse lung does
not activate these smoke condensates; yet another tissue that
is not believed to be at risk to smoke, the liver, can activate the smoke condensates. In vitro activation of such a
complex mixture as smoke condensate is obviously quite difficult to interpret, yet the studies presented here do suggest
some approaches that may be able to show direction for the
eventual understanding of the biological effects of tobacco-
related chemicals.

MATERIALS AND METHODS

Bacterial Strains

     The Salmonella typhimurium strains used, TA1538 and TA98,
were obtained from Dr. B. Ames (Biochemistry Department,
University of California, Berkeley, CA) and have been described
previously (5).

Compounds

     NADH, NADPH, 6-aminochrysene (6-AC), and BP were obtained
from Sigma Chemical Company, St. Louis, MO; 2-AF was from
Aldrich Chemical Company, Milwaukee, WI; Aflatoxin B_1 (AfB_1)
was from Calbiochem, La Jolla, CA; and 7,8-benzoflavone (7,8-
BF) was from Eastman Organic Chemical Company, Rochester,
NY. 7,8-dihydro, dihydroxy-BP (7,8-diol-BP) was provided by
Dr. D. Jerina (NIH-NIAID). Aroclor 1254 was from Analabs,
North Haven, CT; 2,3,7,8-tetrachlorodibenzo-p-dioxin (TCDD)
was provided by Dr. A. Poland (McArdle Laboratories, Univer-
sity of Wisconsin). The CSC fractions were generated by
Meloy Laboratories according to the methods of Patel et al.
(6). Dimethyl sulfoxide (DMSO) was obtained from Schwarz/
Mann, Inc., Rockville, MD.

Rat Hepatic and Mouse Pulmonary S-9

     Male Sprague-Dawley rats, weighing approximately 200 g
each were treated intraperitoneally (IP) with 0.5 ml of 200
mg Aroclor 1254/ml corn oil, in order to induce hepatic

enzymes. Forty-eight hours after injection, rats were sacrificed and livers excised. C57BL/6Cum mice, 6-8 weeks old,
were treated intratracheally (IT) with 0.02 ml of 6.0 μg
TCDD/ml trioctanoin in order to induce pulmonary and hepatic
enzymes. After 48 hr, mice were sacrificed and their lungs
and liver were excised. The 9000 x g post-mitochondrial supernatant (S-9) fractions from the liver and lung tissues were
prepared as previously described (7,8).

Preparation of S-9 Mix

     The S-9 mix for the suspension assay contained 1.2 mM
NADPH, 1.41 mM NADH, 136.9 mM NaCl, 2.68 mM KCl, 8.1 mM
Na_2HPO_4, 1.47 mM KH_2PO_4, and 3.0 mM $MgCl_2$, pH 7.4; total S-9
varied from 0.002 to 0.2 ml per ml of S-9 mix. The S-9 for
the pour-plate assay contained 3.6 mM NADPH, 4.2 mM NADH,
136.9 mM NaCl, 2.68 mM KCl, 8.1 mM Na_2HPO_4, 1.47 mM KH_2PO_4,
and 3.0 mM $MgCl$, pH 7.4, and varies from 0.2 to 0.3 ml of
S-9 fraction per ml. The liver preparations were sterile,
but the lung S-9 mix contained bacterial contaminants which
were removed by passing the S-9 mix through a sterile Milli-
pore disposable filter unit (0.45 μ pore diameter). Total
protein was determined for each condition with fluorescamine
according to the method of Weigele et al. (9).

Aryl Hydrocarbon Hydroxylase (AHH) Assay

     The assay for AHH activity was done according to proce-
dures outlined by Nebert and Gielen (10) and modified by
Kouri et al. (11).

Mutagenesis Assays

     All pour plate incorporation mutagenesis assays were
performed according to the method of Kier et al. (12). For
suspension assays, 0.1 ml of the bacterial tester strain,
0.5 ml of S-9 mix, and the sample to be tested were incubated
in a 37°C water bath for 35 min. After incubation, samples
were taken from each condition, diluted, and spread on nutrient agar plates to determine the number of bacteria at risk.
Two ml samples of molten top agar containing L-histidine
(0.05 mM) and biotin (0.05 mM) were added to each incubated
sample, mixed and poured onto Spizzizen minimal agar plates.
After 48 hr incubation at 37°C, prototrophic revertant

colonies were counted on an NBS Model C111 Colony Counter
(New Brunswick Scientific, Edison, NJ). DMSO and acetone,
in the amounts used, have no toxic or mutagenic effects on
the tester strains. Preliminary studies were done using the
suspension protocol in order to assess the relationship
between the number of mutant colonies observed relative to
the number of bacteria added per plate when trace amounts
(0.05 mM) of L-histidine and biotin were present. Initial
bacterial concentrations ranging from 5×10^6 to 1×10^8/plate
resulted in only an 0.3-fold increase in the number of mutants
per plate. Thus, when L-histidine and biotin were present,
the number of revertant colonies did not really reflect the
initial number of bacteria that were added because the trace
levels of L-histidine allowed for a certain amount of growth
to occur. Therefore, a mutation frequency was calculated only
when the numbers of surviving bacteria for the various test
groups remained relatively constant. In all other cases, the
mutation data were given just in terms of number of revertant
colonies per plate.

RESULTS

     A summary of the biological effects of the smoke conden-
sate from 1A1 low nicotine, normal tar content cigarettes is
shown in Table 1. The assays used included measurements of:
(a) pulmonary AHH following IT administration of fraction;
(b) competitive inhibition of BP metabolism in vitro; (c)
mutagenesis at the his locus in S. typhimurium strains TA1538;
and (d) neoplastic transformation of C3H 10T½ cells in culture. The whole condensate and reconstituted fractions were
weak inducers of pulmonary AHH, weak competitive inhibitors
of BP metabolism, mutagenic to TA1538, and transforming to
the 10T½ cells. Mutagenesis required the presence of an exogenous metabolic activation system in the form of Aroclor
1254-induced rat hepatic S-9 preparation. Fraction $B_I b$ (for
discussion of fraction nomenclature, see 13) contained chemicals that were potent inducers and inhibitors of AHH, could
be metabolized to forms highly mutagenic to TA1538, and could
transform the 10T½ cells. Fractions $B_I a$, B_E, and WA_I also
were active in most of these systems. Fractions N_{MeOH} and
N_{NM} were inducers of pulmonary AHH and could competitively
inhibit BP metabolism in vitro, but had low mutagenic potential and did not transform the 10T½ cells. The N_{NM} fraction
accounts for most of the BP content of the smoke condensate
(see footnotes in Table 1). The strong acid fractions (SA_I,

Table 1

Effects of Fractions of 1A1 CSC in Various Model Systems

Fraction[1]	mg/Cig.	AHH Ind.[2]	[X]/[BP]to Give 50% Inhibition[3]	Mutants/ Plate[4]	Transfor- mation[5]
Whole CSC	23.50	1.7	5.0	+++	+
Reconsti- tuted CSC	23.00	1.8	5.2	+++	+
$B_I a$	0.81	3.6	0.8	++	−
$B_I b$	0.29	2.5	0.5	+++	+
B_E	0.95	1.5	3.0	++	−
B_W	0.36	0.5	>10.0	−	−
WA_I	2.27	1.6	5.0	++	+
WA_E	1.98	1.1	2.0	$\pm$	−
SA_I	0.39	0.5	>10.0	+	−
SA_E	0.78	0.3	>10.0	−	−
SA_W	8.69	0.4	>10.0	−	−
N_{MeOH}	1.19	2.5	3.0	$\pm$	−
N_{CH}	4.58	1.2	ND	−	−
N_{NM}	0.70	3.2	1.0	$\pm$	−

[1] Whole cigarette smoke condensate (CSC) has 21.0 mg nicotine, 5.70 mg phenols, 0.98 µg BP/g. Reconstituted CSC has 22.0 mg nicotine, 5.51 mg phenols, 0.90 µg BP/g. B_E has 31.0 mg nicotine/g. WA_E has 41 mg phenols/g. N_{NM} has 13.1 µg BP/g.

[2] Aryl hydrocarbon hydroxylase (AHH) inducibility = Effect of fractions of 1A1 CSC on pulmonary AHH activity of C57BL/6Cum mice relative to a corn oil control (11).

[3] BP inhibition = Competitive in vitro effect of CSC fractions on BP metabolism by hepatic microsomes from 3-MC-treated C57BL/6Cum mice (14).

[4] Mutagenesis = Mutagenic activity of 1A1 CSC fractions in the Ames assay with S. typhimurium TA1538 in the presence of liver microsomal S-9 mix (12).

[5] Transformation = Malignant transformation frequency in C3H 10T½ Cl. eight cells treated with CSC fractions (15).

SA_E, and SA_W) actually inhibited pulmonary AHH activity and only the SA_I fraction had an effect in any of the in vitro bioassays.

A repeat (using a blind protocol) of the mutation studies, this time using the 2A1 CSC and fractions derived from this condensate, is shown in Table 2. Experiments 1 and 2 are results from studies completed one year apart. The condensate and fractions were stored at -70°C during the interim. The total tar content was higher in this cigarette condensate relative to the 1A1 condensate. However, on a per cigarette basis, these data are very similar to those of Kier et al. (12) using the 1A1 cigarette condensate. The mutagenic activity of the whole condensate, reconstituted fractions and the 12 fractions were very similar to that of 1A1 condensate. The most active fractions were B_Ib, B_E, B_Ia, WA_I and WA_E. The only discrepancies relative to the 1A1 condensate were the higher activity of the nicotine-containing B·· fraction and the slightly lower activity of the WA_I fraction. The mutagenic activity was stable at -70°C for at least one year since both experiments yielded quite similar results.

The use of TCDD-induced mouse pulmonary tissue as an in vitro activation system for the 2A1 condensate is shown in Table 3. Under conditions in which the pulmonary S-9's could efficiently metabolize BP to 3-hydroxybenzo(a)pyrene (3-OH-BP) (the basis for the AHH assay), no metabolism of the 2A1 condensate to a mutagenic form could be observed. Use of pulmonary S-9's from 3-methylcholanthrene (3-MC) treated or control mice also did not activate the 2A1 condensate material (data not shown). Comparison of the in vitro metabolic capacity of TCDD-induced pulmonary S-9 with mouse or rat hepatic S-9's in either a pour plate or suspension assay is shown in Table 4. Under conditions in which both the rat and mouse hepatic S-9's activated 6-AC or the 2A1 condensate to mutagenic forms, the mouse pulmonary S-9 failed to activate either of these chemicals. Also, addition of similar levels of total AHH activity for both pulmonary and hepatic S-9's (by adjusting total protein concentration) yielded conditions in which only the hepatic S-9 activated 2A1 condensate to forms mutagenic to strain TA98 (data not shown).

There was the possibility that even though these pulmonary S-9's were capable of metabolizing BP to 3-OH-BP, some sort of inhibitor of bacterial mutagenesis was functioning

Table 2

Mutagenesis of TA98 with 2A1 CSC Fractions[1]

Fraction	Sample No.	mg/Cig.	Mutants/Plate					
			Per 250 μg of Sample		Per Cigarette		% Activity	
			Exp. 1	Exp. 2	Exp. 1	Exp. 2	Exp. 1	Exp. 2
Whole CSC	5	40.00	133	122	21,280	19,520		
Reconstituted CSC	14	39.50	159	148	25,122	23,384		
B$_I$a	3	0.60	987	1,149	2,369	2,758	11	10
B$_I$b	11	0.24	3,510	5,266	3,370	5,055	16	18
B$_E$	4	1.04	1,258	2,065	5,233	8,590	25	30
B$_W$	6	0.52	102	151	212	314	1	1
WA$_I$	10	3.63	289	394	4,196	5,721	20	20
WA$_E$	7	2.66	141	206	1,500	2,192	7	8
SA$_I$	12	1.44	53	43	305	248	2	1
SA$_E$	2	0.88	8	14	28	49	0	0
SA$_W$	13	15.60	8	20	499	1,248	2	4
N$_{MeOH}$	9	2.16	65	93	562	804	3	3
N$_{CH}$	8	9.54	57	24	2,175	916	10	3
N$_{NM}$	1	1.24	80	139	397	689	2	2

[1]The 2A1 (low nicotine) CSC fractions were generated by Meloy Laboratories according to the methods of Patel et al. (6) and the pour plate incorporation mutagenesis assay was performed according to the methods of Kier et al. (12).

Table 3

Activation of 2A1 CSC
by Pulmonary S-9 Using a Suspension Assay[1]

Pulmonary S-9[2] (mg protein)	2A1 Condensate (μg/Tube)	AHH[3]	BAR (x 10^7)	Mutants/ Plate	MF[4] (x 10^{-7})
0.72	1300	547.4	0.78	15	19.23
0.72	650		0.76	13	17.11
0.72	260		0.79	12	15.19
0.72	0		0.86	13	15.12
1.44	650	1071.0	0.81	21	25.92
1.44	260		0.88	19	21.59
1.44	0		0.72	13	18.06
TA98 (alone)			0.78	14	17.95

[1] For suspension assays, 0.1 ml of the bacterial tester strain, 0.5 ml of S-9 mix, and the sample to be tested were incubated in a 37°C water bath for 35 min. After incubation, samples were taken from each condition, diluted, and spread on nutrient agar plates to determine the number of bacteria at risk (BAR). Two ml samples of molten top agar containing L-histidine (0.05 mM) and biotin (0.05 mM) were added to each incubated sample, mixed, and poured on Spizzizen's minimal agar plates. After 48 hr incubation at 37°C, prototrophic revertant colonies were counted on an NBS Model C111 Colony Counter.

[2] S-9 was derived from pulmonary tissue of C57BL/6Cum ♀ mice induced by IT installation of 120 ng TCDD/0.02 ml trioctanoin 48 hr prior to sacrifice.

[3] AHH = pMoles 3-OH-BP formed per 35 min incubation in separate tubes which contained 25 μg BP/ml as substrate and which were assayed under the same conditions and at the same time as those tubes containing 2A1 condensate.

[4] MF = Mutation frequency, i.e., the number of his$^+$ revertant bacterial colonies per BAR.

Table 4

Comparison of C57BL/6Cum TCDD-Induced Hepatic
and Pulmonary S-9 Mediated Metabolism of 2A1 Whole CSC
to Form(s) Mutagenic to S. typhimurium TA98

Source of S-9	Compound (μg/Plate)		mg Protein	AHH[1]	BAR[2] (x 10^7)	Mutants/ Plate	MF[3] (x 10^{-7})
C57BL/6Cum Hepatic:							
Pour plate	2A1	(650)	3.14	4929		132.0	
	AfB$_1$	(1)	3.14	4929		189.0	
Suspension	2A1	(260)	3.14	4929	0.84	75.7	90
	6-AC	(5)	3.14	4929	0.74	131.7	178
C57BL/6Cum Pulmonary:							
Pour plate	2A1	(650)	1.44	1071		12.0	
	AfB$_1$	(1)	1.44	1071		20.7	
Suspension	2A1	(650)	1.44	1071	0.81	21.0	26
	6-AC	(5)	1.44	1071	0.78	39.7	51
Rat Hepatic:							
Pour plate	2A1	(1300)	4.43	9625		202.3	
	AfB$_1$	(1)	4.43	9625		775.0	
Suspension	2A1	(650)	2.95	7371	0.82	97.7	120
	6-AC	(0.5)	0.03	190	0.63	329.0	522
TA98 alone:							
Pour plate						14.7	
Suspension					0.77	14.7	19

[1]AHH = pMoles 3-OH-BP formed per assay tube; total time was 35 min; 25 μg BP/ml
was substrate.

[2]BAR = Number of bacteria at risk.

[3]MF = Mutation frequency, i.e. the number of his$^+$ revertant mutant bacterial
colonies per number of BAR.

in these S-9's. Tables 5 and 6 show that these pulmonary
S-9's are capable of activating the 7,8-diol-BP and 2-AF to
mutagenc forms, respectively. Thus, these S-9's were capable
of activating at least some PAH and some aromatic amines to
forms mutagenic to tester strain TA98. The activation of
2-AF was dependent on the integrity of the mixed-function
oxidase system because inhibition of AHH by the inhibitor
7,8-BF resulted in concomitant inhibition of 2-AF-induced
mutagenesis (Table 7).

DISCUSSION

     Cigarette smoke contains chemicals that have been shown
to be biologically active in a variety of model systems both
in vitro (see Tables 1 and 2; 12,14,15) and in vivo (3,9,16,
17). Of prime importance is the fact that either whole smoke
(14,18) or smoke condensate material is capable of interacting
with those microsomal monooxygenases known to play a major
role in the activation of many chemical carcinogens to their
cytotoxic (19-21), mutagenic (5,22,23), or carcinogenic (24-27)
forms. In this paper, we show that both the 1A1 and 2A1 reference cigarettes contain chemicals that are substrates for
hepatic monooxygenases and as a result of metabolism by these
hepatic tissue preparations, intermediates are generated which
are mutagenic to S. typhimurium tester strains TA1538 and TA98
(see Tables 1 and 2). Two interesting facts emerge from these
studies: (a) the tester strain TA98 is selectively more sensitive to mutagenesis induced by smoke condensate; and (b) the
fractions that contain most of the mutagenic activity are not
those known to contain the PAH, but rather should contain such
base-soluble chemicals as aromatic amines. Thus, the data
suggest that the majority (approximately 58%) of the total
mutagenic activity of these condensates is in the basic fractions, and not in those fractions containing the PAH.

     Another main issue of concern is the fact that mouse
pulmonary tissue fails to activate the 2A1 smoke condensates
to mutagenic forms (see Tables 3 and 4). That is, under
conditions in which these pulmonary S-9's metabolize BP to
3-OH-BP and metabolize both 7,8-diol-BP (Table 5) and 2-AF
(Table 6) to mutagenic forms, these S-9's fail to activate
either 6-AC or the 2A1 condensate (see Table 4). Thus, mouse
pulmonary tissues seem to be capable of activating certain
PAH and aromatic amines, but not others. If the 2A1 smoke
condensate does contain aromatic amines and these chemicals
are responsible for the high mutagenic activity of these condensates when metabolically activated by hepatic S-9's, then

Table 5

Activation of 7,8-diol-BP by Pulmonary S-9

Pulmonary S-9 (mg Protein)	7,8-diol-BP (μg/Tube)	AHH[1]	BAR[2] (x 10^7)	Mutants/ Plate
6.57	0.5	1928	0.44	544
3.28	0.5	1680	0.14	338
0	1.0	–	0.68	76
0	0.1	–	0.76	84
TA98 alone	–	–	0.77	20

[1]AHH = pMoles 3-OH-BP formed per 35 min incubation in separate tubes containing 25.0 μg BP/ml as substrate.

[2]BAR = Number of bacteria at risk. Because of large variation in BAR, no mutation frequency is given.

Table 6

Activation of 2-AF by Pulmonary S-9

Pulmonary S-9 (mg Protein)	2-AF (μg/Tube)	AHH[1]	BAR[2] (x 10^7)	Mutants/ Plate
6.57	25	1904	0.37	536
	10	1925	0.62	668
	10	1904	0.47	521
3.28	25	1452	0.40	458
	10	1452	0.40	580
	5	1680	0.78	627
0	25	–	0.63	39
	10	–	0.65	52
TA98 alone	–	–	0.77	20

[1]AHH = pMoles 3-OH-BP formed per 35 min incubation in separate tubes containing 25.0 μg BP/ml as substrate.

[2]BAR = Number of bacteria at risk. Because of large variation in BAR, no mutation frequency is given.

Table 7

Effect of 7,8-BF on Pulmonary and Hepatic S-9
Mediated Activation of 2-AF to Forms
Mutagenic to S. typhimurium TA98

S-9[1]	2-AF (µg)	7,8-BF (µg)	AHH[2]	BAR[3] (x 10^7)	Mutants/ Plate
Rat Hepatic	10	0	856.8	0.31	564
(0.15 mg protein)	10	10	630.7	0.64	210
	10	25	202.3	0.52	187
Mouse Pulmonary	–	–	1701.3	0.68	28
(2.9 mg protein)	10	0	1701.7	0.46	538
	10	10	261.8	0.71	125
	10	25	66.5	0.54	68
TA98 (alone)	–	–	–	1.07	16

[1]S-9's were derived from: (1) hepatic tissue from Fischer
334 ♂ rates (200-250 g) induced by IP administration of
500 mg Aroclor-1254/kg body weight 48 hr prior to sacrifice, or (2) pulmonary tissue of C57BL/6Cum ♀ mice (approximately 20 g) induced by IT instillation of 120 ng TCDD/
0.02 ml trioctanoin 48 hr prior to sacrifice.

[2]AHH = pMoles 3-OH-BP formed per 35 min incubation in
separate tubes containing 25 µg BP/ml and, when necessary,
the indicated levels of 7,8-BF.

[3]BAR = Number of bacteria at risk; because of variation in
BAR, no mutation frequency is given.

the aromatic amines would seem to mimic the effects of 6-AC
more nearly than those of 2-AF. This is likely since 6-AC is
activated by hepatic tissue, but not by mouse pulmonary tissue
whereas 2-AF can be activated by pulmonary tissue (see RESULTS
and Table 4).

Whether or not the inability to metabolically activate
smoke condensate is unique to the mouse pulmonary tissue
cannot be answered at this time. Kier et al. (12) reported

that rat pulmonary S-9's gave only slight increases in number of mutants with the 1A1 smoke condensate and its fractions. Hutton and Hackney (28) reported different results using the 1R1 tobacco smoke condensate fractions and induced rat and normal human pulmonary S-9. These authors observed no statistically significant increase in mutagenicity of these condensates with either of these lung-derived activation systems. We are presently comparing pulmonary tissue from mouse, rat, and human sources for their ability to metabolically activate CSC material to biologically active forms.

REFERENCES

1. Wakeham H. Recent trends in tobacco and tobacco smoke research. In: The Chemistry of Tobacco and Tobacco Smoke (Schmeltz I, ed.), New York, Plenum Press, 1972, pp 1-20

2. Weisburger JH, Cohen LA, Wynder EL. On the etiology and metabolic epidemiology of the main human cancers. In: Origins of Human Cancer (Hiatt H, Watson JD, and Winsten JA, eds.), Cold Spring Harbor, New York, Cold Spring Harbor Laboratory, 1977, pp 567-602

3. Bock FG, Swain AP, Stedman RL. Bioassay of major fractions of cigarette smoke condensate by an accelerated technic. Cancer Res 29:584-587, 1969

4. Van Duuren B, Katz C, Goldschmidt BM. Co-carcinogenic agents in tobacco carcinogenesis. J Natl Cancer Inst 51:703-705, 1973

5. McCann J, Spingarn NE, Kobori J, Ames BN. Detection of carcinogens as mutagens: Bacterial tester strains with R factor plasmids. Proc Natl Acad Sci US 72:979-983, 1975

6. Patel AR, Haq MZ, Innerarity CI, Innerarity LJ, Weisgraber K. Fraction studies of smoke condensate samples from Kentucky reference cigarettes. Tobacco 176:61-62, 1974

7. Schechtman LM, Kouri RE. Control of benzo(a)pyrene-induced mammalian cell cytotoxicity, mutagenesis and transformation by exogenous enzyme fractions. In: Progress in Genetic Toxicology (Scott D, Bridges BA, Sobels FH, eds.), New York, Elsevier/North Holland Biomedical Press, 1977, pp 307-316

8.   Kouri RE, Schechtman LM. In vitro metabolic activation
     systems. In: Short-Term In Vitro Testing for Carcino-
     genesis, Mutagenesis and Toxicity (Berky J, Sherrod PC,
     eds.), Philadelphia, Franklin Inst. Press, 1978, pp 423-
     430

9.   Weigele M, DeBernardo S, Tenji J, Leimgruber W. A
     novel reagent for the fluorometric assay of primary
     amines. J Amer Chem Soc 94:5927-5931, 1972

10. Nebert DW, Gielen JE. Genetic regulation of aryl
     hydrocarbon hydroxylase induction in the mouse. Fed
     Proc 31:1315-1324, 197?

11. Kouri RE, Rude T, Thomas PE, Whitmire CE. Studies on
     bred strains of mice. Chem-Biol Interactions 13:317-
     331, 1976

12. Kier LD, Yamasaki E, Ames BN. Detection of mutagenic
     Acad Sci US 71:4159-4163, 1974

13. Swain AP, Cooper JE, Stedman RL. Large scale fraction-
     ation of cigarette smoke condensate for chemical and
     biological investigations. Cancer Res 29:579-583,
     1969

14. Kouri RE, Demoise CF, Whitmire CE. The significance
     of aryl hydrocarbon hydroxylase enzyme systems in the
     selection of model systems for respiratory carcinogens.
     In: Experimental Lung Cancer, Carcinogenesis and Bio-
     assays (Karbe E, Park J, eds.), New York, Springer-
     Verlag, 1974, pp 48-61

15. Benedict WF, Rucker N, Faust J, Kouri RE. Malignant
     transformation of mouse cells by cigarette smoke con-
     densate. Cancer Res 35:857-860, 1975

16. Lazer P, Chouroulinkov I, Izard C, Moree-Testa P,
     Hemon D. Bioassays of carcinogenicity after fraction-
     ation of cigarette smoke condensate. Biomedicine
     20:214-222, 1974

17. Stanton MF, Miller E, Wrench C, Blackwell R. Experimental induction of epidermoid carcinoma in the lungs of rats by cigarette smoke condensate. J Natl Cancer Inst 49:867-877, 1972

18. Gielen JE, Van Cantfort J. Organ selectivity and biochemical characteristics of aryl hydrocarbon hydroxylase induction by cigarette smoke in rats and mice. IARC, Scientific Publication No. 12:275-291, 1975

19. Gelboin HV, Huberman E, Sachs L. Enzymatic hydroxylation of benz(a)pyrene and its relationship to cytotoxicity. Proc Natl Acad Sci USA 64:1188-1195, 1969

20. Somogyi A, Kovacs K, Solymoss R, Kuntzman R, Conney AH. Suppression of 7,12-dimethylbenz(a)anthracene produced adrenal necroses by steroids capable of inducing aryl hydrocarbon hydroxylase. Life Sci 10:1261-1271, 1971

21. Lubet RA, Brown DQ, Kouri RE. The role of 3-OH benzo-(a)pyrene in mediating benzo(a)pyrene induced toxicity and transformation in cell culture. Res Commun Chem Path Pharm 6:929-952, 1973

22. Ames BN, Lee FE, Durston WE. An improved bacterial test system for the detection and classification of mutagens and carcinogens. Proc Natl Acad Sci USA 70: 782-785, 1973

23. Umeda M, Saito M. Mutagenicity of demethylnitrosamine to mammalian cells as determined by the use of mouse liver microsomes. Mutat Res 30:249-254, 1975

24. Gelboin HV, Wiebel FW, Diamond L. Dimethylbenzanthracene tumorigenesis and aryl hydrocarbon hydroxylase in mouse skin: inhibition by 7,8-benzoflavone. Science 170:169-170, 1970

25. Kouri RE, Ratrie H, Whitmire CE. Evidence of a genetic relationship between susceptibility to 3-methylcholanthrene-induced subcutaneous tumors and inducibility of aryl hydrocarbon hydroxylase. J Natl Cancer Inst 51: 197-200, 1973

26. Kouri RE, Ratrie H, Whitmire CE. Genetic control of susceptibility to 3-methylcholanthrene-induced subcutaneous sarcomas. Int J Cancer 13:714-720, 1974

27. Kouri RE, Nebert DW. Genetic regulation of suscepti-
     bility to polycyclic hydrocarbon-induced tumors in the
     mouse. In: Origins of Human Cancer (Hiatt HH, Watson
     JD, Winsten JA, eds.), Cold Spring Harbor, New York,
     Cold Spring Harbor Laboratory, 1977, pp 811-835

28. Hutton JJ, Hackney C. Metabolism of cigarette smoke
     condensates by human and rat homogenates to form
     mutagens detectable by Salmonella typhimurium TA1538.
     Cancer Res 35:2461-2468, 1975

Note added in proof:

    We have recently found that a pulmonary S-9 preparation
from Aroclor 1254-induced mice is capable of weakly activating
6-AC to a bacterial mutagen using a pour plate assay (~2-3
fold over background). We have still not observed an increase
in bacterial mutations using this S-9 preparation and 2A1 cigarette smoke condensate.

MUTAGENIC, CARCINOGENIC, AND TOXIC EFFECTS OF RESIDUAL ORGANICS IN DRINKING WATER

John C. Loper and Dennis R. Lang
Department of Microbiology
College of Medicine
University of Cincinnati
Cincinnati, Ohio

Epidemiologic studies have indicated a possible correlation between pollution of drinking water and incidence of
cancer. Much of the data for these analyses was collected
during the period 1950-1969. Considering the latency period
for clinical cancers, the findings contribute to the general
concern about long term exposure to the myriad pollutants
in our environment.

Associations have been drawn between enhanced carcinogenesis and trihalomethane content in drinking water (2). A
complication in documenting such associations as cause-and-
effect relationships is our ignorance of the effects of most
organic compounds. In water, volatile organic compounds
including the trihalomethanes represent only about 10% of
the weight of total organic material. Of the remaining 90%,
it is estimated that 90-95% of the compounds are yet to be
identified (3). Constituent chemicals are present in very
low amounts; identification and toxicological assessment of
even a minority of the total number is a practical impossi-
bility. Moreover, such complex mixtures raise the prospect
of additive, synergistic, antagonistic, or promoter effects
similar to those discussed by others at this symposium.

Toxicological analysis of such mixtures requires some
type of initial concentration procedure. This study was
begun to test the applicability of reverse osmosis in the
concentration of drinking water residue organics. Sequential samples have been prepared from drinking water of
cities representative of United States municipal water
sources (6,12). Our studies have examined the use of two

in vitro systems for analysis of such complex mixtures: the
Salmonella/microsome system and BALB/3T3 cell transformation
(8,9). In this paper some of our results will be used to
emphasize problems and current directions in the study.

TEST SAMPLES

     Residual organics were prepared for USEPA by Gulf South
Research Institute. The procedure as described by Kopfler
et al. (6) is presented briefly here (see Figure 1). Solutes
are concentrated from repeated 200 l samples of tap water,
maintained at pH 5.5 with the addition of HCl, by reverse
osmosis at 15°C using a cellulose acetate membrane (CA); a
Donnan softening loop is included to avoid precipitation of
salts rejected by the membrane. The CA permeate is treated
with NaOH to pH 10 and its solutes are concentrated by a
similar process using a nylon membrane, the nylon permeate
being discarded. Both the CA and the nylon concentrates are
then adjusted to pH 7 and extracted sequentially using pentane and methylene chloride. The aqueous phases are adjusted
to pH <2 by addition of HCl and methylene chloride extraction
is repeated. Twenty percent of each of the organic fractions
is saved for chemical analysis, while the bulk of the material
is concentrated and combined to generate the reverse osmosis
concentrate-organic extract fraction (ROC-OE).

     The remaining concentrates are purged of excess solvent
by bubbling with N_2 and are passed through columns of XAD-2
resin. After column rinses of 1M HCl and of distilled H_2O,
the organics are eluted using 95% ethanol. Eluent solutions
are dried with sodium sulfate, concentrated by vacuum distillation of solvent, and pooled to generate the XAD eluate fraction (XAD eluate). Both ROC-OE and XAD eluate fractions were
stored at 4°C in sealed containers before delivery and have
been maintained similarly in the dark between samplings. A
portion of each ROC-OE and XAD eluate was further fractionated
by sequential extraction with hexane, ethyl ether, and acetone
according to a method of R.G. Melton (USEPA report, Cincinnati, 1976). For later samples the pentane and methylene
chloride extracts of ROC were obtained as discrete fractions.

     It was intended at the outset that the ROC-OE residues
should be obtained in 1 g amounts. A comparison of the initial water volumes used and yields of the samples provided
for our examination appear in Table 1. For most samples
total organic carbon in the water ranged from 6.4 to 1.7 ppm,
and volumes of 2000 to 8000 liters were sufficient to generate

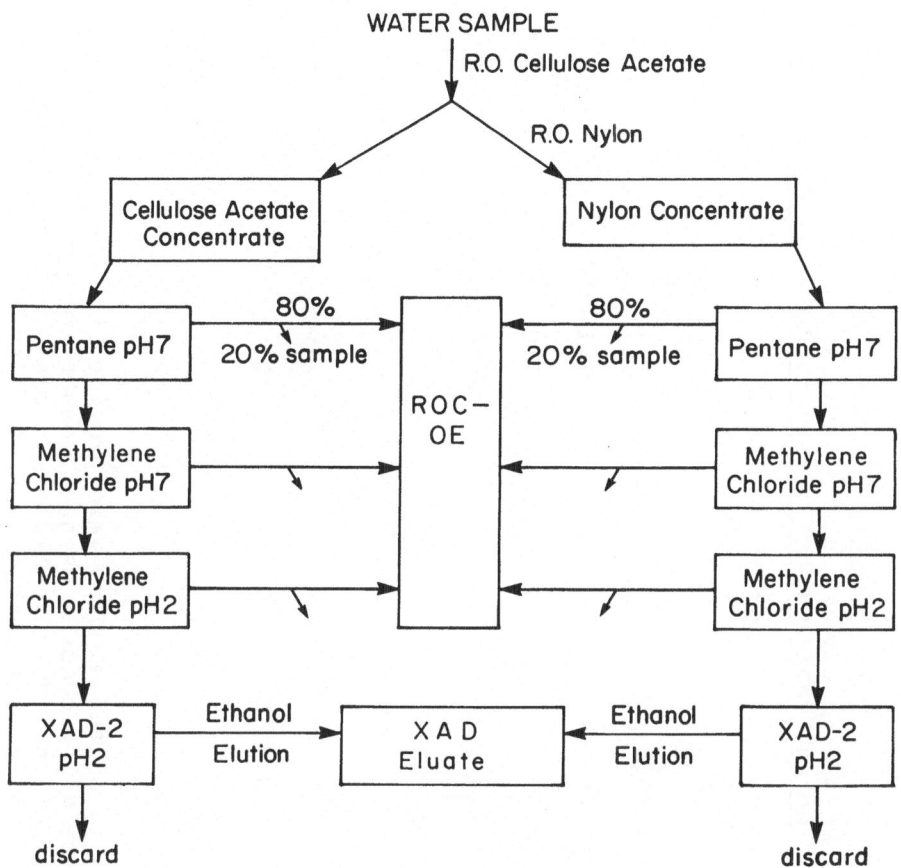

Figure 1. The origin of reverse osmosis concentrates—organic
extract (ROC-OE) and XAD eluate fractions is shown in a diagram of the procedure of Kopfler et al. (6). Twenty percent
of each organic extract, indicated by the short arrows, is
removed and stored for chemical analysis, and the remaining
portions are pooled and concentrated to constant weight to
form the ROC-OE. The remaining aqueous solutions are purged
of excess solvent using N_2 and are passed through columns of
XAD-2. Organics eluated in 95% ethanol are dried, pooled,
and concentrated under vacuum to yield the XAD eluate.

about 1 g of ROE-OE. Seattle and Tucson drinking water yielded less than 0.4 g amounts of ROC-OE from much larger sample
volumes.

Table 1

Yield of Residue Organics from Drinking Water*

City Sample	Tap Water Processed (liters)	Data Collected	Total Organic Carbon in Water (ppm)	Reverse Osmosis Concentrate, Organic Extract (ROC-OE) (g)	XAD Eluate (g)
New Orleans 1a	7,994	10/75	2.0	1.2	4.6
New Orleans 1b	7,812	10/75	2.0	1.0	5.6
New Orleans 2	6,624	1/76	1.7	1.4	6.4
Miami 1a	1,999	11/75	6.4	1.1	7.2
Miami 2	2,271	2/76	6.4	1.0	8.8
Philadelphia 1	5,814	12/75	1.7	0.8	3.9
Philadelphia 2	5,814	2/76	1.7	0.6	4.9
Ottumwa 1	5,961	6/76	3.0	0.9	6.4
Ottumwa 2	5,450	9/76	N.A.	1.0	6.85
Seattle 1	12,604	7/76	1.0	0.38	3.0
Seattle 2	11,752	11/76	N.A.	0.34	2.4
Tucson 1	27,631	10/77	-	0.25	0.7

*Residue fractions and this data on the samples were provided by EPA. Except for those labeled 1a, samples were all processed using a Donnan softening unit for exchange of sodium ions with cations in the cellulose acetate membrane concentrate. For the Tucson sample an anion exchange loop was also included. N.A.: not available.

SALMONELLA/MICROSOME TESTING

The assay system has been described by Ames et al. (1),
who provided the strains TA1535, TA1538, TA98, and TA100.
Promutagen activation was conducted in soft agar overlays
using S-9 mixtures prepared from livers of rats induced with
a PCB mixture Aroclor 1254; characteristic activation potential of each homogenate preparation was verified using known
promutagens (9), and activation of 2-aminoanthracene served
as a positive control during tests of the unknowns. Samples
were dissolved in dimethyl sulfoxide (DMSO) and were delivered
in volumes of 0.01 to 0.3 ml/plate. Tests involved duplicate
platings of 5 doses over a 30-fold dose range for mutagenesis
of TA98 and TA100 in the absence of S-9 mix. This assay was
then repeated, with dose adjustments as appropriate, in an
expanded protocol including the addition of the activation
system, and optionally the strains TA1535 and TA1538. For
assays in which little or no cell killing was evident, dose
responses of net revertant colonies/mg of sample were determined from linear regression plots generated with a computer-
plotter; otherwise initial rates were used. In nearly all
cases mutagenesis for a tester strain was determined from
data which included experimental colony counts which were
at least twice those obtained from the spontaneous control
plates. Presence of characteristic pinpoint histidine-
requiring colonies and appearance of less than spontaneous
colony counts were recorded as having apparent lethal toxicity. Bioassays for histidine in samples were determined
turbido-metrically using strain hisDC129, a stable deletion
histidine auxotroph of Salmonella typhimurium.

BALB/3T3 TRANSFORMATION AND TOXICITY TESTING

We obtained clone 1-13 BALB/3T3 cells from Dr. Takeo
Kakunaga of the National Cancer Institute, Bethesda, Maryland.
Cells were routinely maintained at sub-confluence in anti-
biotic-free Eagle's minimum essential medium (MEM) which was
supplemented with 10% heat inactivated fetal calf serum.
Cells were incubated in a humidified atmosphere of 5% CO_2 in
air.

The experimental conditions were essentially those
described by Kakunaga (5). Cells were plated at a concentration of 10^4 per 60 mm cell culture dish in 5 ml media and
incubated overnight. Appropriate concentrations of carcinogen or water sample were then added in 0.01 ml DMSO. Control
plates received 0.01 ml DMSO alone. Cultures were incubated

for 72 hours after which time the media was removed, cells
rinsed once with phosphate buffered saline (PBS), and refed
with fresh media. Cultures were maintained for an additional four weeks on a bi-weekly feeding schedule. Cells
were then rinsed with PBS, fixed with methanol, and stained
with Giemsa. Areas of piled up cells growing in a disorgan-
ized, criss-cross pattern were quantitated as foci. Prior
to fixing, cells from foci and from normal appearing areas
were cloned for isolation and storage in liquid nitrogen for
eventual testing of their in vivo tumorigenicity. Cytotoxicity was assayed by determining the plating efficiency of
200 cells plated in 5 ml MEM per 60 mm dish with exposure to
test compounds as described for the transformation assay.

RESULTS AND DISCUSSION

Aspects of Salmonella Mutagenesis Testing

     Tests of residues and residue subfractions from each of
the samples listed in Table 1 have been conducted using two
or more strains of the Salmonella testing system; mutagenesis
was induced by residues from each drinking water sample.
Where possible we have tabulated and compared results of
repeat samples from a given city and among cities, giving
attention to (a) the amount of mutagenesis for a strain, as
expressed in terms of net revertant colonies/mg of residue
material tested; (b) the relative mutagenicity of the test
material for TA98 and TA100; and (c) the distribution of
mutagenic activity among ROC-OE and XAD eluate fractions and
their subfractions. Repeat samples have exhibited consistencies of mutagenic patterns that were characteristic for
that city.

     The data and our analysis have been presented elsewhere
in detail (8,9). In this paper we describe some of our
general findings in assay of these complex mixtures. All
the Salmonella mutagenesis measured to date has been direct
acting, with little or no enhancement due to the presence
of the microsome activation system. Direct mutagenic activity for TA1538 was usually similar to that seen using TA98,
while TA1535 often was unaffected in cases where TA100
showed a response. An example of these patterns appears in
Figure 2.

     Many of the fractions tested gave linear dose response,
and with most of these we were able to test amounts of
material which yielded colony counts from responding strains

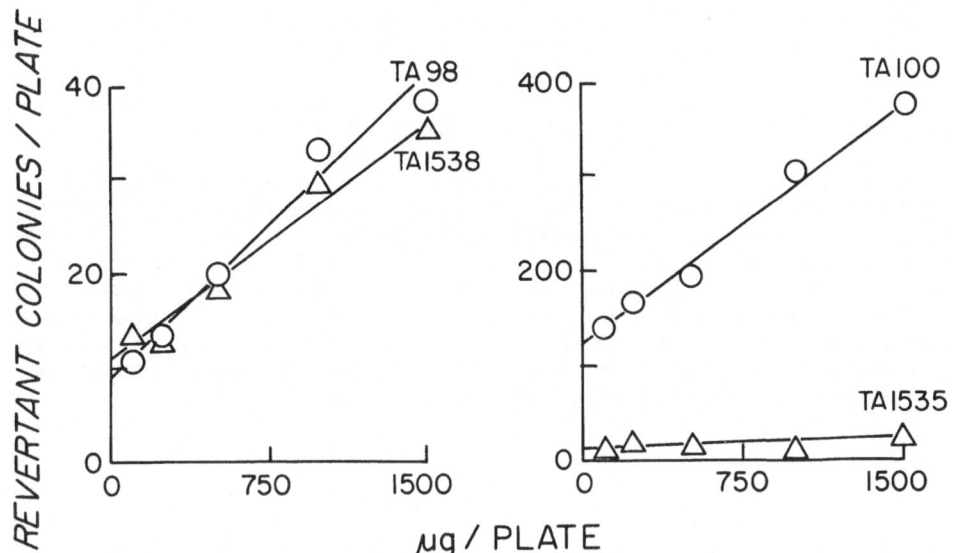

Figure 2. Strain specific mutagenic effects of Miami 2 ROC-OE. Each point is the average of colony counts from 2 plates. Lines were drawn as linear regressions of the original data. All assays presented in this and in the following figures were conducted in the absence of S-9.

of two-fold spontaneous or better. Dose response data involving colony counts of this magnitude are convenient since normal appearing dose dependent increases of less than two-fold could be due to histidine enrichment. We considered this a possibility since large amounts of water were processed to generate these residues. Salmonella typhimurium strain hisDC129 is an organism which grows well in histidine enriched media but is stably dependent upon the presence of the amino acid for growth; the strain is thus convenient for use in turbidometric microbiological assays of histidine. In cases where test fractions showed marginally two-fold mutagenesis, use of this histidine bioassay showed negligible histidine in the samples.

In certain instances assay for mutagens in these mixtures was complicated by antagonistic or toxic effects. With some fractions the lethal effects simply precluded assay for mutagens; with others the mutagenic dose responses were non-linear, some showing a masking of further mutagenic responses, and some showing clear toxicity at higher dose. In such cases the mutagenesis was scored as present or was calculated from the initial rate of response. Plots of data from samples representative of such mixtures appear in Figures 3-5.

For samples from the first 5 cities listed in Table 1 cell killing effects were determined by suspending TA100 cells in a fixed concentration of sample and establishing the decrease in colony forming cells as a function of minutes of exposure. By this method ROC-OE fractions showed two- to five-fold greater toxicity per mg than did the corresponding XAD eluates (unpublished observations). As noted below this trend was even more pronounced in determinations of cellular toxicity for clone 1-13 cells.

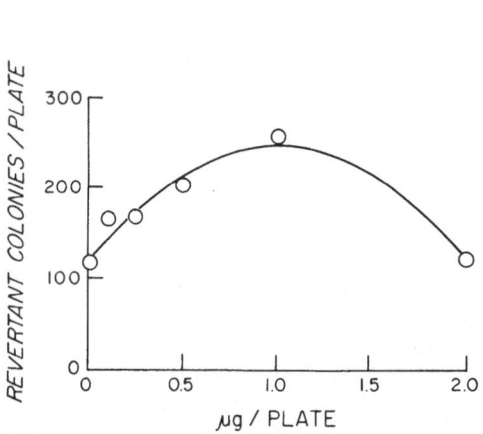

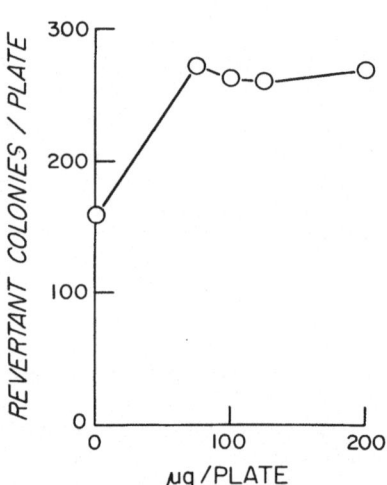

Figure 3. Typical non-linear dose response curve of mutagenic effects of Philadelphia 2 ROC-OE on strain TA100.

Figure 4. Dose response curve representative of apparent mutagenic plus antagonistic effects. The data show effects of Ottumwa 1 XAD eluate on strain TA100. No calculations of net revertant colonies/mg were attempted for such responses.

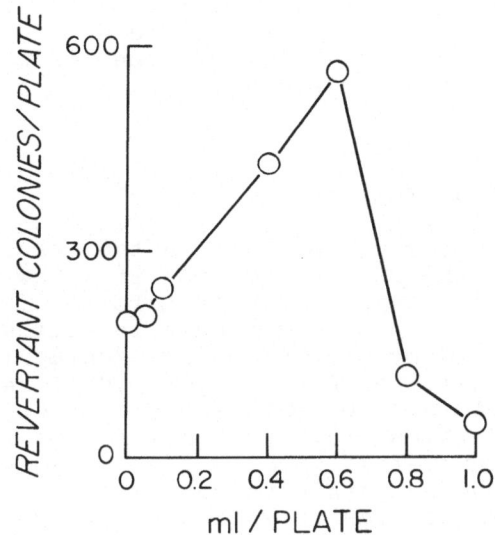

Figure 5. Mutagenic and toxic response of strain TA100 to increasing volumes of aqueous concentrate from Philadelphia drinking water. A volume of concentrate obtained by reverse osmosis using the cellulose acetate membrane, provided to us by EPA, was concentrated ten-fold further by lyophilization. Test volumes were incorporated directly into the soft agar in the standard assay procedure.

Little information is available as to the chemical agents causing this toxicity. As reported elsewhere (9) for various residue fractions from the first 5 cities, oxidation of the organics by refluxing in nitric acid or in a mixture of nitric and sulfuric acids removed bacterial toxicity. In contrast, the ROC-OE and XAD eluate fractions obtained from the Tucson sample showed unusually high bacterial toxicity, and the toxicity of that XAD eluate fraction was stable to oxidation. Spectrographic examination of the Tucson XAD eluate fraction showed 1800 ± 200 ppm of Hg, together with lesser amounts of other metals. Control toxicity experiments using reagent

HgCl$_2$ showed that, should the mercury content of the XAD eluate be present as Hg^{++}, it could account for all the toxicity of that sample. A relatively large sample of drinking water, 27,631 liters, was processed to yield the Tucson fractions and this concentration of mercury ion, calculated per liter of original drinking water, would be well below acceptable levels (9).

Cell Transformation and Toxicity

We have previously reported data on the transformation of BALB/3T3 cells by the ROC-OE fraction from New Orleans 1b sample (8,9). Clones obtained from transformed foci have demonstrated enhanced plating efficiency in soft agar using the technique described by MacPherson and Montagnier (10). We have also shown that the BALB/3T3 cells differentiate between ROC-OE and XAD eluates on the basis of cellular toxicity. For four cities examined, toxicity on a weight basis was ten- to twenty-fold greater for the ROC-OE samples than for the corresponding XAD eluates (9, and unpublished observations). Tests of transformation activity of additional samples are in progress.

We have been attempting recently to develop a mutagenesis assay with this same clone of 3T3 cells using ouabain resistance as a marker. Huberman et al. have shown about a 20:1 ratio of transformation to mutation frequency occurred when both were measured in hamster embryo cells (4). We are attempting to see if similar measurements can be made with BALB/3T3 cells using the focus assay for transformation rather than the colony assay employed by Huberman. Preliminary data indicate that these cells may lend themselves to studies of mutagenesis at the locus for the Na$^+$K$^+$ATPase. If the assay can be developed, it would be of obvious utility in describing the relative carcinogenic and mutagenic activities present in the complex mixtures obtained from drinking water.

We are also attempting to develop a tumor promotor assay with these BALB/3T3 cells. It has been shown by Mondal et al. (11) that the tumor promotor tetradecanoylphorbol acetate (TPA) can have stimulatory effects on transformation of 10T½ cells by known carcinogens under conditions where there is no transformation by either carcinogen or TPA alone. If we are successful in extending this observation to BALB/3T3 cells, we will be able to test the promoting activity of ROC-OE or XAD eluate fractions on transformation initiated by

3-methylcholanthrene or other carcinogens known to be active in this system.

Relationship to Drinking Water

Many problems remain in relating the available data to the frequency and variety of mutagens/carcinogens in the original water samples, and several approaches are in progress or are planned to address these.

1.   Although residues with specific mutagenic properties are reproducibly isolated, we do not know how representative these residues are of the total organics in water. Mutagens may be preferentially concentrated or preferentially lost. Some compounds may have been chemically altered during concentration, extraction, or storage. In one study, the mutagenic potential of New Orleans 2 ROC-OE was equally stable over a one week period when sealed in serum vials and stored at room temperature in DMSO, or when stored at -70°C in either DMSO or dimethyl formamide. But more attention to limiting oxidation throughout the procedure may be important.

From the data of Gulf South Research Institute (6), nearly all of the drinking water TOC is retained during concentration to the reverse osmosis membrane reject volume. Aqueous concentrates from the cellulose acetate membrane were provided by EPA from New Orleans, Miami, and Philadelphia. No mutagenesis was detected in assays of the first two of these samples, but a further 10-fold concentration of the Philadelphia sample gave the dose dependent mutagenic-toxic response presented in Figure 5. We have initiated tests of the transforming activity of this high salt material, and we have begun examining alternate methods of extraction of the mutagenic activity. If we are successful in establishing the TPA promotion assay in BALB/3T3 cells, we will test the promoting activity of these aqueous concentrates as well.

2.   The complexity of these mixtures may lead to a variety of additive and antagonistic effects. Some fractions are too toxic for reliable determination of mutagenicity. Fractions yielding dose responses of the types shown in Figures 3-5 may contain components that prevent accurate measure of the mutagens present. Microsomal activation has not been required for the mutagenesis we have detected to date, but here too antagonistic effects of compounds in certain of these mixtures may mask detection of mutagens requiring activation. The mutagenesis/mg that we measure could be

due to a broad range of chemicals of different specific mutagenic activity.

    An initial fractionation was included in the survey
study of samples provided from the six cities, by which ROC-
OE and XAD eluates were sequentially extracted using hexane,
ethyl ether, and acetone. Mutagenic assays on these subfractions helped identify differences among samples from separate
cities and also revealed some common patterns of distribution
of active components (9). However, even these subfractions
contain a great number of components, and the identification
of the active species by direct analytical methods of GC-MS
will be impossible; chemical analysis in progress on one subfraction of a Cincinnati water ROC-OE so far has revealed
several hundred compounds (E. Coleman, personal communication).
By combining selected solvents with acid, neutral, or basic
aqueous phases, and through application of HPLC, our group
will attempt separation of the bulk of the components into
smaller subfractions, monitoring progress in fractionation
using the Salmonella/microsome test. Active fractions may
be obtained sufficiently free of inactive and toxic components to facilitate peak-to-peak identification by GC-MS.

    In addition, a number of pooled component studies are
possible. Kraybill et al. have compiled a list of direct
acting mutagenic compounds known to be in finished or raw
water (7), and some of these are sufficiently non-volatile
as to be retained in ROC residues. We plan to characterize
mutagenic separation properties of such mutagens in prepared
mixtures. Using known mixtures, and available mutagenic
organic residues, a number of water reconstitution-reconcen-
tration experiments can be initiated. These studies should
allow us to define more clearly the significance of mutagenicity and in vitro carcinogenicity found for reverse osmosis-
derived residue organics of drinking water.

    This work was supported by research grant R804202 from
the USEPA.

REFERENCES

1.   Ames BN, McCann J, Yamasaki E: Methods for detecting
     carcinogens and mutagens with the Salmonella/mammalian
     microsome mutagenicity test. Mutat Res 31:347-363, 1975

2. Control of organic chemical contaminants in drinking water, Environmental Protection Agency Interim Primary Drinking Water Regulations, U.S. Federal Register 43: 5756-5780, 1978

3. Drinking Water and Health, Report of the National Research Council Safe Drinking Water Committee, National Academy of Science, p 492, 1977

4. Huberman E, Mager R, Sachs L: Mutagenesis and transformation of normal cells by chemical carcinogenesis. Nature 264:360-361, 1976

5. Kakunaga T: A quantitative system for assay of malignant transformation by chemical carcinogens using a clone derived from BALB/3T3. Int J Cancer 12:463-473, 1973

6. Kopfler FC, Coleman WE, Melton RG, Tardiff RG, Lynch SC, Smith JK: Extraction and identification of organic micropollutants: Reverse osmosis method. Ann NY Acad Sci 298:20-30, 1977

7. Kraybill HF, Helmes CT, Sigman CC: Biomedical aspects of biorefractories in water. In: Proceedings Second International Symposium on Aquatic Pollutants, Oxford, England: Pergamon Press, Ltd., in press

8. Loper JC, Lang DR, Smith CC: Mutagenicity of complex mixtures from drinking water. In: Proceedings of the Conference on Water Chlorination Environmental Impact and Health Effects, Chapter 33. Ann Arbor Science Publishers, Inc., pp 433-450, 1978

9. Loper JC, Lang DR, Schoeny RS, Richmond BB, Gallagher PM, Smith CC: Residue organic mixtures from drinking water show in vitro mutagenic and transforming activity. Submitted for publication

10. MacPherson I, Montagnier L: Agar suspension culture for the selective assay of cells transformed by polyoma virus. Virology 23:291-294, 1964

11. Mondal S, Brankow DW, Heidelberger C: Two-stage chemical oncogenesis in cultures of C3H/10T cells. Cancer Res 36:2254-2260, 1976

12.   Tardiff RG, Carlson GP, Simmon V: Halogenated organics
      in tap water: a toxicological evaluation. In: Pro-
      ceedings Conference on the Environmental Impact of Water
      Chlorination, pp 213-227, 1975

MUTAGENIC ANALYSIS OF COMPLEX SAMPLES OF AQUEOUS EFFLUENTS, AIR PARTICULATES, AND FOODS

Barry Commoner, Anthony J. Vithayathil,
and Piero Dolara
Center for the Biology of Natural Systems
Washington University
St. Louis, Missouri

INTRODUCTION

Opportunities and problems arise when the Ames mutagenesis technique is applied to the analysis of samples, such
as those derived from the environment, which are mixtures of
unknown compounds that may or may not include mutagens. The
chief advantage of this application of the method is well
known: one can use it as a rapid, inexpensive, biological
screen capable of detecting mutagens by their biological
effect. This makes it possible to avoid the very difficult
task of detecting and identifying all of the numerous organic
compounds that may occur in such a sample in order to compare
them with a list of known mutagens. The chief disadvantage
of this approach is that one is "flying blind," so to speak,
unaware in advance of what types of compounds are present,
their concentrations, and their possible interference with
the test.

In order to appreciate these difficulties and to devise
strategies for overcoming them, it is useful to recall certain characteristics of the Ames system:

● The various Ames strains of <u>Salmonella</u> are designed
specifically to respond to different classes of
organic mutagens. Therefore, in dealing with samples containing unknown mutagens one cannot know
in advance which strains the mutagens will act
upon. For the same reason there is no a <u>priori</u>
basis for quantitative comparisons of mutation
rates obtained with different strains.

- The dose-response curves that relate a given
  strain's response to various concentrations of a
  given mutagen are almost always decidedly nonlinear,
  in some cases falling to a zero response at high
  concentrations. This means that a test designed
  to determine whether or not mutagens are present,
  if carried out at only one concentration, may
  readily give a false negative result. For the
  same reason, a positive value obtained at a single
  concentration is insufficient to estimate the level
  of mutagenic activity.

- Certain mutagens are inherently active in the
  system, while other needs to be "activated" by
  the "S-9" microsome preparation. However, the
  latter is a complex system of related enzymes and
  there is no way of knowing in advance whether the
  microsome preparation will convert a particular
  substance to an active mutagen and whether, on the
  contrary, it will convert an inherently active
  mutagen into an inactive substance.

- There is a certain inherent biological variability,
  from time to time, in the background rate of mutation of each of the Ames strains. At the same
  time, as in any experimental procedure, there are
  certain sources of imprecision (e.g., in volume
  measurements) that also affect this value. This
  raises the question of how these two sources of
  variability are to enter into the computation of
  the experimental results.

- Despite its considerable value, the system is
  still something of a "black box" because certain
  features are poorly understood. These features
  include, in addition to the unresolved properties
  of the microsome preparation, synergistic and/or
  inhibitory interactions between mutagens concurrently present in the system, and possible transformations of test substances by enzymes associated
  with Salmonella.

In our present circumstance it is useful, while using the
Ames test, to remain alert to anomalies that may provide
useful clues for learning more about how the test works.

We will consider how the foregoing features of the Ames system may affect the results obtained from complex, unknown samples, and suggest some procedures which may offset the resultant difficulties. We will present specific examples of analyses of unknown mutagens present in samples of water, air particulates, and food. The analyses presented are concerned with one or more of the following general aims, which commonly arise in applying the Salmonella system as a screen to environmental and other complex samples:

- Isolation and identification of active mutagens from complex unknown samples.

- Evaluation of the level of mutagenic activity associated with a complex unknown sample, especially in relation to relevant environmental parameters.

- Characterization of such samples with respect to the presence of inherently active mutagens, mutagens capable of being active, and mutagens that are inactivated by the microsome preparation.

- Comparison of mutagens detected in complex samples with known ones.

- Application of techniques for studying the formation of mutagens in experimental systems.

The following specific examples are discussed: (a) detection and isolation of mutagens in the aqueous effluents of petrochemical plants along the Houston Ship Channel; (b) analysis of the mutagenic activity of Chicago air particulates; (c) analysis of a minor anomaly in the Ames test that has led to the discovery of situations in which mutagens are produced during conventional cooking of certain foods.

DETECTION AND ISOLATION OF MUTAGENS IN THE AQUEOUS EFFLUENTS OF PETROCHEMICAL PLANTS ALONG THE HOUSTON SHIP CHANNEL

These studies have been carred out under a collaborative arrangement with the Harris County Pollution Control Department (Pasadena, Texas). First, water samples (two gallons each) were collected directly from the effluent pipes of the various chemical plants under the joint supervision of the Pollution Control Department staff and the plants' personnel. The samples were stored in our laboratory at 4°C.

A total of 24 effluent samples were collected from 16
different industrial plants (see Table 1). At two locations
samples were collected from the same outflow pipe on a series
of dates. Initially, benzene/isopropanol extracts of each
sample (usually 2 liters of water extracted successively at
pH 2.5 and pH 11) were dried, then dissolved in DMSO. Aliquots representing varying amounts of the original water
samples were tested against strain TA1538 with and without
the liver microsome preparation, in keeping with the procedures described by Ames et al. (2). Throughout the work
described in this paper, the microsome preparation used was
the standard S-9 preparation from the livers of PCB-induced
rats. All plate counts reported are the averages of duplicate plates. These techniques of sample preparation and
mutagenesis testing are not suitable for volatile compounds,
and such compounds are not involved in our studies.

The results of some initial tests of the acid extracts
of the samples are shown in Table 1 (alkaline extracts were
uniformly negative). In interpreting the significance of
these results, we have employed an approach developed earlier,
based on the comparison of 50 known organic noncarcinogens
and 50 organic compounds that previously had been shown to
be carcinogenic toward laboratory animals (3). In this comparison we computed a "mutagenic activity ratio" from the
quotient $\frac{E-C}{C_{Av}}$, where E is the number of mutant colonies
obtained from the experimental sample; C is the control
value (i.e., the number of mutant colonies observed when the
experimental material is not included) obtained on the day
of analysis; and C_{Av} is the "historical" control value, or
the average control value for all runs carried out during
the course of the study. The rationale for this procedure,
which was described earlier (3), is intended to take into
account daily variations in the background mutation rate as
well as those variations inherent in the method itself.

As we have shown previously, in the test of equal numbers of known carcinogens and noncarcinogens, the reliability
with which the two classes of compounds can be distinguished
depends on the value of $\frac{E-C}{C_{Av}}$ which is chosen as the cut-off
point. Thus 82% of the noncarcinogens yield a mutagenic
activity ratio below 2, and 82% of the carcinogens yield a
ratio above that value. If a higher reliability of detecting
carcinogens is desired, a somewhat lower cut-off ratio is
chosen, at the risk of increasing the chance of falsely

Table 1

Analysis of pH 2.5 extracts of samples from industrial
sources in Houston Ship Channel area

Plant	Type of Plant	Type of Sample	Date Collected	Date Analyzed	Equivalent Amount of Sample/Plate (ml)	No. of Colonies/Plate (TA1538, Liver S9)		Mutagenic Activity Ratio $\frac{E-C}{C_{Av}}$
						Control	Experimental	
A	Pulp mill	Black liquor	10/3/75	10/30/75	1	31	36	0.2
A	Pulp mill	Water	10/3/75	10/20/75	50	41	32	-0.4
A	Pulp mill	Water	10/3/75	10/20/75	125	41	47	0.3
A	Pulp mill	Water	10/3/75	10/20/75	250	41	63	1.0
B	Steel mill	Water	9/5/75	10/16/75	25	42	41	0
B	Steel mill	Water	9/5/75	10/16/75	62.5	42	64	1.0
B	Steel mill	Water	9/5/75	10/16/75	125	42	17	-1.1
C	Chemical	Water	9/23/75	10/29/75	125	26	86	2.6
C	Chemical	Water	9/23/75	10/29/75	250	26	82	2.4
D	Chemical	Water	6/26/75	10/16/75	100	42	41	0
D	Chemical	Water	1/5/76	6/15/76	250	27	86	2.7
E	Chemical	Water	6/19/75	10/23/75	25	22	11	0.5
F	Chemical	Sludge	6/26/75	10/25/75	5	17	35	0.8
F	Chemical	Water*	1/6/76	6/22/76	250	21	29	0.4
F	Chemical	Water*	1/6/76	6/22/76	250	21	40	0.9
F	Chemical	Water*	1/6/76	6/22/76	250	21	35	0.6
F	Chemical	Water*	1/6/76	6/22/76	250	21	38	0.8
G	Industrial waste treatment	Water	9/23/75	10/30/75	50	31	127	4.2
G	Industrial waste treatment	Water	9/23/75	10/30/75	125	31	182	6.6
G	Industrial waste treatment	Water	1/5/76	1/13/76	100	17*	259	24.2
G	Industrial waste treatment	Water	1/5/76	6/15/76	250	27	568	24.6
H	Chemical	Water	1/5/76	6/22/76	250	24	48	1.2
I	Chemical	Water	1/5/76	6/22/76	250	21	36	0.7
J	Chemical	Water	1/5/76	6/22/76	250	21	35	0.6
K	Chemical	Water	1/5/76	6/22/76	250	21	59	1.7
L	Industrial waste treatment	Water	1/5/76	6/22/76	250	21	28	0.3
L	Industrial waste treatment	Sludge	1/5/76	6/22/76	250	21	29	0.3
M	Chemical	Water	1/5/76	6/22/76	250	21	32	0.5
N	Chemical	Water	1/6/76	6/15/76	250	27	37	0.7
O	Chemical	Water	1/6/76	6/15/76	250	27	131	4.7
P	Chemical	Water	1/6/76	6/15/76	250	27	63	1.6

*These samples were collected from 4 different effluent outlets from the same plant.

identifying noncarcinogens. In more general terms, ratios
of 2-3 should be regarded as at least suggestive of the
presence of mutagenic activity; values above the range of
3-5 are clearly indicative of the presence of mutagenic
acitivity.

From the values shown in Table 1 it appears that effluents from industrial waste treatment Plant G consistently
yielded significant levels of mutagenic activity. We have
carried out a systematic analysis of this effluent designed
to isolate and identify the substances responsible for the
observed activity. Figure 1 shows the procedures we applied
to a 20-gallon sample of effluent from Plant G.

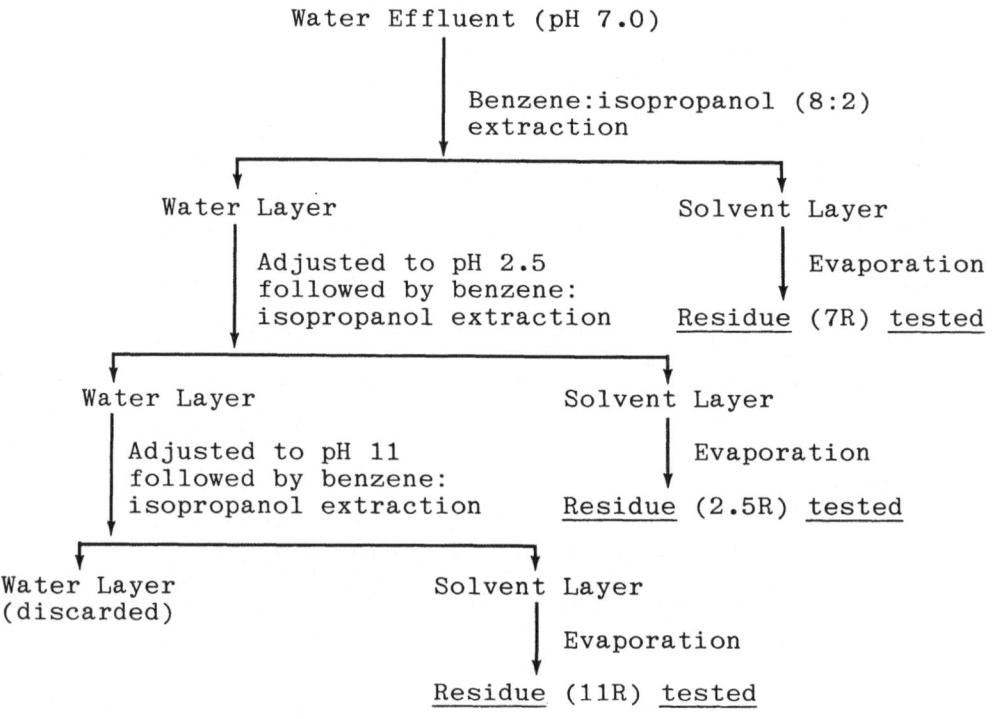

Figure 1. Procedures applied to a 20-gallon sample of
effluent from Plant G.

Aliquots of the residues yielded by this scheme (7R, 2.5R, and 11R), equivalent to 200 ml of the original effluent were tested for mutagenic activity in the usual way, with strain TA1538, in the presence of the microsome preparation. The results are shown in Table 2. It is apparent that the neutral and acidic fractions are clearly active, while the activity of the alkaline fraction is marginal.

Table 2

Mutagenic Activity of Industrial Waste Treatment
Plant Effluent Extracted with Benzene:
Isopropanol at Different pH

Sample Number	pH of Extraction	Number of Revertant Colonies/Plate* (TA1538, Liver Microsomes)	Mutagenic Activity Ratio $\dfrac{E-C}{C_{Av}}$
7R	7	2116	95.2
2.5R	2.5	839	37.1
11R	11	87	3.0

*Colonies/plate for equivalent of 400 ml of the water sample.

In the next step, aliquots (representing 200 ml of the original effluent) of the neutral (7R) and acidic (2.5R) residues were subjected to thin-layer chromatographic (TLC) fractionation using silica gel paper and a benzene:hexane (1:1) solvent. A series of 1 cm sequential zones were then cut from the developed chromatogram, each extracted in 10 percent methanol in chloroform and allowed to dry. The successive zonal samples were then taken up in DMSO and tested in the usual way against TA1538 in the presence of the microsome preparation. From the numbers of mutant colonies produced by each zonal sample it was possible to characterize the chromatographic behavior of the mutagenically active constituent(s).

As shown in Figure 2, following this chromatographic
procedure the mutagenic activity of both of the fractions was
found predominantly lodged at the origin. However, UV scans
of the chromatogram showed that several mutagenically inactive
components had moved away from the origin, so that this chromatographic system was a useful means of initial purification of
the sample.

The zones located at the origins of the foregoing chromatograms were eluted with 10 percent methanol in chloroform and
were rechromatographed using methanol:ethyl acetate:benzene
(1:10:89) as the solvent system. The results of this second
fractionation step are shown in Figure 3 for the 2.5R (acidic
fraction).

How shall we interpret this result? The most obvious
interpretation is that the material at the origin of the
first chromatogram (Figure 2) was heterogeneous, and in the
second chromatogram, resolved into three peaks (at the origin,
at RF = 0.8, and RF = 1.0). The two peaks at RF = 0.8 and
at RF = 1.0 presumably represent two different mutagens. But
this interpretation holds only if the dose-response curve is
linear. If instead the dose-response curve goes through a
maximum and falls to zero at higher concentrations, the appar-
rently double peak in Figure 3 may actually represent a single
substance. For example, Zone 8 might represent a relatively
low concentration of the mutagen, which lies on the rising arm
of the dose-response curve, Zone 9, a _higher_ concentration
which is on the falling arm of the dose-response curve, and
Zone 10, once again a relatively low concentration which lies
on the rising arm of the dose-response curve. Thus, in actu-
ality the mutation rate values for Zones 8, 9, and 10 might
represent a _single_ chromatographic peak centered at Zone 9.

This example is cited only to provide an illustration
of the impact that the possible non-linearity of mutagenic
response to a particular unknown substance may have on the
otherwise simple problem of interpreting chromatographic
peaks. It emphasizes once more the importance of actually
measuring dose-response curves in dealing with such samples.

MUTAGENIC ANALYSIS OF CHICAGO AIR PARTICULATES

A number of organic compounds that include carcinogens
have been found to be associated with urban air particulates.
Accordingly, analysis of such material represents another

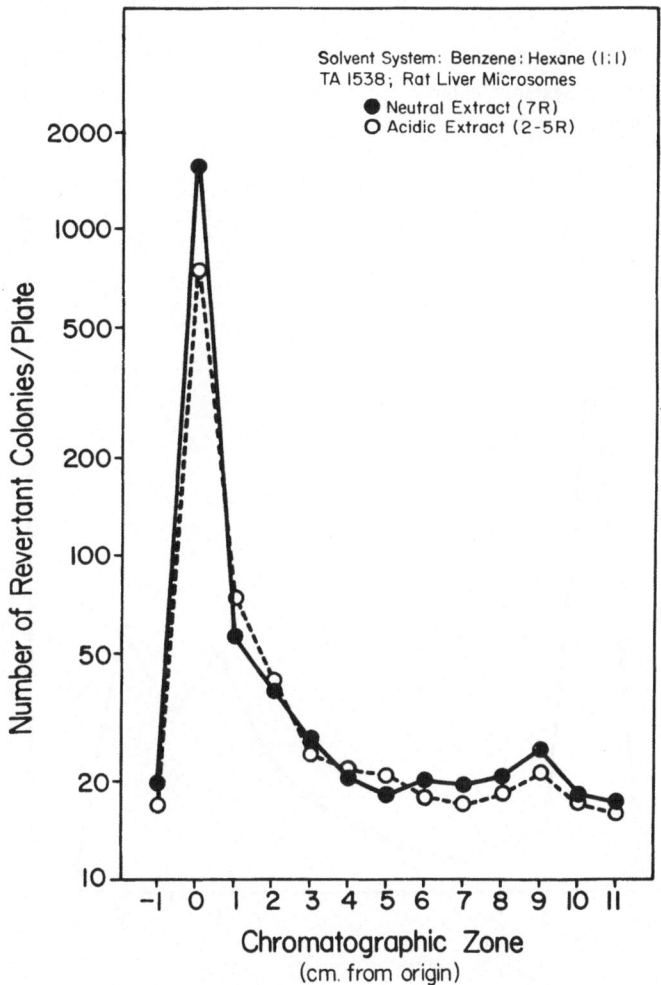

Figure 2. Thin layer chromatographic fractionation of the
neutral (solid lines) and acidic (broken lines) extracts of
a sample of effluent from industrial plant G. The chromatographic solvent system was benzene:hexane (50:50). Chromatographic fractions were tested using TA1538 with microsome
preparation present.

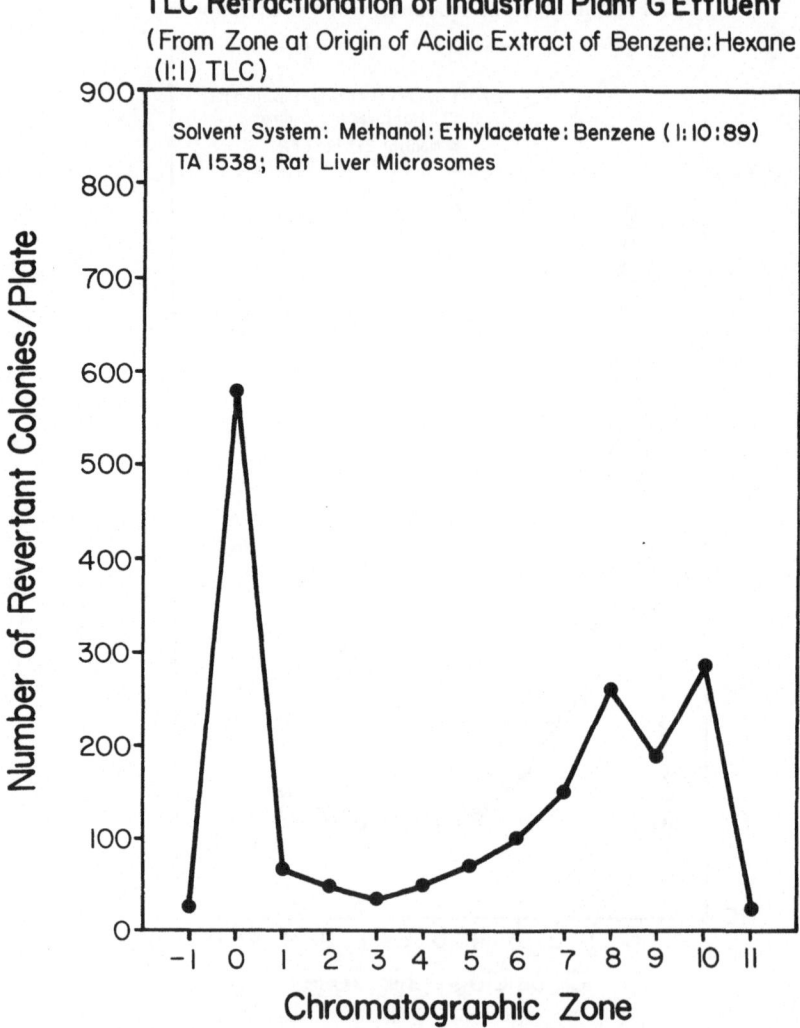

TLC Refractionation of Industrial Plant G Effluent
(From Zone at Origin of Acidic Extract of Benzene:Hexane (1:1) TLC)

Solvent System: Methanol:Ethylacetate:Benzene (1:10:89)
TA 1538; Rat Liver Microsomes

Figure 3. Thin layer chromatographic fractionation of the
zone at the origin of the acidic extraction shown in Figure 2.
The chromatographic solvent system was methanol:ethylacetate:
benzene (1:10:89). Chromatographic fractions were tested
using TA1538 with microsome preparation present.

test of the research strategy for employing the Salmonella test as a means of detecting and identifying environmental carcinogens.

We have established a cooperative arrangement with the City of Chicago Department of Environmental Control to carry out mutagenic analyses on the high volume air particulate samples that they collect daily at 25 stations in that city. Samples are provided for us, together with data on the weight of the collected particulates and associated meteorological information.

As a preliminary step, analyses were made of benzene: hexane (1:1) extracts of two square-inch samples of filters collected concurrently from a series of different stations in the City of Chicago air pollution system. The results, which are shown in Figure 4, revealed a general proportionality between particulate concentration and the numbers of revertant colonies, and identified the Washington School in South Chicago as a site considerably more active than the rest. In an effort to improve the efficiency of extraction, it was then found that extracts obtained with benzene:hexane:isopropanol (70:10:20) yielded somewhat higher revertant colony counts than benzene:hexane extracts, and the former solvent was used thereafter. On the basis of these results we have concentrated our studies on the analysis of samples from the Washington School station, using the revised extraction system. Also, for the reasons cited earlier, in these studies we have relied heavily on data based on dose-response curves.

Analyses of air particulate samples collected at intervals during 1975 from the Washington School site have been carried out. Dose-response curves were obtained for each air filter with and without the presence of the microsome preparation from each of the following: (a) the benzene: hexane:isopropanol extract; (b) the benzene-soluble fraction of the benzene:hexane:isopropanol extract; and (c) the water-soluble fraction of the benzene:hexane:isopropanol extract.

Using this procedure we have determined dose-response curves (with strain TA1538) for samples collected at the Washington School site for 15 days during 1975. Figure 5 shows six of the 15 dose-response curves obtained from these samples for, respectively, the benzene:hexane:isopropanol extracts, the benzene fractions, and the water fractions. In each case the results obtained with microsomes present (solid line) and microsomes absent (broken line) are shown.

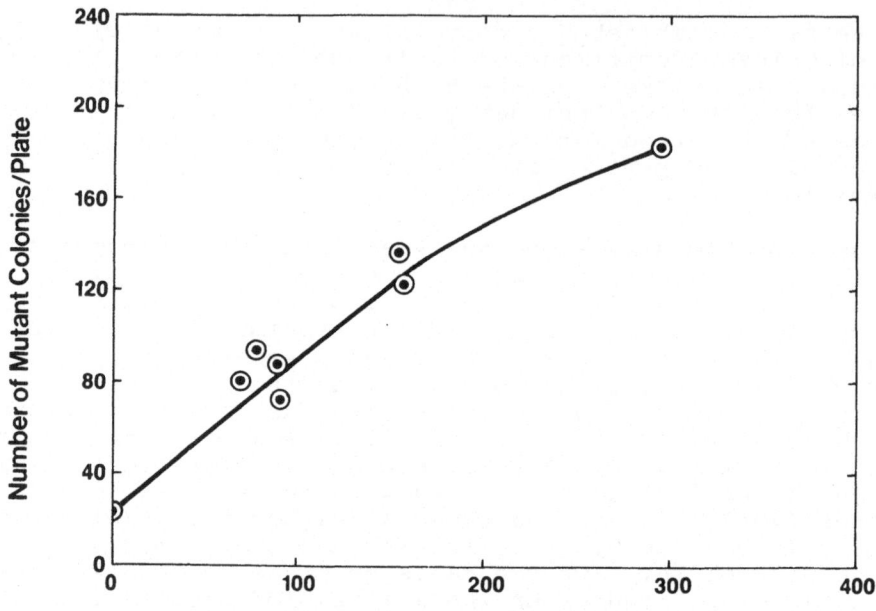

Figure 4. Number of revertant colonies produced per test
plate by benzene:hexane extracts of 2 in² aliquots of air
particulate filters from different Chicago collection sites,
tested on strain TA1538, with the microsome preparation.
The highest value is from the Washington School site.

Approximately linear dose-response curves are exemplified by
those obtained from the February 26 and July 31 benzene fractions, with microsomes present. Many of the curves exhibit
slopes that decline at higher concentrations. Instances of
toxicity at higher concentrations can be seen in the July 31
sample, benzene:hexane:isopropanol extract, without microsomes present. This curve also illustrates the inactivating
effect of microsomes; at two of the lower sample concentrations, the numbers of colonies produced when microsomes are
present are <u>lower</u> than those observed in their absence.

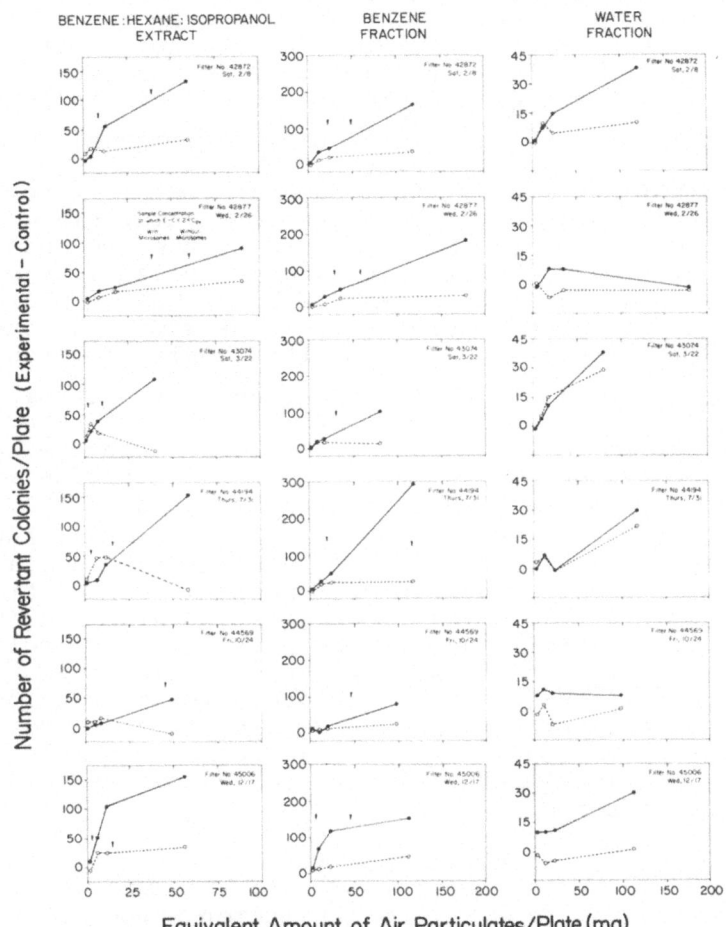

Figure 5. Number of revertant colonies (less control values) produced per plate by increasing amounts of air particulate extracts collected on six different dates in 1975 at Washington School site. Tested on strain TA1538, with (solid line) and without (broken line) microsome preparation. The arrows mark the sample concentration at which experimental minus control values = 2 x C_{Av} (where C_{Av} is the historic control value). The reciprocal of the indicated value is the relative mutagenic activity of the sample.

In order to devise a procedure capable of comparing the mutagenic activities of different samples that takes into account the variable shapes of the dose-response curves, we have adopted the following procedure. To begin with, we note, on the basis of our earlier statistical comparison of the mutagenic activities of noncarcinogens and carcinogens, that there is a minimum value of $\frac{E-C}{C_{Av}}$ which determines, with the stated reliability figure, that the material is carcinogenic. In the present analyses, we may regard a mutagenic activity ratio of 2.0 or greater as indicative of the presence of active substances in the sample, with a reliability of about 98% if microsomes are absent and of about 93% if microsomes are present. We then determine from the sample's dose-response curve the lowest concentration of the sample at which the $\frac{E-C}{C_{Av}} = 2.0.$ This value, which is marked by the arrow shown in Figure 5, can be obtained from the dose-response curve by interpolation to determine the sample concentration at which $E-C = 2.0 \times C_{Av}$. The value can be determined in this way regardless of the shape of the dose-response curve (specifically whether a maximum occurs, or whether the initial slope is different from that at higher concentrations). Finally, the reciprocal of the sample concentration at which the mutagenic activity ratio is 2.0 may be defined as the relative mutagenic activity of the sample. While this procedure does not take into account possible synergistic interactions among separate mutagens present in the sample, it does provide, as a first approximation, relative measures of the mutagenic activities of samples even if they yield dose-response curves that differ in shape.

The relative mutagenic activities computed in this way for the benzene:hexane:isopropanol extracts and the benzene fractions obtained from all 15 Washington School samples are plotted, as a function of sample date, in Figure 6. (The corresponding plot for the water fraction is not shown since in every sample the mutagenic activity ratios are zero.) The reported wind direction at each date is also indicated.

The data of Figure 6 support several conclusions. First, it is evident that in the presence of microsomes the level of activity of the benzene fraction generally parallels that of the original extract from which it is derived, providing that there is little or no inherently active material present. This situation occurs in the latter half of the year. However, the activities of the benzene fraction are generally

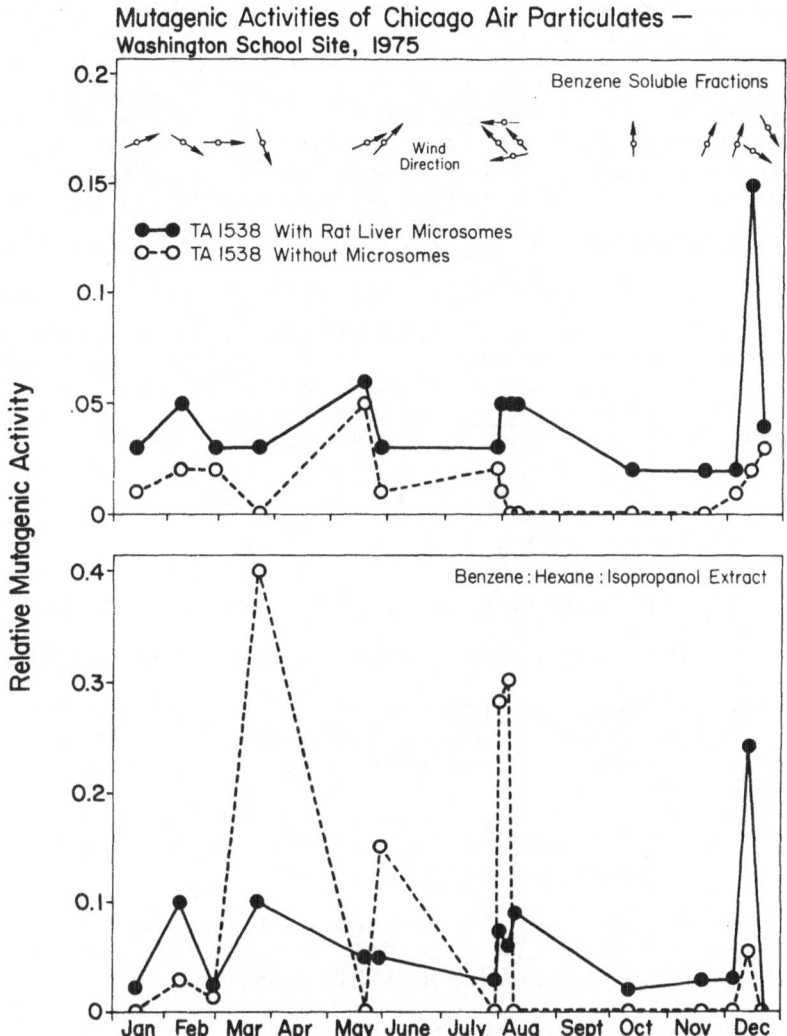

Figure 6. Relative mutagenic activities (computed as indicated in the examples shown in Figure 5) of different Chicago air particulate extracts collected on different dates in 1975 at the Washington School site.

about half of those exhibited by the comparable original extracts, suggesting that active material is lost during the fractionation procedure. Second, it is evident that in the first half of the year several instances occur in which the samples exhibit considerable inherent mutagenic activity, and that at least a good part of this activity is lost when the microsome preparation is present. This means that some of the mutagens that are inherently active are <u>inactivated</u> by the microsome preparation. This situation, often encountered in complex, unknown samples, creates important constraints on the interpretation of the data. This can be seen from the following considerations.

The basic difficulty is that measurements are made under <u>two</u> different conditions relative to the microsome preparation (i.e., with the preparation either present or absent), while the sample may contain <u>three</u> different classes of mutagens relative to their response to the microsome preparation.

Thus:

$$R_{WO} \sim a + b$$

$$\text{and } R_W \sim a + c$$

where:

R_{WO} = revertant rate without microsome preparation.

R_W = revertant rate with microsome preparation.

a = the concentration of compounds which are inherently mutagenic and not inactivated by the microsome preparation.

b = the concentration of compounds which are inherently mutagenic but inactivated by the microsome preparation.

c = the concentration of compounds which are not inherently mutagenic but are activatable by the microsome preparation.

It is evident from these relationships that it is impossible, from only the two measurements of revertant rate (i.e., R_{WO} and R_W) to determine the concentration of any one of the classes of mutagens, except in the special case in which R_{WO} is zero, nearly zero, or at least very much smaller than R_W.

In the example shown in Figure 6, except for several scattered points, the latter condition occurs only in August-December, so in that period the measurements made with the microsome preparation present (i.e., R_w) are indicative of the concentration of activable mutagens. Most of the other measurements made with the microsome preparation cannot be interpreted quantitatively since there is no way of knowing what part of the value is due to inherently active mutagens of class (a), which also contribute to the value of R_w. On the other hand, the values obtained in the absence of the microsome preparation (R_{wo}) are interpreted as representative of the activity of both classes of inherently active mutagens (i.e., classes a and b).

Finally, it is evident that the inherently mutagenic substances which are inactivated by the microsome preparation (i.e., class b) are largely lost when the benzene fraction is prepared. It is possible that this material passes into the water fraction in the second step of the procedure, since several samples (e.g., March 22 and July 31) exhibit a consistently rising trend with sample concentration, even though at the highest concentrations the value of E-C does not reach the statistical criterion of 2.0 x C_{Av}. This suggests that water-soluble active material is in fact present, which would become statistically significant if larger samples were analyzed.

Although it is premature to relate these observations to the general data regarding meteorological conditions, it is perhaps worth noting that most of the high concentrations of inherently mutagenic material observed in the original extract occurred when winds were generally from the northeast quadrant (see Figure 7).

The foregoing observations are indicative of the expected complexity of the mutagenic materials that occur in association with urban airborne particulates. We have further analyzed a particularly active sample, that for December 17, in order to test the feasibility of using the Salmonella technique as a means of isolating and identifying the responsible substances. About 56 square inches of the air filter was extracted in benzene:hexane:isopropanol. The extract was dried, taken up in chloroform, and aliquots were subjected to thin-layer chromatography according to the procedures described earlier. The extracts of successive chromatographic zones were then tested on strain TA1538 in the presence of microsomes. When the original extract was fractionated in a benzene:hexane (1:1) solvent system, two mutagenically active components with RF values of 0 and 0.9 were detected. The zones at RF = 0.9 and 1.0 were then combined, extracted, dried, and rechromatographed using n-hexane as the solvent

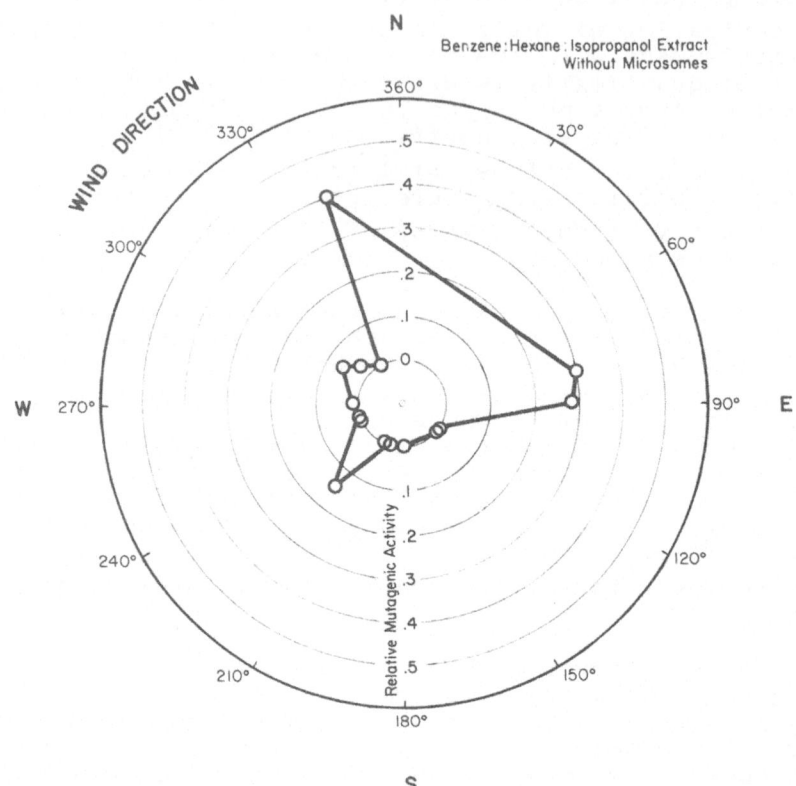

Figure 7. Relative mutagenic activities of benzene:hexane: isopropanol extracts of air particulate samples collected from Washington School site on different dates in 1975, as a function of concurrent wind direction (data of Fig. 6).

system. This procedure yielded a major mutagenically active zone with an RF value of 0.8 and a minor one at the origin. When the former was further chromatographed using iso-octane as the solvent, as shown in Figure 8, a single mutagenically active zone with an RF value of 0.7 was obtained. Under ultraviolet light this zone exhibited a strong fluorescence typical of certain polycyclic hydrocarbons. When preparations of pure benzo(a)pyrene and benzo(e)pyrene were chromatographed in the iso-octane solvent system, they yielded the same RF value as the mutagenically active component, 0.7.

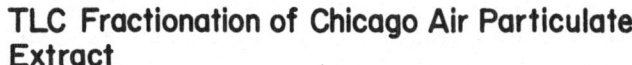

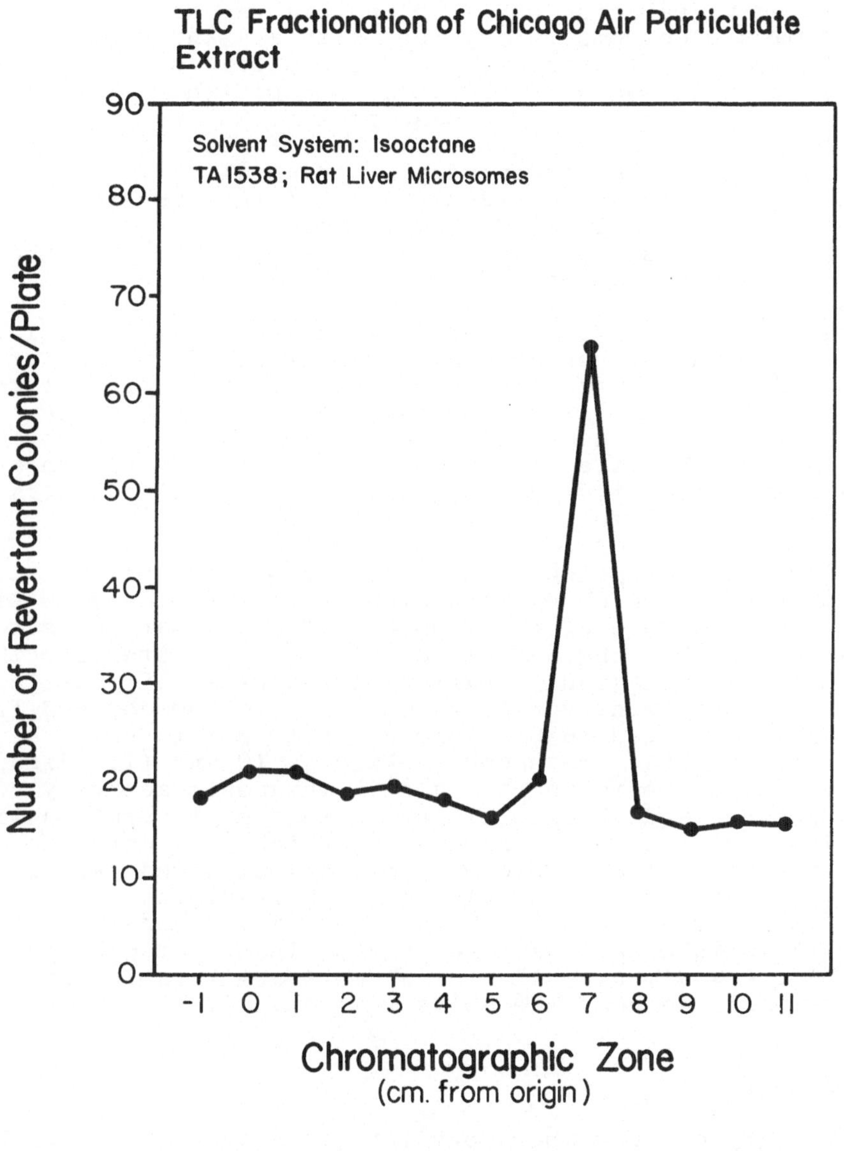
TLC Fractionation of Chicago Air Particulate Extract
Solvent System: Isooctane
TA 1538; Rat Liver Microsomes
Number of Revertant Colonies/Plate
0
10
20
30
40
50
60
70
80
90
-1 0 1 2 3 4 5 6 7 8 9 10 11
Chromatographic Zone
(cm. from origin)

Figure 8. Final TLC fractionation of mutagenic activity of material from Chicago air particulate sample (Washington School: December 7, 1975). See text for fractionation steps.

The purified preparation obtained in this way was analyzed by
means of mass spectrometry together with a standard sample of
benzo(a) pyrene (both isomers yield identifcal spectra in such
an analysis). As shown in Figure 9 the spectrum of the active
component exhibits the strong mass peak at 252 which corre-
sponds to the mass of both the (a) and (e) isomers of benzopyrene, as well as the fragmentation peaks which according
to a standard atlas are characteristic of this substance.
The presence of additional peaks, for example, at 266 and
270, 238 and 248, suggest that a small amount of some other
compound is present as well.

These results indicate that the active material isolated
by successive thin-layer chromatograms is largely a mixture of
benzo(a)pyrene and benzo(e)pyrene. Both isomers are mutagenic
toward strain TA1538 in the presence of the standard microsome
preparation (5). Consequently, benzo(a)pyrene and benzo(a)-
pyrene can be identified as two of the substances responsible
for the mutagenic activity exhibited by the original extract
of the air particulate sample.

All of the foregoing data are based on conventional high
volume samples in which particulates that vary widely in size
are trapped. Because of the tendency of small particles to
be retained in the lungs, it is of interest to determine the
distribution of mutagens in various sized urban air particulates. Some preliminary results on this problem derived from
experiments conducted in Los Angeles (provided to us by Dr.
David Coffin of EPA) are shown in Figures 10 and 11. Figure
10 shows that there is an increase in mutagenic activity both
in the presence and absence of the microsome preparation with
decreasing particle size. Figure 11, which is a chromatographic analysis of a sample of the smallest sized particles,
illustrates once again the value of such fractionation procedures. It shows, for example, that one of the constituents
(RF = 0.1) is inherently active and not inactivated by the
microsome preparation, while one or more activatable constituents is localized near the solvent front.

MUTAGENS IN FOODS

The entry of our laboratory into this area of research
illustrates one of the "black box" aspects of the Ames test,
and emphasizes the importance of paying attention to anomalies
that may arise. One such anomaly has been recognized in certain of the controls used in the test. Among the controls are
determinations of the number of revertant colonies that occur

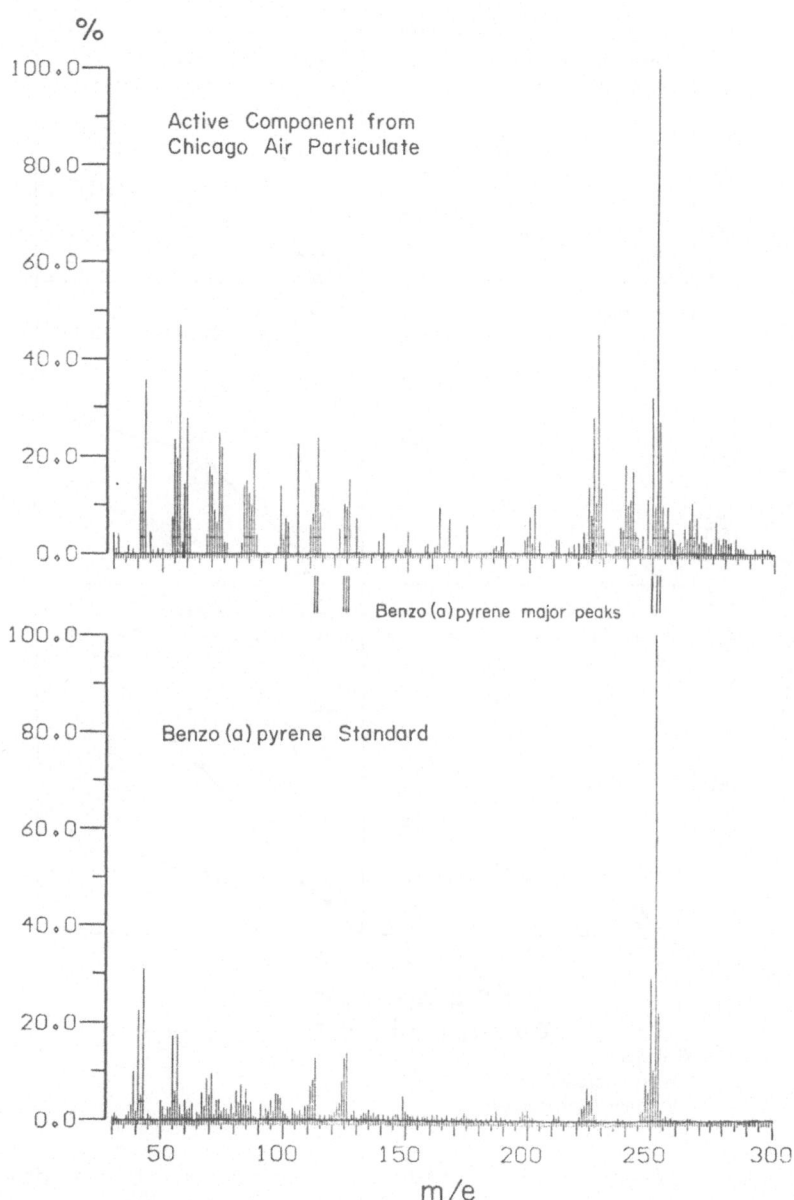

Figure 9. Mass spectra of material from zone (RF = 0.7) of chromatogram shown in Figure 8, and of benzo(a)pyrene.

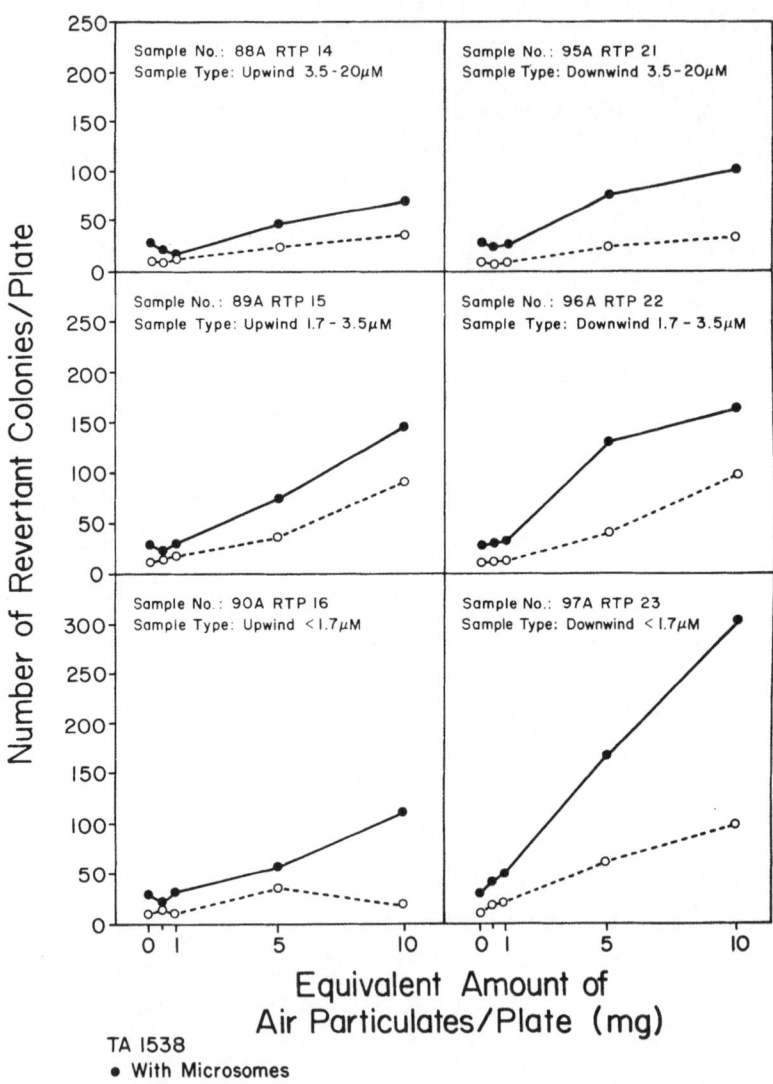

Figure 10. Dose-response curves of extracts of air particulates on strain TA1538 with (solid line) and without (broken lines) microsome preparation. Varying particle size ranges are shown.

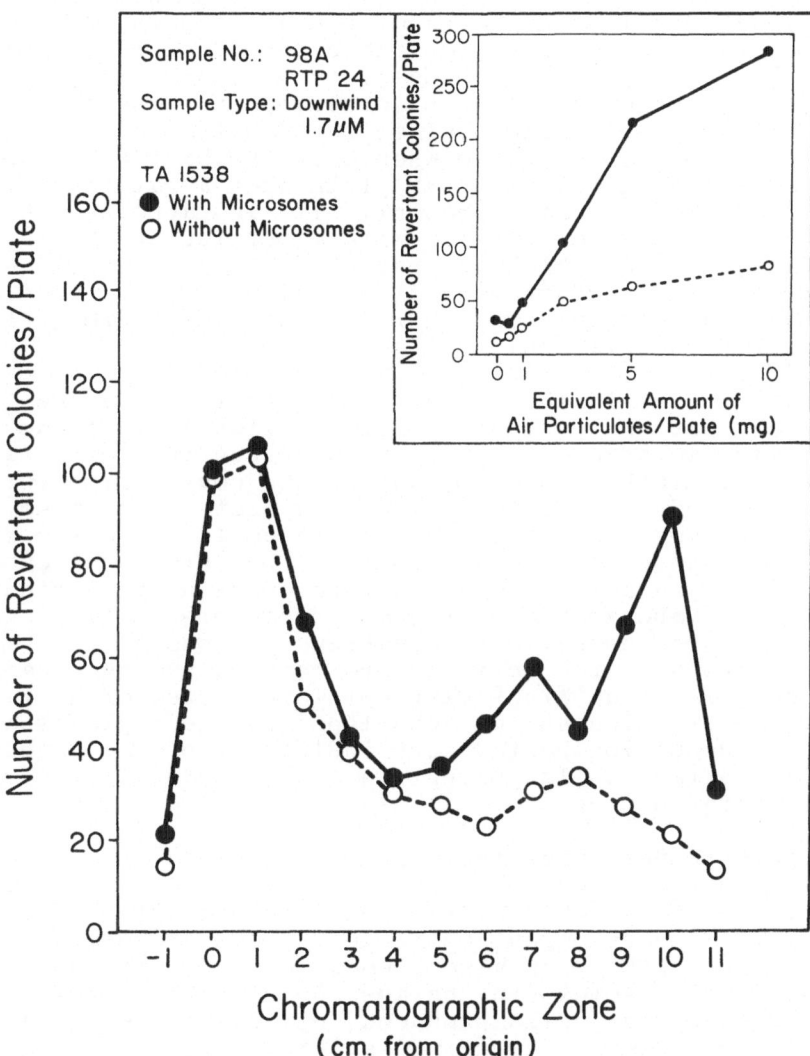

Figure 11. Chromatographic fractionation and dose-response curve for extract of air particulate sample 98A RTP 24. Particles were in the size range <1.7 µM. Chromatographic solvent was benzene:hexane (1:1). Samples were incubated with (solid lines) and without (broken lines) microsome preparation.

on plates that contain only the bacterial inoculum and on
plates that contain the microsome preparation in addition to
the inoculum. In the course of an extended series of tests
of a number of organic compounds, we noticed a small difference between the mutation rates observed in these two controls
(3). For example, average values for 200 test plates were 13
revertant colonies per plate when only the bacterial inoculum
was present and 22 colonies per plate when a rat-liver microsome preparation was also present. It also appeared that the
effect occurred preferentially with a particular strain of
Salmonella, TA1538, which according to Ames is sensitive to
substances that cause frameshift mutations. This effect,
which has been observed in other laboratories as well (1,6),
has remained unexplained.

As it occurs in the standard Ames test, the effect is so
small as to have no influence on the reliability of the test,
since active substances usually produce hundreds of mutant
colonies per plate. However, during the course of experiments
with a modified form of the Salmonella test, also based on the
Ames strains, we found that the effect could be considerably
amplified. These modified tests were conducted by incubating
Salmonella in an aerobic 5 ml culture containing nutrient
broth (Difco Laboratories), the microsome preparation, and the
substance to be tested. After various periods of incubation,
0.1 ml aliquots of the culture were removed and inoculated on
plates containing nutrient agar completely free of histidine.
The numbers of colonies that developed on these plates after
a 48-hour incubation period were indicative of the concentration of revertant cells present in the culture after various
periods of incubation.

Figure 12 describes typical data obtained from such a
liquid-culture test system when bacteria of strain TA1538
were present alone, when microsomes were present as well, and
when a typical carcinogen, activated by a microsome, 2-acetyl-
aminofluorene (AAF), was also present. It is evident that the
presence of microsomes (in the absence of AAF) increases the
number of revertant cells produced in the culture by an order
of magnitude. Similar experiments carried out with a series
of Salmonella strains, using several different types of microsome preparations, showed that the effect occurs only in
strains TA1538 and TA98 (which is similar to TA1538 in its
response to different mutagens). Thus, the phenomenon originally observed in standard plate tests, i.e., the specific
enhancement of the rate of mutation of strain TA1538, also
occurs in the liquid-culture system, but the effect is much
larger and therefore more capable of analysis.

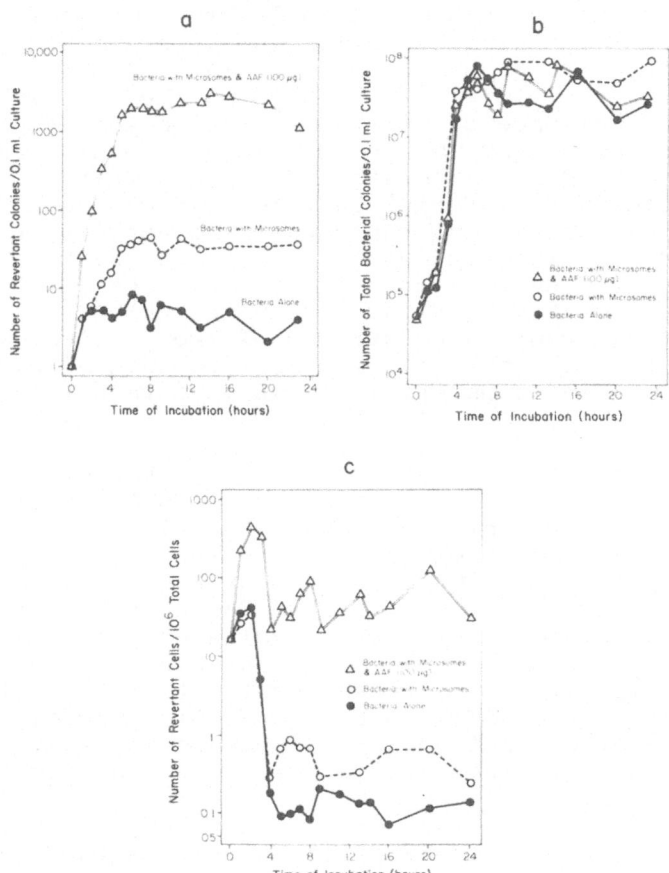

Figure 12. Number of revertant colonies and total number of
colonies produced from 0.1 ml inocula of a culture of strain
TA1538 obtained after increasing periods of incubation of:
bacteria alone (closed circles); bacteria with the microsome
preparation (open circles); bacteria with microsome preparation and 100 µgm of 2-acetylaminofluorene (triangles).
a: Numbers of revertant colonies, as obtained from counts
of culture aliquots inoculated on histidine-free synthetic
medium plates. b: Numbers of total colonies, as obtained
from counts of culture aliquots inoculated on plates of synthetic medium supplemented with histidine. c: Ratio of
revertant to total number of colonies, computed from the
data of a and b.

As a first step in such an analysis, we undertook to
determine the functional basis for the apparent mutagenic
effect of microsome preparations on strain TA1538. These
studies showed that material which is mutagenic toward strain
TA1538 in the presence of microsomes can be extracted by
benzene:isopropanol (80:20) and similar solvents from "Bacto
nutrient broth" (Difco Laboratories), whether fresh or following incubation in a bacterial culture. It can be concluded,
therefore, that the effect represents the conversion of a
substance present in nutrient broth into an active mutagenic
metabolite by the enzymatic activity of microsomes. This is
confirmed by the data of Figure 13, which shows, from dose-
response curves, that such extracts of two samples of commercial nutrient broth contain comparable amounts of microsome-
activatable mutagenic material, to which strain TA1538 readily
responds.

In a survey of a number of commercial bacterial nutrients,
we found that those nutrients which contain "beef extract" or
beef heart infusion contain active material, yielding from 308
to 2789 revertant colonies per gram in the presence of microsomes as compared with 10-36 colonies when microsomes are
absent. Comparison of the several Difco nutrients tested suggests that the number of revertant colonies produced per gram
is roughly proportional to the nutrient's content of beef
extract. It appeared from these results that the mutagen is
a constituent of the beef tissue (generally muscle) used to
produce the beef extract employed in these nutrient preparations, or is derived from such a constituent during the preparation process.

Beef extract used in bacterial nutrients is produced in
abbatoirs, by first preparing beef broth from beef tissue
which has been boiled for about 30 minutes in an equal volume
of water and then defatted. To prepare beef extract this
broth is then boiled down to 20 percent or less of its original volume. The result is a dark brown paste which is used
in the manufacture of bacterial nutrients and in various
foods, such as beef bouillon cubes. "Bacto Beef Extract"
(Difco Laboratories) was tested for mutagenic activity in the
following way: samples were homogenized in distilled water
and then acidified (to pH 2.0) with HCl. Protein was then
precipitated by adding ammonium sulfate to saturation. The
samples were then filtered through glass wool, the filtrate
adjusted to pH 10 with ammonium hydroxide, extracted three
times with methylene chloride and the extract evaporated to
dryness. Aliquots representing varying amounts of the original sample were taken up in DMSO and tested on strain TA1538

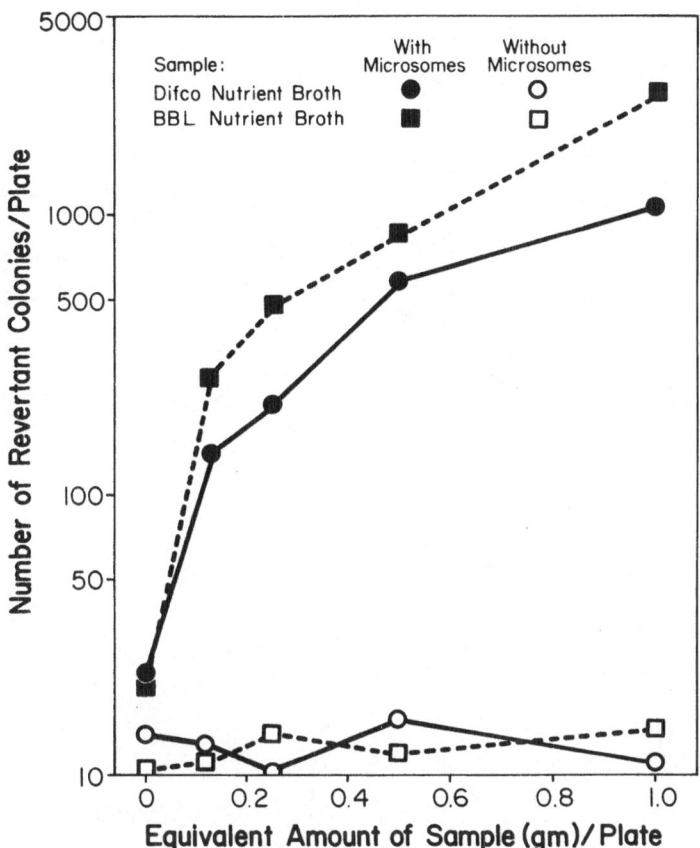

Figure 13. Number of revertant colonies (of strain TA1538) produced per standard test plate by benzene:isopropanol (80:20) extracts of increasing quantities of Difco (circles) and BBL (squares) nutrient broth. Solid lines: microsomes present; broken lines: microsomes absent.

with and without microsomes. The dose-response obtained is clearly indicative of mutagenic activity in the presence of the microsome preparation (see Figure 14). Dose-response curves obtained with other strains show that strain TA98 is equally active, TA1537 about one-fourth as active, while strains TA100 and TA1535 are inactive. In all cases there was no activity when microsomes were absent. A series of

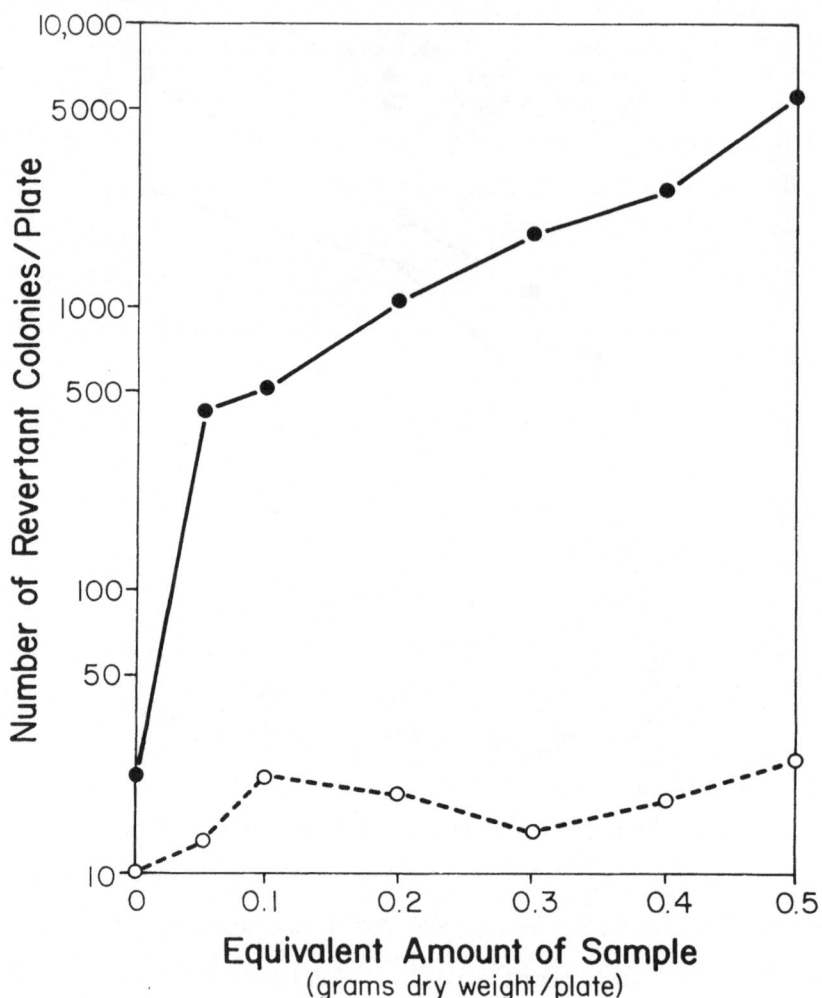

Figure 14.   Dose-response curves of methylene chloride
extracts of "Bacto Beef Extract" (Difco Laboratories) tested
on strain TA1538.   Ordinate: number of histidine-positive
revertants per plate.   Abscissa: amount of sample used to
prepare the methylene chloride extract added per plate.
Solid lines represent plates to which microsome preparation
was added.   Broken lines represent plates to which the microsome preparation was not added.

chromatographic analyses of Difco beef extract and of Difco nutrient broth were carried out with hexane-acetone and with benzene-methanol as solvents. As shown in Figure 15, the chromatrographic mobility of the mutagenically active material from beef extract and from nutrient broth, in benzene:methanol (95:5) was similar. Comparable results occurred in the other solvent system. Thus, the mutagen originally discovered in bacterial nutrient is present in the beef extract itself.

Two commercial preparations, purchased in local stores, "Maggi Beef Bouillon Cubes" and "B.V. Broth & Sauce Concentrate," which according to their labels contain beef extract, have been tested with methods comparable to those described for beef extract. From the dose-response curves against various Salmonella strains, in the presence of a microsome preparation and from chromatographic analysis (see Figure 15), it is evident that these preparations contain mutagens with the characteristics of those found in bacterial nutrients and in beef extract.

Beef broth contains no detectable mutagens whereas beef extract, which is prepared from the broth by extensive boiling does. Accordingly, we have studied the conversion process by testing beef broth for mutagenicity at 30-minute intervals during extensive boiling. The results, which are reported in Figure 16, show that the mutagens are absent from beef stock and are produced during the boiling process, especially when the preparation is reduced to a paste, at which time the mutagenic activity rises sharply to 1572 revertants per plate per 0.69 gm dry weight. It is apparent, then, that the mutagens do not occur as such in beef tissue or in beef broth, but are formed during the heating and evaporation that occurs in the conversion of beef broth to beef extract.

Given these results, it was of obvious interest to determine whether these mutagens are formed when beef is cooked by conventional procedures. Lean ground beef (in 100 gm, dry weight, portions) was cooked in an electrically-heated (plate temperature 200°C) home hamburger cooking appliance for 1.5 minutes ("rare"), 3.0 minutes ("medium"), and 5.5 minutes ("well-done"), respectively. The cooked samples and an uncooked control were homogenized in twice their volume of distilled water in a Waring blender and were treated in the same way as the beef extract described earlier. Aliquots of the final methylene chloride extracts representing 5 and 25 gm dry weight of the cooked beef (in the case of the uncooked control, aliquots represents 5, 10, and 35 gm were tested) were dried, taken up in DMSO, and tested in the usual way against strain TA1538 in the presence and absence of the

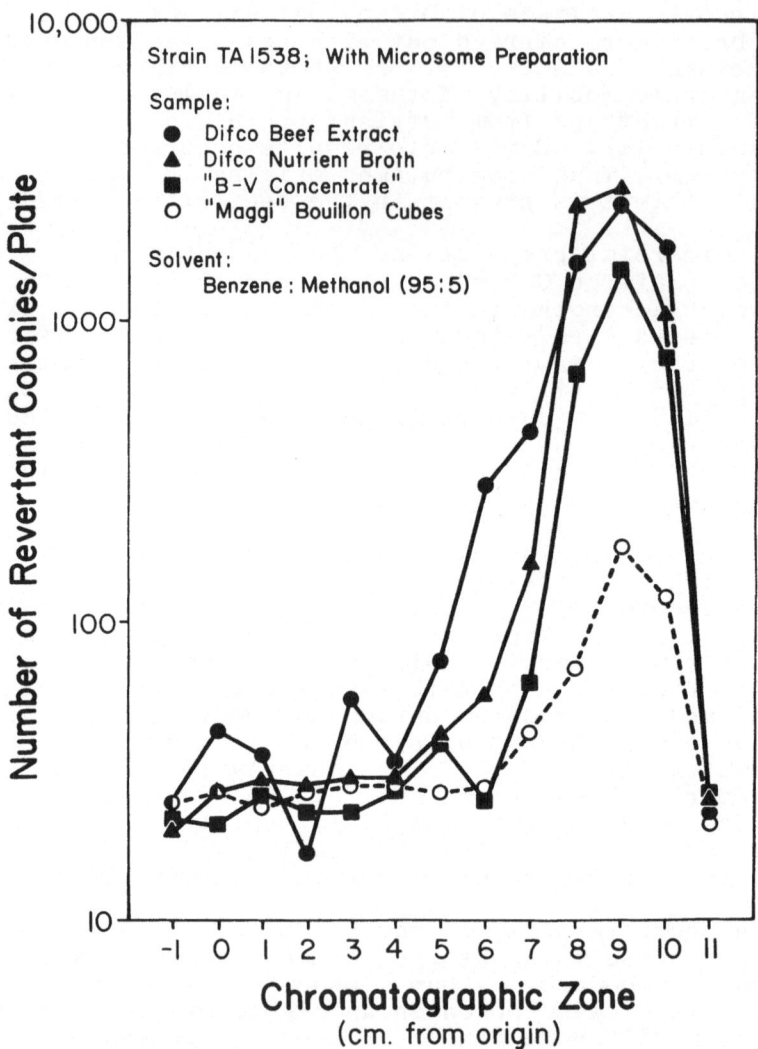

Figure 15. Thin layer chromatographic fractionation in benzene:methanol (95:5) of the mutagenic material of Difco beef extract, Difco nutrient broth, "B-V concentrate," and "Maggi" bouillon cubes. Tested on strain TA1538 with microsome preparation present.

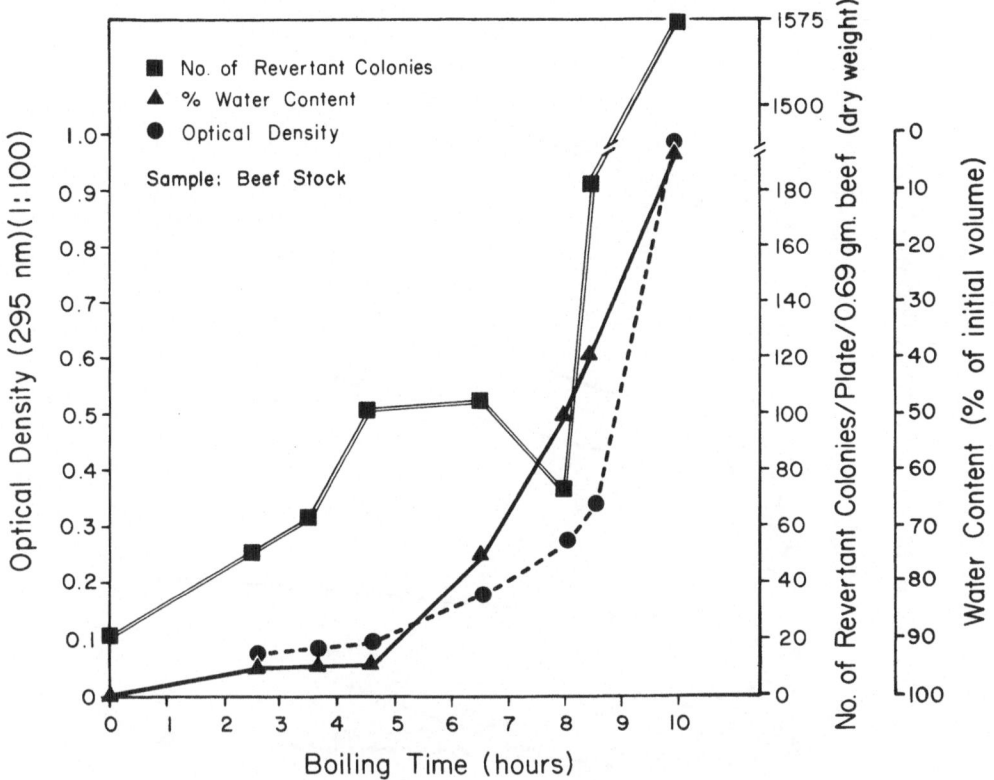

Figure 16. Variation of mutagenic activity (determined on strain TA1538; microsome preparation present), optical density at 295 nm (dilution 1:100), and water content of beef stock as a function of boiling time.

microsome preparation. The results are shown in Figure 17. The mutagenic activities of all samples tested in the absence of the microsome preparation fall within the range of control values. The values obtained from the uncooked sample in the presence of the microsome preparation are slightly above the control value, but the increase is of doubtful significance. However, all the cooked samples yielded substantial levels of mutagenic activity. The values increase with cooking time, the "well-done" samples yielding the highest values: 954 revertants per plate with a 5 gm sample and 3388 revertants per plate with a 25 gm sample.

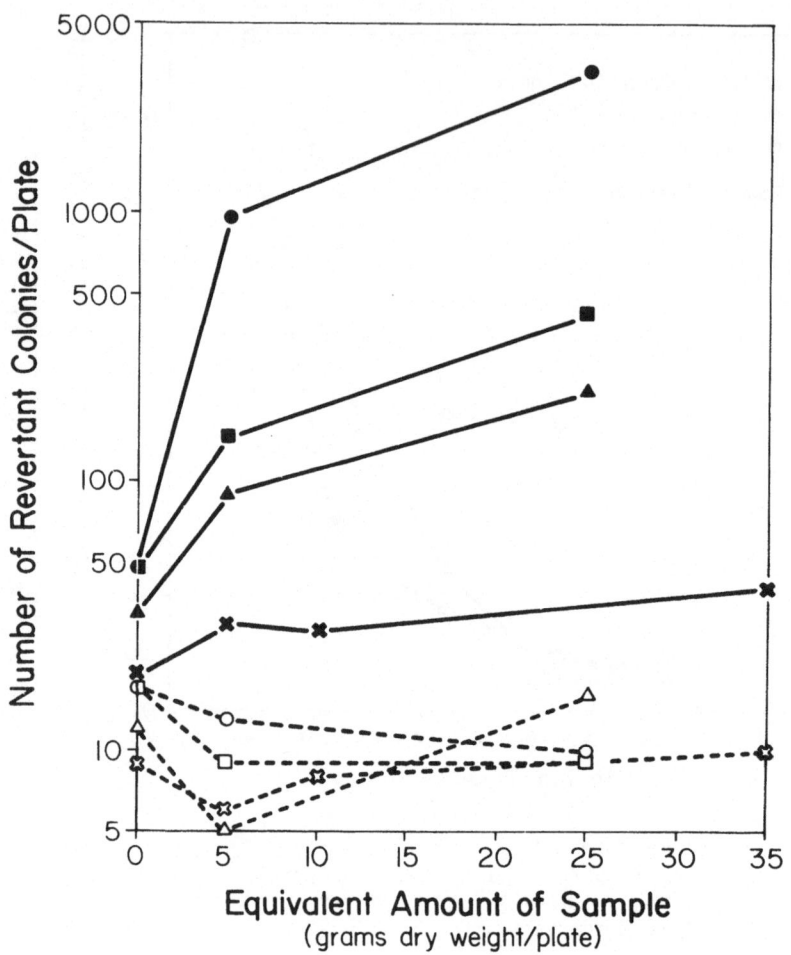

Figure 17. Dose-response curves for methylene chloride extracts of uncooked and cooked lean ground beef. Tests were carried out on strain TA1538 with the microsome preparation present (solid lines) and in its absence (broken lines). 100 gm (wet weight) samples of lean ground beef were tested before cooking (data points indicated by crosses) and after cooking in an electrically-heated home hamburger cooking appliance for the following times: 1.5 minutes ("rare"; data points indicated by triangles), 3.0 minutes ("medium"; indicated by squares) and 5.5 minutes ("well-done"; indicated by circles).

These data suggested a possible relation between our observations and earlier evidence that mutagens, including known carcinogens such as benzo(a)pyrene, are formed in meat and fish during certain cooking procedures. Thus, Sugimura et al. (7) report that condensed smoke from meat and fish broiled over an open gas or charcoal flame contains material that is mutagenic toward strain TA98, usually only in the presence of the microsome preparation. They report that the mutagenic activity levels are much too high to be accounted for by the amounts of benzo(a)pyrene present in the smoke condensates and suggest that other mutagens may arise from pyrolysis of tissue protein and amino acids. This suggestion is based on their observation that pyrolysis (at temperatures of 300°-600°C) of proteins and certain amino acids produces mutagens similar in their effects in the Ames test to those observed in the smoke condensates (4). They also report similar activity in material obtained from the charred surface of a broiled beef steak.

In view of the foregoing results, it was of interest to compare the mutagens that occur in beef extract and cooked beef with those formed by pyrolysis of amino acids, and with benzo(a)pyrene. For this purpose methylene chloride extracts of beef extract, cooked beef, cooked beef with added benzo(a)-pyrene, and a pyrolyzed mixture of amino acids were chromatographed, using a silica-gel impregnated glass fiber sheet (Gelman ITLC-SG) in a suitable solvent. Successive 1 cm zones of the developed chromatograms were extracted in chloroform:methanol (90:10), dried, taken up in DMSO, and tested on strain TA1538 in the usual way. Figure 18 reports such analyses of methylene chloride extracts of "Bacto Beef Extract" and of a beef patty cooked for ten minutes on a ceramic hot plate, using benzene:methanol (95:5) as the chromatographic solvent. From thermocouples at the surface of a patty and in its interior, it was determined that the maximum temperature (at the end of the cooking period) at the surface of the patty was 200°C and in the interior 80°C. The mutagens present in the two samples exhibit identical chromatographic behavior, with a major peak at an RF = 0.5 and a slight shoulder at RF = 0.3. Figure 19(a) reports the results of a similar analysis (using 100 percent hexane as the chromatographic solvent) of methylene chloride extracts of "Bacto Beef Extract," of a hot-plate cooked beef patty, and of such a patty to which 25 μgm per kgm (wet weight) of benzo(a)pyrene had been added (after cooking and extraction). All of the mutagenic activity associated with "Bacto Beef Extract," and cooked beef remains at the origin, while the sample of the latter in which benzo(a)-pyrene had been added exhibits an additional peak at RF = 0.85.

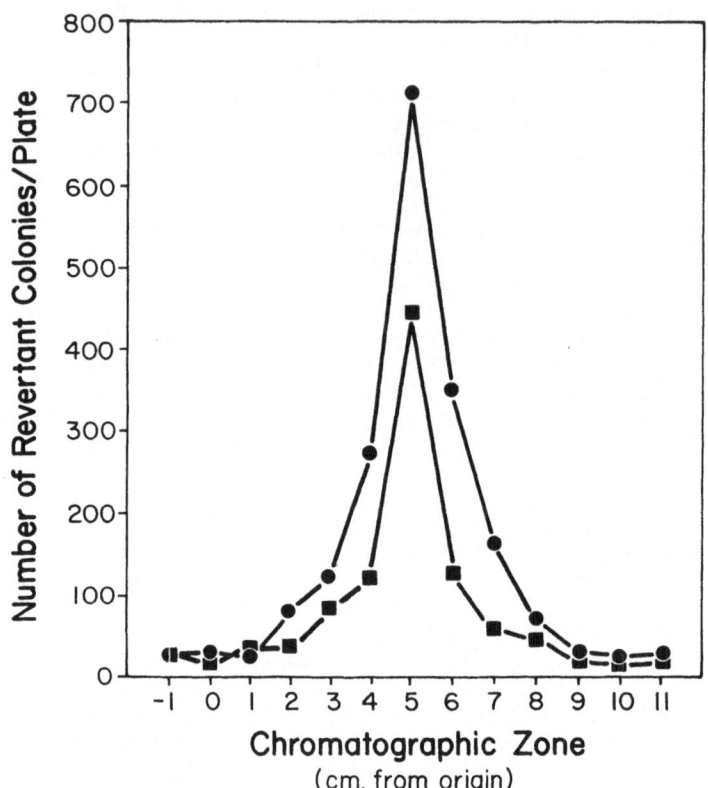

Figure 18. Thin-layer chromatographic fractionation of the
methylene chloride extracts of "Bacto Beef Extract" (● — ●)
and hot plate-cooked lean ground beef (■ — ■). Gelman ITLC-
SG sheets were used with benzene:methanol (95:5) as the solvent system. Four ground beef patties (each approximately
120 gm wet weight) were wrapped in aluminum foil and cooked
on a 350°C ceramic hot plate for 10 to 12 minutes. Thermocouples at the surface of a patty and in its interior recorded
temperatures of 200°C and 80°C, respectively, at the end of
the cooking period. Extracts equivalent to approximately 0.2
grams of beef extract and 26 grams (dry weight) of ground
beef were applied to the chromatogram. One-centimeter zones
of the developed chromatogram were extracted with chloroform:
methanol (90:10). Aliquots were taken to dryness, resuspended
in DMSO and tested on strain TA1538 in the presence of the
microsome preparation.

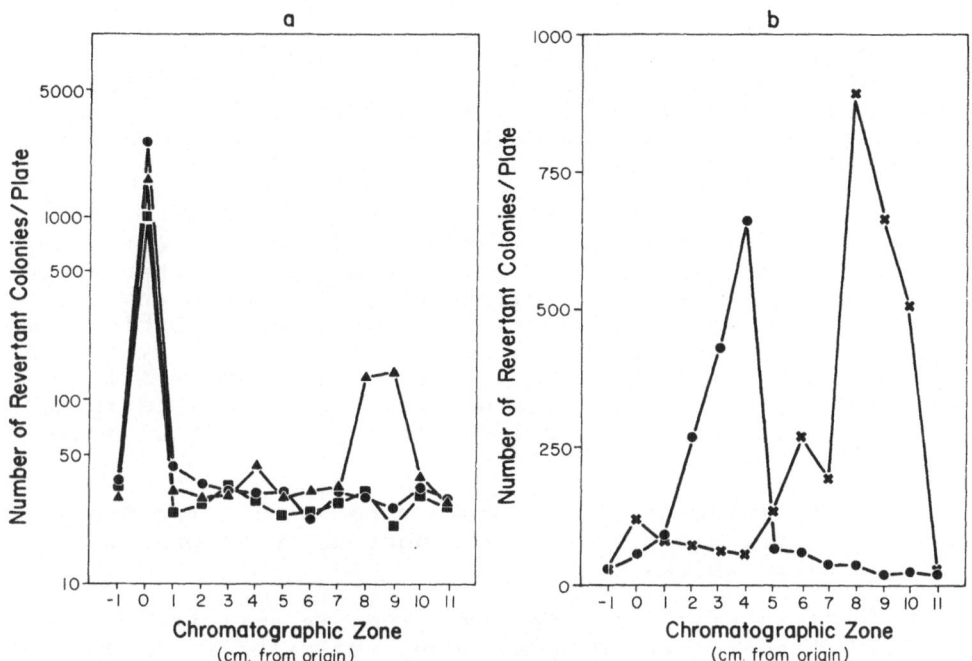

Figure 19. Thin-layer chromatographic fraction of: (a) the methylene chloride extracts of "Bacto Beef Extract" (● — ●), hot plate-cooked beef (■ — ■) and hot plate-cooked beef to which benzo(a)pyrene was added (25 μg/Kg wet weight) (▲ — ▲). Extracts equivalent to approximately 0.2 grams of beef extract and 26 grams of ground beef (with and without benzo(a)pyrene) were applied to the chromatogram. Chromatographic solvent: 100% hexane; and (b) the methylene chloride extracts of "Bacto Beef Extract" (● — ●) and of a mixture of 18 amino acids pyrolyzed at 350°C (✳ — ✳). Extracts equivalent to approximately 0.2 grams of beef extract and 9 mg of amino acids (equal weights of each) were applied to the chromatogram. Chromatographic solvent: hexane:acetone (50:50). Microsome preparation was present.

Figure 19(b) reports a similar chromatographic analysis
[using hexane:acetone (50:50) as the chromatographic solvent]
of methylene chloride extracts of "Difco Beef Extract" and
the pyrolsis product (pyrolysis temperature 350°C) of a mixture of 2 mg of each of the 18 amino acids, which according
to Matsumoto et al., (4) yield mutagenic material when
pyrolyzed. The material from "Bacto Beef Extract" exhibits
a peak at RF = 0.4, while the material from the pyrolyzed
amino acids exhibits a main peak at RF = 0.8, a minor peak
at RF = 0.6, and some residual activity at the origin.

These analyses indicate (a) that the mutagens produced
when beef stock is heated to form beef extract are chromatographically indistinguishable from those produced when ground
beef is cooked on an electrically-heated hot-plate and (b)
that the former are chromatographically distinguishable from
both benzo(a)pyrene and the mutagens produced from pyrolyzed
amino acids. Further studies of the mutagenic material
extractable by methylene chloride from "Bacto Beef Extract,"
partially purified by successive thin-layer chromatographic
separations, show the following:

● The mutagen(s) is a basic substance, extractable
   by organic solvents from aqueous solutions at
   alkaline pH.

● It is unaffected in its mutagenic activity or
   chromatographic behavior by refluxing in 6N HCl
   for six hours.

● On treatment with nitrous acid, the material becomes
   inherently mutagenic (i.e., in the absence of the
   micromsome preparation), suggesting the possible
   formation of a nitroso group. The conditions in
   which these mutagens are formed are similar to those
   characteristics of the Maillard or "Browning" reactions in which amino acids and sugars react to produce a variety of complex substances (8).

The foregoing experiments show that one or more substances
which are mutagenic in the Ames system (in the presence of the
microsomal preparation) are produced when beef stock is heated
and condensed to form beef extract and when ground beef is
cooked (at temperatures not exceeding 200°C) on an electric
hot-plate or a home hamburger cooking appliance. These mutagens are neither benzo(a)pyrene nor the mutagenic substances
produced when amino acids are pyrolyzed. This is indicated
by the chromatographic analyses reported above. Moreover,

according to Matsumoto et al. (4), the mutagenic pyrolysis products are formed only at temperatures in excess of 300°C, which can readily occur in foods cooked over open flames. In contrast, the mutagens we have detected in beef extract are produced at temperatures that do not exceed 105°C, while those detected in cooked ground beef are produced at temperatures that do not exceed 200°C. Thus, these mutagens are produced in conditions that occur in common cooking procedures, including the preparation of hamburgers on electrically-heated hot-plates at conventional cooking temperatures and times.

The mutagens found in beef extract and cooked beef are rather active, as compared with a typical mutagen which is also active toward strains TA1538 and TA98, 2-acetylamino-fluorene (AAF). Tested on strain TA1538, 50 µgm of AAF (which is in the linear portion of the dose-response curve) yields about 4800 revertants per plate. Active material prepared from a bacterial nutrient containing 37 percent beef extract yielded 1367 revertants per plate containing 3.5 µgm (in the linear part of the dose-response curve) of a preparation partially purified, by successive chromatographic fractionation, from the original methylene chloride extract. Accordingly, the specific activity of the beef extract mutagen(s) is a minimum of about 350 revertants per plate per µgm, as compared with 96 revertants per plate per µgm for AAF. Based on the estimate of 350 revertants per plate per µgm, a 3.6 gm beef bouillon cube contains a minimum of approximately 0.3 µgm of mutagen and a 100 gm wet weight lean-beef hamburger contains approximately 1 to 14 µgm of mutagen, depending on the extent of cooking. These figures correspond to concentrations, on a wet weight basis, of 0.1 ppm of mutagen in beef bouillon cubes and from .01 to .14 ppm in cooked hamburgers.

If, as indicated by the observed correlation between mutagenicity in the Ames test and carcinogenicity, these mutagens--once purified and tested on laboratory animals--are found to be carcinogens, their apparent concentration in some foods may represent an appreciable risk to certain populations. The relatively ordinary circumstances in which these mutagens are formed suggest that they may arise during the course of certain conventional cooking procedures, in addition to the preparation of hamburgers, such as the braising of beef and the evaporation of beef stock in the preparation of stews. However, the sensitivity of the effect to cooking times, which is evident in the results shown in Figure 17, suggests that it may be possible to modify cooking procedures in ways that reduce the formation of the mutagens.

DISCUSSION

The substantive conclusion of the foregoing results is
that mutagens occur in the effluents of certain petrochemical
plants, in Chicago air particulates, in beef extracts, and in
hamburgers. Clearly, the Ames test is a very useful means of
detecting the occurrence of such environmental carcinogens.
It is also evident from these results that, combined with
chromatographic techniques, the method can be used to isolate
and ultimately identify mutagens which occur in such samples.
However, such qualitative conclusions--for example, the determination of whether or not a given environmental sample contains a significant amount of mutagenic material--depend on
certain quantitative procedures. Specifically, the appropriate procedure is to determine, from a dose-response curve,
whether at any sample concentration the mutagenic activity
ratio, $\frac{E-C}{C_{Av}}$, exceeds the statistical criterion previously
established from test of standard substances. Such determinations must be made separately with microsomes present
and absent. Constraints on this type of determination include the following:

- The determination relates only to substances that
  are active on the particular strain of <u>Samonella</u>
  that is used.

- A false negative result may be obtained if the
  sample contains sufficient toxic or bacteriostatic
  material to suppress the growth of mutants.

Subject to these constraints and to the previously stated
limits of the reliability of the test system, the <u>Salmonella</u>
technique can readily be used for the rapid, qualitative detection of organic carcinogens in environmental samples.

It is also evident that, subject to additional constraints,
<u>quantitative</u> estimation of the level of mutagenic activity is
possible, based on the analytical procedures described above.
In these procedures one determines by interpolation from the
dose-response curve the lowest sample concentration at which
the mutagenic activity ratio that is representative of statistically significant mutagenic activity occurs. The sample's
mutagenic activity is expressed, in relative terms, by the
reciprocal of this sample concentration. A major constraint
on this procedure is that it is not applicable to data obtained
in the presence of microsomes, unless the sample's mutagenic

activity in the absence of microsomes can be shown to be zero, or small relative to the value obtained when microsomes are present. Where an initial extract of the sample does not conform to this requirement, it would be necessary to introduce a fractionation procedure that separates inherently active mutagens from those requiring microsomal activation before quantitative estimates of the latter are made.

While the emphasis of this paper is on the methodological aspects of these results, certain substantive aspects of the results are worth noting. The results of studies of air particulates from the Washington School site are probably related to the fact that this site, which appears to yield the highest concentrations of carcinogens in air particulates from the Chicago area, is located within a heavily industrialized neighborhood. Steel mills, including coke-oven operations, are present. Since these operations are known to produce high concentrations of benzo(a)pyrene and other carcinogens, the high levels of mutagenic activity that we have observed in air particulates, and direct evidence that benzopyrene isomers occur in them, is not suprising. While the data obtained from this site are insufficient to establish firm correlations with wind direction, they do suggest that with more detailed analyses it will be possible to define the origins of the particulate-associated carcinogens. It would appear, therefore, that screening procedures based on the Salmonella mutagenesis technique can be used to determine how the environmental distribution of the detectable carcinogens may be associated with the local epidemiology of cancer incidence, and with the activities of possible sources of the relevant substances.

In the same way, the studies of the formation of mutagens in cooked beef and in beef extract, together with earlier studies in Japanese laboratories, show that the technique can be a very useful means of monitoring the role of cooking practices on the formation of mutagens.

The Ames technique, suitably applied and subject to certain constraints, is a valuable means of screening environmental samples for mutagens. Given the established correlation between mutagenicity in this test and carcinogenicity toward laboratory animals, those procedures form the basis for an analysis of the role of environmental agents in the incidence of cancer.

REFERENCES

1. Ames BN, Durton WE, Yamasaki E, Lee FD: Carcinogens
    are mutagens: A simple test system combining liver
    homogenates for activation and bacteria for detection.
    Proc Natl Acad Sci USA 70:2281-2285, 1973

2. Ames BN, McCann J, Yamasaki E: Methods for detecting
    carcinogens and mutagens with the Salmonella/mammalian-
    microsome mutagenicity test. Mutat Res 31:347-364, 1975

3. Commoner B: Reliability of bacterial mutagenesis
    techniques to distinguish carcinogenic and noncarcin-
    ogenic chemicals. Washington DC, US Environmental
    Protection Agency Publ. No. EPA-600/1-76-022, p 104,
    1976

4. Matsumoto T, Yoshida D, Mizusaki S, Okamoto H: Muta-
    genic activity of amino acid pyrolyzates in Salmonella
    typhimurium TA98. Mutat Res 48:279-286, 1977

5. McCann J, Choi E, Yamasaki E, Ames BN: Detection of
    carcinogens as mutagens in the Salmonella/microsome
    test: Assay of 300 chemicals. Proc Natl Acad Sci USA
    72:5135-5139, 1975

6. Nebert DW, Feton JS: Evidence for the activation of
    3-methylcholanthrene as a carcinogen in vivo and as a
    mutagen in vitro by P_1-450 from inbred strains of mice.
    In: Cytochromes P-450 and b_5, Structure, Function and
    Interaction (Cooper DW, Rosenthall O, Snyder R, Witmer
    C, eds.). New York, Plenum, pp 127-149, 1975

7. Sugimura T, Nagao M, Kawachi T, Honda M, Yahagi T,
    Seino Y, Sato S, Matsukura N, Matsushima T, Shirai A,
    Sawamura M, Matsumoto H: Mutagen-carcinogens in food
    with special reference to highly mutagenic pyrolitic
    products in broiled foods. In: Origins of Human
    Cancer (Hiatt HH, Watson JD, Winsten JA, eds.). Book
    B, Cold Spring Harbor, NY, Cold Spring Harbor Labora-
    tory, pp 1561-1577, 1977

8. Tarr H: Ribose and the Maillard reaction in fish
    muscle. Nature 171:344-345, 1953

Index